Antimicrobial Resistance and Virulence Mechanisms

Antimicrobial Resistance and Virulence Mechanisms

Editor

Manuela Oliveira

MDPI • Basel • Beijing • Wuhan • Barcelona • Belgrade • Manchester • Tokyo • Cluj • Tianjin

Editor
Manuela Oliveira
Department of Animal Health,
Faculty of Veterinary Medicine
University of Lisbon
Lisbon
Portugal

Editorial Office
MDPI
St. Alban-Anlage 66
4052 Basel, Switzerland

This is a reprint of articles from the Special Issue published online in the open access journal *Antibiotics* (ISSN 2079-6382) (available at: www.mdpi.com/journal/antibiotics/special_issues/epidemiology_antibiotics).

For citation purposes, cite each article independently as indicated on the article page online and as indicated below:

LastName, A.A.; LastName, B.B.; LastName, C.C. Article Title. *Journal Name* **Year**, *Volume Number*, Page Range.

ISBN 978-3-0365-3906-5 (Hbk)
ISBN 978-3-0365-3905-8 (PDF)

Contents

About the Editor

Manuela Oliveira

Manuela Oliveira received her PhD degree from the Faculty of Veterinary University of Lisbon (FMV-ULisbon) in 2005 and is a Diplomate of the European College of Veterinary Microbiology (DipECVM). She is an Associate Professor of the Department of Animal Health of FMV-ULisbon, lecturing on Microbiology and other subjects in the Integrated Master of Veterinary Medicine and on Food Microbiology in other Master's courses. Her research interests include One Health, clinical veterinary bacteriology, bacterial biofilms, antimicrobial resistance, food safety, wildlife diseases and mycology. She has supervised and participated in several national and international research projects in these areas, authoring more than 100 peer-reviewed publications (articles, abstracts and book chapters) and 200 scientific communications in congresses (oral and poster communications). She also has experience in student supervision at several levels (PhD, MSc and LSc/BSc). Currently, she is responsible for two diagnostic laboratories of FMV-ULisbon, the Laboratory of Bacteriology and the Laboratory of Mycology.

Preface to "Antimicrobial Resistance and Virulence Mechanisms"

The worldwide emergence of antimicrobial-resistant bacteria, specially those resistant to last-resource antibiotics, is now a common problem being defined as one of three priorities for the safeguarding of One Health by the Tripartite Alliance, which includes the World Health Organization (WHO), the Food and Agriculture Organization (FAO) and the Office International des Epizooties (OIE). Bacteria resistance profiles, together with the expression of specific virulence markers, have a major influence on the outcomes of infectious diseases. These bacterial traits are interconnected, since not only the presence of antibiotics may influence bacterial virulence gene expression and consequently infection pathogenesis, but some virulence factors may also contribute to an increased bacterial resistance ability, as observed in biofilm-producing strains. The surveillance of important resistant and virulent clones and associated mobile genetic elements is essential for decision making in terms of mitigation measures to be applied for the prevention of such infections in both human and veterinary medicine. However, the role of natural environments as important components of the dissemination cycle of these strains has not been consider until recently. This Special Issue aims to publish manuscripts that contribute to the understanding of the impact of bacterial antimicrobial resistance and virulence in the three areas of the One Health triad–i.e., animal, human and environmental health.

Manuela Oliveira
Editor

 *antibiotics* 

Article

Characterization of Fosfomycin and Nitrofurantoin Resistance Mechanisms in *Escherichia coli* Isolated in Clinical Urine Samples

Antonio Sorlozano-Puerto [1] , Isaac Lopez-Machado [1], Maria Albertuz-Crespo [1],
Luis Javier Martinez-Gonzalez [2] and Jose Gutierrez-Fernandez [1,3,*]

[1] Department of Microbiology, School of Medicine and PhD Program in Clinical Medicine and Public Health, University of Granada-ibs, 18016 Granada, Spain; asp@ugr.es (A.S.-P.); isloma@correo.ugr.es (I.L.-M.); albertuzmaria@correo.ugr.es (M.A.-C.)

[2] Pfizer-University of Granada-Junta de Andalucía Centre for Genomics and Oncological Research (GENYO), 18016 Granada, Spain; luisjavier.martinez@genyo.es

[3] Laboratory of Microbiology, Virgen de las Nieves University Hospital-ibs, 18014 Granada, Spain

* Correspondence: josegf@ugr.es

Received: 17 July 2020; Accepted: 24 August 2020; Published: 24 August 2020

Abstract: Fosfomycin and nitrofurantoin are antibiotics of choice to orally treat non-complicated urinary tract infections (UTIs) of community origin because they remain active against bacteria resistant to other antibiotics. However, epidemiologic surveillance studies have detected a reduced susceptibility to these drugs. The objective of this study was to determine possible mechanisms of resistance to these antibiotics in clinical isolates of fosfomycin- and/or nitrofurantoin-resistant UTI-producing *Escherichia coli*. We amplified and sequenced *murA*, *glpT*, *uhpT*, *uhpA*, *ptsI*, *cyaA*, *nfsA*, *nfsB*, and *ribE* genes, and screened plasmid-borne fosfomycin-resistance genes *fosA3*, *fosA4*, *fosA5*, *fosA6*, and *fosC2* and nitrofurantoin-resistance genes *oqxA* and *oqxB* by polymerase chain reaction. Among 29 isolates studied, 22 were resistant to fosfomycin due to deletion of *uhpT* and/or *uhpA* genes, and 2 also possessed the *fosA3* gene. Some modifications detected in sequences of NfsA (His11Tyr, Ser33Arg, Gln67Leu, Cys80Arg, Gly126Arg, Gly154Glu, Arg203Cys), NfsB (Gln44His, Phe84Ser, Arg107Cys, Gly192Ser, Arg207His), and RibE (Pro55His), and the production of truncated NfsA (Gln67 and Gln147) and NfsB (Glu54), were associated with nitrofurantoin resistance in 15/29 isolates; however, the presence of *oqxAB* plasmid genes was not detected in any isolate. Resistance to fosfomycin was associated with the absence of transporter UhpT expression and/or the presence of antibiotic-modifying enzymes encoded by *fosA3* plasmid-mediated gene. Resistance to nitrofurantoin was associated with modifications of NfsA, NfsB, and RibE proteins. The emergence and spread of these resistance mechanisms, including transferable resistance, could compromise the future usefulness of fosfomycin and nitrofurantoin against UTIs. Furthermore, knowledge of the genetic mechanisms underlying resistance may lead to rapid DNA-based testing for resistance.

Keywords: *Escherichia coli*; fosfomycin; nitrofurantoin; antimicrobial resistance

1. Introduction

The high incidence of urinary tract infections (UTIs) and their usually mild character means that most patients receive empirical antibiotic treatment. However, clinicians are now faced with major challenges due to multiple factors, including population aging, the presence of allergies or adverse reactions to antibiotics, an increased number of immunodepressed patients, and, especially, high rates of multi-resistant pathogens, which can cause therapeutic failure. A good alternative option may be to return to antibiotics such as fosfomycin and nitrofurantoin [1].

The characteristics of fosfomycin and nitrofurantoin make them especially useful for UTI treatment, including their rapid oral absorption, high urine concentration, and bactericide activity against a wide range of Gram-negative and Gram-positive bacteria. Both are first-line treatments for non-complicated UTIs of community origin [2]. They have also been reported to preserve their activity against multi-resistant microorganisms, especially uropathogenic enterobacteria such as *Escherichia coli* and extended-spectrum beta-lactamase-producing isolates [3], although these are usually less susceptible to fosfomycin and nitrofurantoin than are non-producers [4,5].

Currently, resistance to fosfomycin or nitrofurantoin is not common in our setting, and >85% of bacteria isolated in UTIs are susceptible to these antibiotics. Nonetheless, the gradual decrease in susceptibility to these drugs may lead to their contraindication as an empirical treatment in the future [1]. Any expansion of their clinical utilization would, therefore, require the adoption of epidemiological surveillance measures to detect the possible emergence of resistance [6]. With this background, the objective of this study was to explore possible molecular mechanisms underlying the resistance of clinical isolates of UTI-producing *E. coli* to fosfomycin and nitrofurantoin in our setting.

2. Methods

2.1. Bacterial Isolates

The study included clinical isolates of fosfomycin- and/or nitrofurantoin-resistant *E. coli* with significant bacterial count selected from among urine cultures conducted for UTI analysis in the Microbiology Laboratory of Virgen de las Nieves University Hospital (Granada, Spain). They were identified by matrix-assisted laser desorption ionization-time of flight (MALDI-TOF) mass spectrometry (Bruker Daltonics, Bremen, Germany) as part of the routine microbiology laboratory workup [1]. A disk diffusion procedure (Kirby–Bauer) was also conducted on agar Mueller-Hinton plates, using McFarland 0.5 bacterial inoculum and disks with 200 μg fosfomycin supplemented with 50 μg glucose-6-phosphate (G6P) or disks with 300 μg nitrofurantoin. Each isolate was defined as "susceptible," "intermediate," or "resistant" according to the Clinical and Laboratory Standards Institute (CLSI) breakpoints [7]. For fosfomycin, an inhibition zone diameter ≥16 mm was considered susceptible, 13–15 mm intermediate, and ≤12 mm resistant; for nitrofurantoin, an inhibition zone diameter of ≥17 mm was considered susceptible, 15–16 mm intermediate and ≤14 mm resistant. *E. coli* ATCC 25922 (American Type Culture Collection, Manassas, VA, USA) was used as the control strain in the susceptibility assays.

Furthermore, in order to identify *E. coli* isolates producing fosfomycin resistance-mediating glutathione S-transferases, 20 μL sodium phosphonoformate (PPF) (Sigma-Aldrich, Madrid, Spain) was added at a concentration of 50 mg/mL on a second disk with 200 μg fosfomycin supplemented with 50 μg G6P, located at a distance of 30–35 mm from the first. After overnight incubation at 36 ± 1 °C, diameters of the growth inhibition zone were compared between the first disk (with PPF) and the second (without PPF). A difference of ≥5 mm between diameters was considered to confirm the phenotypic presence of the enzyme [8]. All assays were performed in duplicate.

2.2. Carbohydrate Utilization Test

All isolates were studied to determine the capacity for bacterial growth in the presence of a single source of carbon, *sn*-glycerol 3-phosphate (G3P), or G6P, using a previously described procedure [9]. After incubating bacteria in Mueller–Hinton broth for 24 h at 36 ± 1 °C in agitation, they were collected by centrifugation and resuspended in normal saline solution (0.9% NaCl). After five washes (to remove any remains that may act as carbon source), bacterial suspensions were then streaked onto M9 minimal medium agar supplemented with glucose (as a positive growth control), with G3P or G6P at 0.2% (w/v). Bacterial growth was determined after incubation at 36 ± 1 °C for 48 h. All assays were performed in duplicate. The absence of bacterial growth or poor growth with no colony formation in media supplemented with G3P or G6P was considered to indicate GlpT or UhpT function deficiency, respectively [9].

2.3. PCR Amplification

Polymerase chain reaction (PCR) was used to amplify *murA*, *glpT*, *uhpT*, *uhpA*, *ptsI*, *cyaA*, *nfsA*, *nfsB*, and *ribE* genes of *E. coli*, using previously reported procedures [5,9], separately amplifying two fragments (*cyaA1* and *cyaA2*) for the *cyaA* gene. The primer pairs used are listed in Table 1. DNA was obtained from clinical isolates and *E. coli* ATCC 25922 (used as control strain) using the PureLink Microbiome DNA Purification Kit (Thermo Fisher Scientific, Waltham, MA, USA). One microliter of the purified DNA was added to a master mix containing PCR buffer (1×), $MgCl_2$ (2 mM), dNTPs (0.4 mM), primers (0.4 μM), and Taq polymerase (1.25 U).

Table 1. Primers used for amplification and sequencing of the *Escherichia coli* genes involved in fosfomycin or nitrofurantoin resistance.

Gene	Forward Primer	Reverse Primer	Amplicon Size (bp)	Reference
murA	5′-AAACAGCAGACGGTCTATGG-3′	5′-CCATGAGTTTATCGACAGAACG-3′	1542	[9]
glpT	5′-GCGAGTCGCGAGTTTTCATTG-3′	5′-GGCAAATATCCACTGGCACC-3′	1785	[9]
uhpT	5′-TTTTTGAACGCCCAGACACC-3′	5′-AGTCAGGGGCTATTTGATGG-3′	1667	[9]
uhpA	5′-GATCGCGGTGTTTTTTCAG-3′	5′-GATACTCCACAGGCAAAACC-3′	771	[9]
ptsI	5′-GAAAGCGGTTGAACATCTGG-3′	5′-TCCTTCTTGTCGTCGGAAAC-3′	1908	[9]
cyaA1	5′-AACCAGGCGCGAAAAGTGG-3′	5′-TGATGGCTGATGATCGACTC-3′	1559	[9]
cyaA2	5′-AAAGCTCAGCCGTGAACGC-3′	5′-ACCTTCTGGGATTTGCTGG-3′	1648	This study
nfsA	5′-ATTTTCTCGGCCAGAAGTGC-3′	5′-AGAATTTCAACCAGGTGACC-3′	1036	[5]
nfsB	5′-CTTCGCGATCTGATCAACG-3′	5′-CAACAGCAGCCTATGATGAC-3′	923	[5]
ribE	5′-AAGGGAAGCAGCGCACGAA-3′	5′-GGACAACTGCCAGGAGTAGA-3′	634	This study
fosA3	5′-GCGTCAAGCCTGGCATTT-3′	5′-GCCGTCAGGGTCGAGAAA-3′	282	[10]
fosA4	5′-CTGGCGTTTTATCAGCGGTT-3′	5′-CTTCGCTGCGGTTGTCTTT-3′	230	[11]
fosA5	5′-TATTAGCGAAGCCGATTTTGCT-3′	5′-CCCCTTATACGGCTGCTCG-3′	177	[11]
fosA6	5′-GCTACGGTTCAGCTTCCAGA-3′	5′-CGAGCGTGGCGTTTTATCAG-3′	242	This study
fosC2	5′-CGTTCCGTGGAGTTCTATAC-3′	5′-CTTGATAGGGTTTAGACTTC-3′	334	[8]
oqxA	5′-GACAGCGTCGCACAGAATG-3′	5′-GGAGACGAGGTTGGTATGGA-3′	339	[12]
oqxB	5′-CGAAGAAAGACCTCCCTACCC-3′	5′-CGCCGCCAATGAGATACA-3′	240	[12]

PCR amplification of *murA*, *glpT*, *uhpT*, *cyaA1*, *cyaA2*, *nfsA*, *nfsB*, and *ribE* genes was performed as follows: 2 min of denaturation at 94 °C, followed by 35 cycles of denaturation at 94 °C for 30 s, annealing at 55 °C for 30 s and extension at 72 °C for 2 min (1 min for *nfsA, nfsB,* and *ribE*), with a final period of extension at 72 °C for 5 min. The same conditions were used for the amplification of *uhpA* and *ptsI* genes except that the annealing temperature was 57 °C.

For the isolates in which *uhpT* or *uhpA* genes could not be detected by the aforementioned procedure, a new PCR was designed using an outer primer pair (*uhpT*-F_2: 5′-GATGTTAATCGGTATGGCGGC-3′; *uhpT*-R_2: 5′-CAGTCGCTGGCGGAACAAAT-3′; *uhpA*-F_2: 5′-CGTAATTCTGGAGCTCACCG-3′; *uhpA*-R_2: 5′-CGCCTGCGTTAGCCAGTAA-3′). Besides re-amplification with outer primers, the amplification specificity was increased by using the forward outer primer with the reverse inner primer and the forward inner primer with the reverse outer primer.

Plasmid-borne fosfomycin resistance genes *fosA3, fosA4, fosA5, fosA6,* and *fosC2* and nitrofurantoin resistance genes *oqxA* and *oqxB* were screened by PCR amplification with the primers listed in Table 1, following previously reported procedures [8,10–12].

All PCR products were separated in 0.8% agarose gel and visualized under UV light after staining with ethidium bromide.

2.4. Nucleotide Sequencing

Pools of 8 and 10 amplicons were established, and each amplicon was equimolarly normalized in the pool. Each pool was tagmented (tagged and fragmented) using the Nextera XT transposome, which fragments the DNA and then tags it with adapter sequences in a single step. The tagmented DNA was amplified with 12 PCR cycles. The PCR step also adds index 1 (i7), index 2 (i5), and full adapter sequences required for cluster formation. Each DNA sample was purified using 30 µL of AMPure XP beads and was resuspended in 50 µL of water. Then, it was quantified using a Qubit® 3.0 Fluorometer (Thermo Fisher Scientific) and normalized. Pools were sequenced in a high cartridge of 300 cycles using a NextSeq platform. Data were mapped against the reference sequence of *E. coli* str. K-12 substr. MG1655 (NCBI Reference Sequence: NC_000913.3). A BAM file was generated, followed by a variant calling, and the most representative variants were recorded. The online Protein Variation Effect Analyzer (PROVEAN) platform (http://provean.jcvi.org/index.php) was used to predict the impact of identified amino acid substitutions on the biological function of each protein [13]. PROVEAN is able to provide predictions for any type of protein sequence variation, including single or multiple amino acid substitutions, insertions, or deletions. The platform introduces a delta alignment score based on the reference and variant versions of a protein query sequence with respect to sequence homologs collected from the NCBI protein database through BLAST. If the PROVEAN score (P-score) was equal to or below a predefined cutoff of −2.5, the protein variant was predicted to have a "deleterious" effect (potential loss of protein structure or function). If the P-score was above the threshold, the variant was predicted to have a "neutral" effect (no alteration in the structure or function of the protein).

3. Results

A total of 29 fosfomycin- and/or nitrofurantoin-resistant clinical isolates were identified: 8 were resistant to both fosfomycin and nitrofurantoin, 14 were resistant to fosfomycin and susceptible to nitrofurantoin, and 7 were susceptible to fosfomycin and resistant to nitrofurantoin. Figure 1 depicts PCR amplification of chromosomal genes *murA*, *glpT*, *uhpT*, *uhpA*, *ptsI*, *cyaA* (cyaA1 and cyaA2), *nfsA*, *nfsB*, and *ribE* in *E. coli* ATCC 25922.

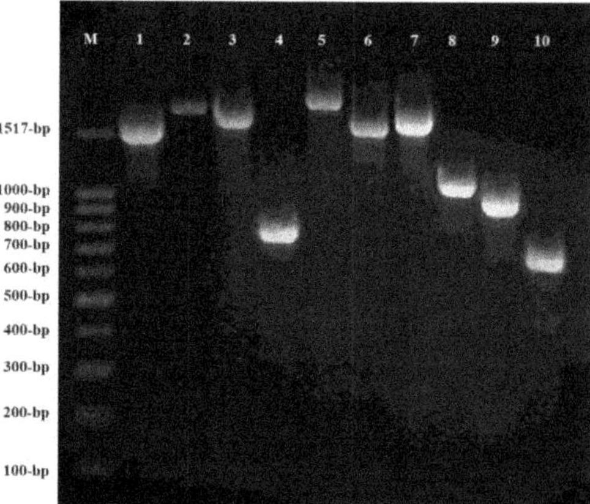

Figure 1. Electrophoresis results of polymerase chain reaction (PCR) products in *Escherichia coli* ATCC 25922 on 0.8% agarose gel. M: Molecular weight. Lines 1 to 10: PCR products of *murA* (1542 bp), *glpT* (1785 bp), *uhpT* (1667 bp), *uhpA* (771 bp), *ptsI* (1908 bp), *cyaA1* (1559 bp), *cyaA2* (1648 bp), *nfsA* (1036 bp), *nfsB* (923 bp), and *ribE* (634 bp), respectively.

3.1. Fosfomycin Resistance

Table 2 summarizes the characteristics of the 22 fosfomycin-resistant (inhibition zone diameter ≤12 mm around the disk with 200 µg fosfomycin supplemented with 50 µg glucose-6-phosphate) and 7 fosfomycin-susceptible (inhibition zone diameter ≥16 mm around the disk with 200 µg fosfomycin supplemented with 50 µg glucose-6-phosphate) clinical isolates of *E. coli* according to the CLSI procedure, displaying the diameter of the bacterial growth inhibition in the presence of PPF, the bacterial growth capacity in the presence of G3P or G6P as sole carbon source, and the amino acid substitutions in MurA, GlpT, UhpT, UhpA, PtsI, and CyaA proteins detected in each isolate.

Three of the twenty-two fosfomycin-resistant isolates (strains 789, 809, and 853) showed a single substitution in the amino acid sequence of MurA (Leu370Ile), categorized as neutral (no alteration in structure or function of the protein) in the PROVEAN analysis (P-score: −1.995).

Twelve amino acid substitutions were detected in GlpT: Glu448Lys (P-score: 0.486, categorized as neutral), in all isolates, both resistant and susceptible; Ala16Thr (P-score: −0.713, categorized as neutral), and Phe133Cys (P-score: −5.549), Gly135Trp (P-score: −7.756), Ala197Val (P-score: −3.472), and Leu373Arg (P-score: −5.328), all categorized as deleterious (potential loss of protein structure or function), in susceptible isolates alone; and Met52Leu (P-score: −1.261), Leu297Phe (P-score: −2.375), Glu443Gln (P-score: 0.014), and Gln444Glu (P-score: −0.106), all four categorized as neutral, and Gly84Asp (P-score: −6.056) and Pro212Leu (P-score: −9.698), both categorized as deleterious, in resistant isolates alone.

No amplification product of the *uhpT* gene was obtained from strains 11 and 26 using the two primer pairs reported above (loss of entire gene). Amino acid substitution in UhpT (Glu350Gln), categorized as neutral (P-score: −0.016), was observed in 16 of the 22 fosfomycin-resistant isolates but in none of the susceptible isolates.

The *uhpA* gene was detected in all fosfomycin-susceptible isolates, and two of these showed substitution of Arg46Cys in the protein sequence, categorized as neutral (P-score: −0.268). By contrast, this gene was detected in only 1 of the 22 fosfomycin-resistant isolates: strain 26 (wild-type).

Three amino acid substitutions were detected in PtsI: Arg367Lys (P-score: 0.842), in all isolates, both resistant and susceptible; Ala306Thr (P-score: 0.030), in two susceptible isolates alone; and Val25Ile (P-score: −0.606), in 10 resistant isolates alone; and all three substitutions were categorized as neutral in the PROVEAN analysis.

Finally, 11 amino acid substitutions were detected in CyaA: Asn142Ser (P-score: 0.016, categorized as neutral), in all isolates; Gly222Ser (P-score: −3.447, categorized as deleterious), in one of the seven susceptible isolates; Ala349Glu (P-score: 2.261), Glu362Asp (P-score: −0.286), Asp837Glu (P-score: 0.123), and Thr840Ala (P-score: −0.314) all four categorized as neutral; and Ser356Leu (P-score: −2.624) and Gly359Glu (P-score: −3.077) both categorized as deleterious, in some susceptible and resistant isolates; and Ala363Ser (P-score: 0.900), Ala363Gly (P-score: −0.251), and Ser352Thr (P-score: −0.645) all three categorized as neutral, in fosfomycin-resistant isolates alone.

The effect of these amino acid substitutions on transporters GlpT and UhpT in resistant and susceptible isolates was evaluated by testing bacterial growth on M9 minimal medium agar supplemented with G3P or G6P (substrates for GlpT or UhpT, respectively). As reported in Table 2, all isolates grew on M9 medium with G3P, indicating no significant loss of GlpT function with any substitution detected in the amino acid sequence of this transporter. However, fosfomycin-resistant isolates did not grow or showed poor growth on the medium containing G6P, because of the loss of function of UhpT due to the complete deletion of *uhpT* (strains 11 and 26) and/or *uhpA* genes (strains 11, 17, 66, 381, 387, 462, 632, 752, 757, 776, 789, 792, 795, 799, 809, 853, 854, 860, 871, 883 and 891).

Table 2. Susceptibility to fosfomycin according to the Clinical and Laboratory Standards Institute (CLSI) procedure and supplemented with sodium phosphonoformate (PPF); bacterial growth on M9 minimal medium agar supplemented with *sn*-glycerol 3-phosphate (G3P) or glucose-6-phosphate (G6P); and amino acid substitutions in MurA, GlpT, UhpT, UhpA, PtsI, and CyaA proteins in 29 clinical isolates of *Escherichia coli*.

Strain	Fosfomycin Disk [1]	Clinical Category [2]	Fosfomycin Disk Plus PPF [3]	G3P [4]	G6P [5]	Amino Acid Substitutions in					
						MurA	GlpT	UhpT	UhpA	PtsI	CyaA
11	6	R	6	+	−	None	Leu297Phe Glu443Gln Gln444Glu Glu448Lys	Not detected	Not detected	Arg367Lys	Asn142Ser Ala349Glu Ser352Thr Ser356Leu Gly359Glu Glu362Asp
17	6	R	6	+	−	None	Glu448Lys	Glu350Gln	Not detected	Val25Ile Arg367Lys	Asn142Ser Asp837Glu Thr840Ala
26	12	R	12	+	− [a]	None	Gly84Asp Glu448Lys	Not detected	None	Arg367Lys	Asn142Ser Ala349Glu Ser356Leu Gly359Glu Glu362Asp Ala363Ser Asp837Glu Thr840Ala
66	6	R	13	+	−	None	Glu448Lys	Glu350Gln	Not detected	Val25Ile Arg367Lys	Asn142Ser Asp837Glu Thr840Ala
302	29	S	30	+	+	None	Glu448Lys	None	None	Arg367Lys	Asn142Ser Gly222Ser
334	31	S	32	+	+	None	Glu448Lys	None	None	Arg367Lys	Asn142Ser

Table 2. *Cont.*

Strain	Fosfomycin Disk [1]	Clinical Category [2]	Fosfomycin Disk Plus PPF [3]	G3P [4]	G6P [5]	Amino Acid Substitutions in					
						MurA	GlpT	UhpT	UhpA	PtsI	CyaA
381	6	R	6	+	−	None	Glu448Lys	Glu350Gln	Not detected	Val25Ile Arg367Lys	Asn142Ser Asp837Glu Thr840Ala
387	6	R	6	+	−	None	Glu448Lys	Glu350Gln	Not detected	Val25Ile Arg367Lys	Asn142Ser Asp837Glu Thr840Ala
462	12	R	13	+	−	None	Glu448Lys	Glu350Gln	Not detected	Val25Ile Arg367Lys	Asn142Ser Asp837Glu Thr840Ala
632	6	R	6	+	−	None	Leu297Phe Glu443Gln Gln444Glu Glu448Lys	Glu350Gln	Not detected	Arg367Lys	Asn142Ser Ala349Glu Ser352Thr Ser356Leu Gly359Glu Glu362Asp Ala363Gly
751	30	S	32	+	+	None	Ala16Thr Glu448Lys	None	Arg46Cys	Ala306Thr Arg367Lys	Asn142Ser Ala349Glu Ser356Leu Gly359Glu Glu362Asp Asp837Glu Thr840Ala
752	11	R	11	+	−	None	Pro212Leu Glu448Lys	Glu350Gln	Not detected	Val25Ile Arg367Lys	Asn142Ser Asp837Glu Thr840Ala

Table 2. *Cont.*

Strain	Fosfomycin Disk [1]	Clinical Category [2]	Fosfomycin Disk Plus PPF [3]	G3P [4]	G6P [5]	Amino Acid Substitutions in					
						MurA	GlpT	UhpT	UhpA	PtsI	CyaA
757	6	R	6	+	− [a]	None	Leu297Phe Glu443Gln Gln444Glu Glu448Lys	None	Not detected	Arg367Lys	Asn142Ser Ala349Glu Ser352Thr Ser356Leu Gly359Glu Glu362Asp
776	12	R	12	+	− [a]	None	Glu448Lys	Glu350Gln	Not detected	Val25Ile Arg367Lys	Asn142Ser Asp837Glu Thr840Ala
789	6	R	6	+	−	Leu370Ile	Leu297Phe Glu443Gln Gln444Glu Glu448Lys	Glu350Gln	Not detected	Arg367Lys	Asn142Ser Ala349Glu Ser352Thr Ser356Leu Gly359Glu Glu362Asp
792	6	R	6	+	−	None	Glu448Lys	Glu350Gln	Not detected	Val25Ile Arg367Lys	Asn142Ser Asp837Glu Thr840Ala
795	6	R	6	+	−	None	Leu297Phe Glu443Gln Gln444Glu Glu448Lys	Glu350Gln	Not detected	Arg367Lys	Asn142Ser Ala349Glu Ser352Thr Ser356Leu Gly359Glu Glu362Asp

Table 2. *Cont.*

Strain	Fosfomycin Disk [1]	Clinical Category [2]	Fosfomycin Disk Plus PPF [3]	G3P [4]	G6P [5]	Amino Acid Substitutions in					
						MurA	GlpT	UhpT	UhpA	PtsI	CyaA
797	20	S	21	+	+	None	Ala16Thr Leu373Arg Glu448Lys	None	Arg46Cys	Ala306Thr Arg367Lys	Asn142Ser Ala349Glu Ser356Leu Gly359Glu Glu362Asp Asp837Glu Thr840Ala
799	6	R	7	+	−	None	Leu297Phe Glu443Gln Gln444Glu Glu448Lys	None	Not detected	Arg367Lys	Asn142Ser Ala349Glu Ser352Thr Ser356Leu Gly359Glu Glu362Asp
802	30	S	30	+	+	None	Phe133Cys Gly135Trp Ala197Val Glu448Lys	None	None	Arg367Lys	Asn142Ser
809	6	R	6	+	−	Leu370Ile	Leu297Phe Glu443Gln Gln444Glu Glu448Lys	Glu350Gln	Not detected	Arg367Lys	Asn142Ser Ala349Glu Ser352Thr Ser356Leu Gly359Glu Glu362Asp
853	6	R	8	+	−	Leu370Ile	Leu297Phe Glu443Gln Gln444Glu Glu448Lys	Glu350Gln	Not detected	Arg367Lys	Asn142Ser Ala349Glu Ser352Thr Ser356Leu Gly359Glu Glu362Asp

Table 2. *Cont.*

Strain	Fosfomycin Disk [1]	Clinical Category [2]	Fosfomycin Disk Plus PPF [3]	G3P [4]	G6P [5]	Amino Acid Substitutions in					
						MurA	GlpT	UhpT	UhpA	PtsI	CyaA
854	6	R	6	+	−	None	Glu448Lys	None	Not detected	Arg367Lys	Asn142Ser
860	6	R	7	+	−	None	Met52Leu Leu297Phe Glu443Gln Gln444Glu Glu448Lys	None	Not detected	Arg367Lys	Asn142Ser Ala349Glu Ser352Thr Ser356Leu Gly359Glu Glu362Asp
871	6	R	14	+	− [a]	None	Leu297Phe Glu443Gln Gln444Glu Glu448Lys	Glu350Gln	Not detected	Arg367Lys	Asn142Ser Ala349Glu Ser352Thr Ser356Leu Gly359Glu Glu362Asp Ala363Gly
872	35	S	35	+	+	None	Glu448Lys	None	None	Arg367Lys	Asn142Ser Asp837Glu Thr840Ala
883	11	R	12	+	−	None	Glu448Lys	Glu350Gln	Not detected	Val25Ile Arg367Lys	Asn142Ser Asp837Glu Thr840Ala
891	11	R	11	+	−	None	Glu448Lys	Glu350Gln	Not detected	Val25Ile Arg367Lys	Asn142Ser Asp837Glu Thr840Ala
892	21	S	21	+	+	None	Glu448Lys	None	None	Arg367Lys	Asn142Ser

[1] Diameter (in mm) of the bacterial growth inhibition halo around the disk with 200 μg fosfomycin supplemented with 50 μg glucose-6-phosphate on Mueller-Hinton agar. [2] Clinical categories of each isolate against fosfomycin according to CLSI breakpoints (S: susceptible; R: resistant). [3] Diameter (in mm) of the bacterial growth inhibition halo around the disk with 200 μg fosfomycin supplemented with 50 μg glucose-6-phosphate and 20 μL sodium phosphonoformate (PPF) in order to identify *E. coli* isolates producing fosfomycin resistance-mediating glutathione S-transferases (between-diameter difference of ≥5 mm considered to confirm the phenotypic presence of the enzyme). [4] Bacterial growth on M9 minimal medium agar supplemented with 0.2% *sn*-glycerol 3-phosphate (all isolates showed growth). [5] Bacterial growth on M9 minimal medium agar supplemented with 0.2% glucose-6-phosphate (+: bacterial growth; −: absence of bacterial growth). [a] Only poor growth was observed after 48 h of incubation. Not detected: gene not detected by PCR after the different combinations of two pairs of primers (loss of the entire gene). None: no amino acid substitutions found.

Furthermore, fosfomycin resistance-mediating glutathione S-transferase was observed in two of the fosfomycin-resistant isolates (strains 66 and 871) due to a significantly increased bacterial growth inhibition halo (≥5 mm) in the presence of PPF (Table 2). This phenotypic finding was confirmed by PCR amplification of the *fosA3* gene in both isolates (Figure 2). No *fosA4*, *fosA5*, *fosA6*, and *fosC2* plasmid genes were detected in any isolate.

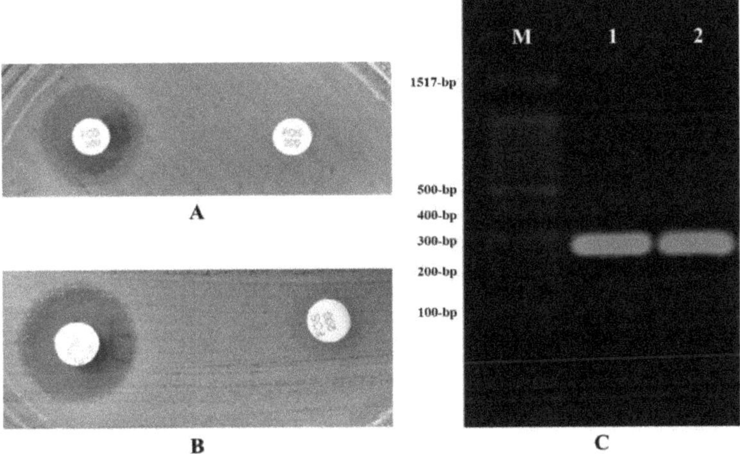

Figure 2. Detection of fosfomycin resistance-mediating glutathione S-transferase (sodium phosphonoformate test) and *fosA3* gene (electrophoresis) in strains 66 and 871. (**A,B**) Phenotypic detection of fosfomycin resistance-mediating glutathione S-transferase in strains 66 and 871, respectively, showing an increase of ≥5 mm in growth inhibition halo around the disk of 200 μg fosfomycin supplemented with 50 μg G6P plus sodium phosphonoformate in comparison to the disk containing 200 μg fosfomycin supplemented with 50 μg G6P alone. All assays were performed in duplicate in all isolates, obtaining the same between-assay results; (**C**) Electrophoresis results for the PCR products of *fosA3* gene (282 bp) on 0.8% agarose gel in strains 66 (line 1) and 871 (line 2). M: molecular weight.

3.2. Nitrofurantoin Resistance

Table 3 summarizes the characteristics of the 15 nitrofurantoin-resistant or intermediate isolates (inhibition zone diameter ≤14 mm or 13–15 mm, respectively, around the disk with 300 μg nitrofurantoin) and the 14 nitrofurantoin-susceptible (inhibition zone diameter ≥17 mm around the disk with 300 μg nitrofurantoin) clinical isolates of *E. coli* according to the CLSI procedure. Amino acid substitutions in NfsA, NfsB, or RibE proteins were detected in all isolates.

Among the 15 nitrofurantoin-resistant or nitrofurantoin-intermediate and 14 nitrofurantoin-susceptible isolates, 14 amino acid substitutions were detected in the NfsA protein: Glu58Asp (P-score: −1.866), Ile117Thr (P-score: −0.634), Lys141Glu (P-score: 1.207), Gln147Arg (P-score: −1.170), and Gly187Asp (P-score: 1.554) all of these categorized as neutral in the PROVEAN analysis (no alteration in structure or function of the protein), in susceptible isolates; Asp19Asn (P-score: −2.091) and Ser180Asn (P-score: 0.071) both categorized as neutral, and His11Tyr (P-score: −5.746), Ser33Arg (P-score: −2.526), Gln67Leu (P-score: −5.860), Cys80Arg (P-score: −11.148), Gly126Arg (P-score: −7.544), Gly154Glu (P-score: −7.608), and Arg203Cys (P-score: −7.090) all of these categorized as deleterious (potential loss of protein structure or function), only in isolates with some level of resistance (resistant or intermediate). In addition, a single nucleotide mutation in the *nfsA* gene was detected in strains 757 and 802, leading to truncation of the NfsA sequence in Gln67 (Gln67stop; CAA to TAA) and Gln147 (Gln147stop; CAG to TAG), respectively. These mutations produced 66 and 146 amino acid long proteins, respectively, instead of a wild-type protein with 240 amino acids.

Table 3. Susceptibility to nitrofurantoin according to the Clinical and Laboratory Standards Institute (CLSI) procedure and amino acid substitutions in NfsA, NfsB, or RibE proteins of 29 clinical isolates of *Escherichia coli*.

Strain	Nitrofurantoin Disk [1]	Clinical Category [2]	Amino Acid Substitutions in		
			NfsA	**NfsB**	**RibE**
11	11	R	Ile117Thr Gly126Arg Lys141Glu Gln147Arg Gly187Asp	Truncated at Glu54	None
17	27	S	Ile117Thr Lys141Glu Gly187Asp	Gly66Asp Val93Ala Ala174Glu	None
26	24	S	Ile117Thr Lys141Glu Gly187Asp	Gly66Asp Val93Ala Ala174Glu	None
66	14	R	Ile117Thr Lys141Glu Gly187Asp	Gly66Asp Val93Ala Ala174Glu Arg207His	None
302	14	R	None	Gly66Asp Met75Ile Val93Ala Ala174Glu Arg207His	None
334	15	I	Gln67Leu	Gly66Asp Met75Ile Val93Ala Arg107Cys	Pro55His
381	25	S	Ile117Thr Lys141Glu Gly187Asp	Gly66Asp Val93Ala Ala174Glu	None
387	21	S	Ile117Thr Lys141Glu Gly187Asp	Gly66Asp Val93Ala Ala174Glu	None
462	14	R	Cys80Arg Ile117Thr Lys141Glu Gly187Asp	Gly66Asp Val93Ala Ala174Glu Gly192Ser	None
632	20	S	Glu58Asp Ile117Thr Lys141Glu Gln147Arg Gly187Asp	Val93Ala	None
751	14	R	Ile117Thr Lys141Glu Gly187Asp Arg203Cys	Gly66Asp Val93Ala Ala174Glu	Val51Ile
752	22	S	Ile117Thr Lys141Glu Gly187Asp	Gly66Asp Val93Ala Ala174Glu	None

Table 3. *Cont.*

Strain	Nitrofurantoin Disk [1]	Clinical Category [2]	Amino Acid Substitutions in		
			NfsA	NfsB	RibE
757	16	I	Glu58Asp Truncated at Gln67	Val93Ala Lys122Arg	None
776	19	S	Ile117Thr Lys141Glu Gly187Asp	Gly66Asp Val93Ala Ala174Glu	None
789	24	S	Glu58Asp Ile117Thr Lys141Glu Gln147Arg Gly187Asp	Val93Ala	None
792	25	S	Ile117Thr Lys141Glu Gly187Asp	Gly66Asp Val93Ala Ala174Glu	None
795	16	I	His11Tyr Glu58Asp Ile117Thr Lys141Glu Gln147Arg Gly187Asp	Val93Ala	None
797	12	R	Ile117Thr Lys141Glu Gly154Glu Gly187Asp	Leu22Ile Gly66Asp Val93Ala Ala174Glu	Val51Ile
799	13	R	Ile117Thr Lys141Glu Gln147Arg Gly187Asp	Val93Ala	None
802	12	R	Truncated at Gln147	Met75Ile Val93Ala	None
809	28	S	Glu58Asp Ile117Thr Lys141Glu Gln147Arg Gly187Asp	Val93Ala	None
853	20	S	Glu58Asp Ile117Thr Lys141Glu Gln147Arg Gly187Asp	Val93Ala	None
854	16	I	None	Gly66Asp Met75Ile Val93Ala	None
860	20	S	Ile117Thr Lys141Glu Gln147Arg Gly187Asp	Val93Ala	None

Table 3. *Cont.*

Strain	Nitrofurantoin Disk [1]	Clinical Category [2]	Amino Acid Substitutions in		
			NfsA	NfsB	RibE
871	24	S	Glu58Asp Ile117Thr Lys141Glu Gln147Arg Gly187Asp	Val93Ala	None
872	15	I	Asp19Asn Ser33Arg Ile117Thr Lys141Glu Gly187Asp	Gln44His Gly66Asp Val93Ala Ala174Glu	None
883	13	R	Ile117Thr Lys141Glu Gly187Asp	Gly66Asp Phe84Ser Val93Ala Ala174Glu	None
891	23	S	Ile117Thr Lys141Glu Gly187Asp	Gly66Asp Val93Ala Ala174Glu	None
892	10	R	None	Gly66Asp Met75Ile Val93Ala	None

[1] Diameter (in mm) of bacterial growth inhibition halo around the disk with 300 µg nitrofurantoin. [2] Clinical categories of each isolate against nitrofurantoin according to CLSI breakpoints (S: susceptible; I: intermediate; R: resistant). None: no amino acid substitutions were found.

Eleven amino acid substitutions were detected in NfsB: Gly66Asp (P-score: −1.775), Val93Ala (P-score: 2.155), and Ala174Glu (P-score: 1.621) all of these categorized as neutral, in susceptible isolates; and Leu22Ile (P-score: 0.334), Met75Ile (P-score: 2.094), and Lys122Arg (P-score: −0.179) all of these categorized as neutral, and Gln44His (P-score: −4.800), Phe84Ser (P-score: −5.862), Arg107Cys (P-score: −7.863), Gly192Ser (P-score: −5.961), and Arg207His (P-score: −4.966) all of these categorized as deleterious, in isolates with some level of resistance (resistant or intermediate). In addition, a single nucleotide mutation in the *nfsB* gene (GAA to TAA) was detected in strain 11, leading to a truncation of the NfsB sequence in Glu54.

Two amino acid substitutions in RibE were detected in three isolates with some level of resistance: Val51Ile (P-score: −0.363, categorized as neutral) and Pro55His (P-score: −8.840, categorized as deleterious). Finally, no *oqxAB* plasmid gene was detected in any isolate.

4. Discussion

4.1. Mechanisms of Resistance to Fosfomycin in E. coli

Mechanisms of resistance to fosfomycin described in various bacteria include the modification or overexpression of target molecule MurA, a reduced permeability, and irreversible antibiotic modification. The first two mechanisms are chromosomal, whereas the third can be chromosomal or encoded in transferable multi-resistance plasmids [14].

4.1.1. Modification or Overexpression of the Target (MurA)

The main action mechanism of fosfomycin is inhibition of the first step of peptidoglycan synthesis. Its chemical structure is analogous to that of phosphoenolpyruvate (PEP), therefore blocking the active center of enzyme UDP-N-acetylglucosamine enolpyruvyl transferase (MurA), covalently binding to the residue of cysteine Cys115 and preventing the binding of the substrate with the enzyme. In *E. coli*, amino acid substitutions in the active center of MurA, specifically Cys115Asp, are related to fosfomycin resistance [15] but are not common in clinical isolates of this species due to a drastic reduction in bacterial cell viability [16]. Only a few reports have associated amino acid substitutions in the MurA sequence of *E. coli* with resistance, especially Asp369Asn and Leu370Ile [9]. The latter was detected in 3 of the 22 fosfomycin-resistant isolates in the present study (strains 789, 809, and 853) but the protein variant was predicted to have a neutral effect in the PROVEAN analysis, with no alteration in the structure or function of the protein. Although previous crystallization studies found that leucine in position 370 of MurA does not interfere with its binding to fosfomycin, the fact that it is a highly preserved residue suggests an important role in the binding of PEP and therefore fosfomycin to the active site of the enzyme [9].

4.1.2. Permeability Reduction

Fosfomycin can use two transport systems to access the bacterial cytoplasm: glycerol-3-phosphate transporter (GlpT) and hexose phosphate transporter (UhpT). They are induced by the presence of their substrates (G3P and G6P, respectively) and require high levels of cyclic AMP (cAMP), whose synthesis depends on the enzyme adenylate cyclase (CyaA) and is regulated by the phosphoenolpyruvate-protein phosphotransferase (PtsI) system. The expression of GlpT is determined by a repressor gene, *glpR*, given that the interaction of GlpR with G3P increases transcription of the *glpT* gene. The expression of UhpT is in turn controlled by various regulating genes (*uhpA*, *uhpB*, and *uhpC*) [14]. This mechanism of action is unique; it does not confer cross-resistance to other antibiotics and it favors additive action with beta-lactams, aminoglycosides, glycopeptides, and fluoroquinolones, among others [17].

GlpT and UhpT are transporters with an extensive amino acid sequence homology that appear in several bacterial species with a high degree of conservation [14]. Various studies of *E. coli* have identified modifications of these proteins and/or proteins that regulate their expression (UhpA, PtsI, and CyaA) due to gene mutations or complete loss [9,18–21]. However, although the most important fosfomycin-resistance mechanism in this bacterium, modifications in chromosomal genes *uhpT*, *glpT*, *uhpA*, *ptsI*, or *cyaA* are reported to carry a high fitness cost, and clinical isolates with this resistance are known to be outcompeted by isolates susceptible to fosfomycin [22].

In the present study, all clinical isolates of *E. coli* presented substitutions in the amino acid sequence of GlpT. Some of them were detected in fosfomycin-susceptible isolates (Ala16Thr, Phe133Cys, Gly135Trp, Ala197Val, Leu373Arg, and Glu448Lys). Hence, these substitutions do not appear to be related per se to an alteration in GlpT function or resistance to the antibiotic. Other substitutions were solely detected in resistant isolates (Met52Leu, Leu297Phe, Glu443Gln, and Gln444Glu) but were classified as neutral in the PROVEAN analysis and would have no impact on the biological function of this protein. According to the PROVEAN analysis, only Gly84Asp and Pro212Leu substitutions could be significantly related to an alteration of GlpT functionality; however, the two isolates with this substitution (strains 26 and 752) proved able to grow in the presence of G3P. Hence, all isolates grew on M9 medium with G3P, indicating no significant loss of GlpT function with any substitution detected in the amino acid sequence of this transporter.

Among the 22 fosfomycin-resistant *E. coli* isolates, 4 showed no amino acid substitution in UhpT, 16 showed one substitution (Glu350Gln) and 2 were defective in UhpT due to gene loss (strains 11 and 26). We highlight that the *uhpA* gene was detected in strain 26 alone and that none of the 22 fosfomycin-resistant isolates were able to grow in the presence of G6P. In the UhpA sequence, the only substitution was Arg46Cys, which was only detected in two fosfomycin-susceptible isolates; therefore, it does not appear to be related per se to an alteration in the function of these proteins or to antibiotic resistance. According to our findings, all of the resistant isolates analyzed were defective in the UhpT transport system due to *uhpT* and/or *uhpA* deletion and showed no growth or only poor growth in a medium containing G6P as sole carbon source. Therefore, this finding supports the hypothesis that fosfomycin resistance in *E. coli* is most frequently attributable to blockage of the entry pathway of the antibiotic into the bacteria, mainly due to modifications in the UhpT transporter or its regulating proteins [18,19].

All of the *E. coli* clinical isolates in the present study showed substitutions in PtsI and CyaA. Given that some of these were detected in fosfomycin susceptible isolates (Ala306Thr and Arg367Lys in PtsI; Asn142Ser, Gly222Ser, Ala349Glu, Ser356Leu, Gly359Glu, Glu362Asp, Asp837Glu, and Thr840Ala in CyaA), they do not appear to be related per se to an alteration in the function of these proteins or to antibiotic resistance. Some other substitutions in these proteins were only detected in resistant isolates (Val25Ile in PtsI; Ser352Thr, Ala363Ser, and Ala363Gly in CyaA), as in previous studies [9]; nevertheless, their contribution to antibiotic resistance in these isolates cannot be affirmed, given that they were categorized as neutral in the PROVEAN analysis and there was no alteration in the function of GlpT, which was permeable to G3P. Therefore, it cannot be affirmed that amino acid substitutions in PtsI and CyaA contributed to resistance to fosfomycin in the clinical isolates of *E. coli* in the present study.

4.1.3. Enzymatic Modification of Fosfomycin

Two mechanisms may underlie fosfomycin resistance due to the action of modifying enzymes: epoxide ring opening, catalyzed by FosA enzymes (glutathione S-transferase), FosB (L-cysteine thiol transferase), or FosX (hydrolase epoxide); or antibiotic phosphorylation by FomA, FomB, or FosC enzymes [23]. Among these enzymes, FosA3 is the most widely described in *E. coli* plasmids, largely in Eastern Asia countries, although its detection is infrequent in Europe [11]. To our knowledge, this is the first time that the *fosA3* gene has been detected in clinical isolates of *E. coli* in Spain (strains 66 and 871). Both isolates were also defective in the UhpT transport system due to *uhpA* deletion; however, the importance of this finding is that this plasmid-mediated gene may accelerate the dissemination of fosfomycin resistance in the near future.

4.2. Mechanisms of Resistance to Nitrofurantoin in E. coli

Nitrofurantoin is a prodrug of the nitrofuran family and exerts its antibiotic activity via multiple mechanisms of action, although none have been fully elucidated. It is known to inhibit: (i) protein synthesis, (ii) aerobic metabolism, (iii) nucleic acid synthesis, and (iv) cell wall synthesis. Its active form is generated within the bacterium by the action of nitroreductase enzymes, which reduce the nitro group coupled to the furan heterocyclic ring, giving rise to active intermediate metabolites that inhibit the synthesis of proteins involved in DNA, RNA, and carbohydrate metabolism.

Various studies have attributed resistance to nitrofurantoin in *E. coli* to the loss of intracellular nitroreductase activity via sequential mutations in *nfsA* and *nfsB* genes, which encode oxygen-insensitive nitroreductases, as well as to deletions affecting the active center of *ribE*, although the latter have not yet been reported in clinical isolates. Mutations in genes encoding oxygen-sensitive nitroreductases have not yet been described [5,24]. However, as in the case of fosfomycin, this nitrofurantoin resistance is reported to confer a high biological cost, and clinical isolates with this resistance are known to be outcompeted by susceptible isolates, reducing the likelihood of its detection in clinical isolates [5].

All *E. coli* clinical isolates in the present study showed substitutions in the amino acid sequence of NfsA and/or NfsB. As reported above, some were detected in nitrofurantoin-susceptible isolates (Glu58Asp, Ile117Thr, Lys141Glu, Gln147Arg, and Gly187Asp in NfsA; and Gly66Asp, Val93Ala, and Ala74Glu in NfsB). Although some of these (positions Ile117 and Lys141 in NfsA; Gly66 and Val93 in NfsB) have been associated with resistance in other studies [5,24,25], they were all classified as neutral in the PROVEAN analysis. Hence, none of these substitutions appear to be related per se to an alteration in the function of these proteins or to resistance to the antibiotic.

Other substitutions were detected in resistant isolates alone (His11Tyr, Asp19Asn, Ser33Arg, Gln67Leu, Cys80Arg, Gly126Arg, Gly154Glu, Ser180Asn, Arg203Cys, and truncation at Gln67 and Gln147 in NfsA; Leu22Ile, Gln44His, Met75Ile, Phe84Ser, Arg107Cys, Lys122Arg, Gly192Ser, Arg207His, and truncation at Glu54 in NfsB; Val51Ile and Pro55His in RibE). Some of these (His11, Ser33, Gln67, Gln147, and Arg203 in NfsA; Gln44, Met75, Arg107, Lys122, Gly192, and Arg207 in NfsB) have been associated with nitrofurantoin resistance in other studies [5,24–26]. According to the PROVEAN analysis, His11Tyr, Ser33Arg, Gln67Leu, Cys80Arg, Gly126Arg, Gly154Glu, and Arg203Cys in NfsA; Gln44His, Phe84Ser, Arg107Cys, Gly192Ser, and Arg207His in NfsB; and Pro55His in RibE were predicted to have a deleterious impact on the protein structure. Production of truncated NfsA (Gln67 and Gln147) or NfsB (Glu54) may have resulted in the inability or reduced ability of nitrofurantoin-resistant isolates to reduce the nitrofurantoin and produce active intermediates from the compound. Hence, these amino acid substitutions and/or truncated proteins would be related to nitrofurantoin resistance.

According to various studies, NfsA inactivation followed by NfsB inactivation is the main mechanism for high-level nitrofurantoin resistance in *E. coli* [5,26]. However, several of our nitrofurantoin-resistant clinical isolates did not show any modification in the NfsA sequence compatible with resistance (strains 66, 302, 799, 854, 883, and 892). Among these six isolates, we only detected substitutions in the NfsB sequence compatible with resistance (Arg207His) in the first two. However, we cannot affirm its association with resistance in the other four, although they presented various amino acid substitutions. Therefore, the mechanism that produces nitrofurantoin resistance in these four isolates is yet to be elucidated. Some authors have affirmed that NfsB inactivation in the presence of a wild-type *nfsA* gene cannot be associated with resistance [26]; in contrast, according to our findings, certain NfsB modifications requiring no previous NfsA alterations may be responsible for the functional alteration of bacterial nitroreductases, as also previously reported [24].

More recently, it has been reported that the presence of OqxAB (a plasmid-encoded multidrug efflux pump that confers reduced susceptibility to quinolones, tigecycline, chloramphenicol, trimethoprim, and disinfectants such as quaternary ammonium compounds) would also enhance nitrofurantoin resistance via an active antibiotic expulsion mechanism in *E. coli* isolates with previous nitroreductase modifications, because it has not been possible to relate the presence of OqxAB per se in the bacterium to antibiotic resistance levels [27]. This plasmid has been widely detected in *E. coli* and other enterobacteria, both in human and animal isolates, mainly in China [25,27]; although its presence has also been reported in Europe [28,29], including Spain [30]. However, this plasmid was not detected in any of our series of isolates, indicating that nitrofurantoin resistance must involve mechanisms other than antibiotic extrusion.

Finally, as in the present study, there have been reports of nitrofurantoin-resistant *E. coli* isolates with no amino acid substitutions in NfsA, NfsB, and RibE, or presence of the *oqxAB* plasmid, indicating the need to identify new mechanisms that explain nitrofurantoin resistance in this bacterium [27].

5. Conclusions

These results suggest that the emergence of fosfomycin resistance in clinical isolates of *E. coli* in our setting is largely attributable to the absence of expression of transporter UhpT due to complete deletion of the *uhpT* and/or *uhpA* regulating genes, reducing the permeability of the bacterium to the antibiotic. To our knowledge, we report for the first time the presence in Spain of the plasmid gene *fosA3*, responsible for the enzyme glutathione S-transferase, which inactivates the antibiotic. We consider this finding to be of major epidemiological importance, given its potential dissemination not only in *E. coli* but also other bacteria. Nitrofurantoin resistance can be explained, at least in part, by the presence of specific modifications in NfsA, NfsB, or RibE proteins. The presence of *oqxAB* plasmid genes does not appear to represent an important resistance mechanism among *E. coli* clinical isolates in our setting at the present time. The emergence and spread of these resistance mechanisms, including transferable resistance, could compromise the future usefulness of fosfomycin and nitrofurantoin against UTIs.

Author Contributions: Conceptualization: A.S.-P. and J.G.-F.; Methodology: A.S.-P., I.L.-M., M.A.-C., L.J.M.-G. and J.G.-F.; Formal Analysis: A.S.-P. and I.L.-M.; Investigation: A.S.-P., I.L.-M. and M.A.-C.; Validation: A.S.-P. and J.G.-F.; Writing—Original Draft Preparation: A.S.-P.; Writing—Review & Editing: A.S.-P., I.L.-M. and J.G.-F. All authors have read and agreed to the published version of the manuscript.

Funding: This research did not receive any specific grant from funding agencies in the public, commercial, or not-for-profit sectors.

Acknowledgments: The authors are grateful to Concepción Gimeno Cardona, Head of the Microbiology Laboratory of the University General Hospital of Valencia, Associate Professor of Microbiology at Valencia University and Coordinator of the External Quality Control Program of the Spanish Society of Infectious Diseases and Clinical Microbiology (Spanish abbreviation: SEIMC), for critically reviewing the article and providing input into the final paper.

Conflicts of Interest: The authors declare no conflict of interest.

References

1. Sorlozano, A.; Jimenez-Pacheco, A.; de Dios Luna Del Castillo, J.; Sampedro, A.; Martinez-Brocal, A.; Miranda-Casas, C.; Navarro-Marí, J.M.; Gutiérrez-Fernández, J. Evolution of the resistance to antibiotics of bacteria involved in urinary tract infections: A 7-year surveillance study. *Am. J. Infect. Control* **2014**, *42*, 1033–1038. [CrossRef] [PubMed]

2. Vardakas, K.Z.; Legakis, N.J.; Triarides, N.; Falagas, M.E. Susceptibility of contemporary isolates to fosfomycin: A systematic review of the literature. *Int. J. Antimicrob. Agents.* **2016**, *47*, 269–285. [CrossRef] [PubMed]

3. Falagas, M.E.; Kastoris, A.C.; Kapaskelis, A.M.; Karageorgopoulos, D.E. Fosfomycin for the treatment of multidrug-resistant, including extended-spectrum beta-lactamase producing, *Enterobacteriaceae* infections: A systematic review. *Lancet Infect. Dis.* **2010**, *10*, 43–50. [CrossRef]

4. Sánchez-García, J.M.; Sorlózano-Puerto, A.; Navarro-Marí, J.M.; Gutiérrez Fernández, J. Evolution of the antibiotic-resistance of microorganisms causing urinary tract infections: A 4-year epidemiological surveillance study in a hospital population. *Rev. Clin. Esp.* **2019**, *219*, 116–123. [CrossRef]

5. Sandegren, L.; Lindqvist, A.; Kahlmeter, G.; Andersson, D.I. Nitrofurantoin resistance mechanism and fitness cost in *Escherichia coli*. *J. Antimicrob. Chemother.* **2008**, *62*, 495–503. [CrossRef]

6. Giske, C.G. Contemporary resistance trends and mechanisms for the old antibiotics colistin, temocillin, fosfomycin, mecillinam and nitrofurantoin. *Clin. Microbiol. Infect.* **2015**, *21*, 899–905. [CrossRef] [PubMed]

7. CLSI. *Performance Standards for Antimicrobial Susceptibility Testing*, 27th ed.; CLSI supplement M100; Clinical and Laboratory Standards Institute: Wayne, PA, USA, 2017.

8. Nakamura, G.; Wachino, J.; Sato, N.; Kimura, K.; Yamada, K.; Jin, W.; Shibayama, K.; Yagi, T.; Kawamura, K.; Arakawa, Y. Practical agar-based disk potentiation test for detection of fosfomycin-nonsusceptible *Escherichia coli* clinical isolates producing glutathione S-transferases. *J. Clin. Microbiol.* **2014**, *52*, 3175–3179. [CrossRef]

9. Takahata, S.; Ida, T.; Hiraishi, T.; Sakakibara, S.; Maebashi, K.; Terada, S.; Muratani, T.; Matsumoto, T.; Nakahama, C.; Tomono, K. Molecular mechanisms of fosfomycin resistance in clinical isolates of *Escherichia coli*. *Int. J. Antimicrob. Agents.* **2010**, *35*, 333–337. [CrossRef]

10. Hou, J.; Huang, X.; Deng, Y.; He, L.; Yang, T.; Zeng, Z.; Chen, Z.; Liu, J.H. Dissemination of the fosfomycin resistance gene *fosA3* with CTX-M β-lactamase genes and rmtB carried on IncFII plasmids among *Escherichia coli* isolates from pets in China. *Antimicrob. Agents Chemother.* **2012**, *56*, 2135–2138. [CrossRef]

11. Benzerara, Y.; Gallah, S.; Hommeril, B.; Genel, N.; Decré, D.; Rottman, M.; Arlet, G. Emergence of plasmid-mediated fosfomycin-resistance genes among *Escherichia coli* isolates, France. *Emerg. Infect. Dis.* **2017**, *23*, 1564–1567. [CrossRef]

12. Chen, X.; Zhang, W.; Pan, W.; Yin, J.; Pan, Z.; Gao, S.; Jiao, X. Prevalence of qnr, aac(6=)-Ib-cr, qepA, and oqxAB in *Escherichia coli* isolates from humans, animals, and the environment. *Antimicrob. Agents Chemother.* **2012**, *56*, 3423–3427. [CrossRef] [PubMed]

13. Choi, Y.; Chan, A.P. PROVEAN web server: A tool to predict the functional effect of amino acid substitutions and indels. *Bioinformatics* **2015**, *31*, 2745–2747. [CrossRef] [PubMed]

14. Castañeda-García, A.; Blázquez, J.; Rodríguez-Rojas, A. Molecular mechanisms and clinical impact of acquired and intrinsic fosfomycin resistance. *Antibiotics* **2013**, *2*, 217–236. [CrossRef] [PubMed]

15. Kim, D.H.; Lees, W.J.; Kempsell, K.E.; Lane, W.S.; Duncan, K.; Walsh, C.T. Characterization of a Cys115 to Asp substitution in the *Escherichia coli* cell wall biosynthetic enzyme UDP-GlcNAc enolpyruvyl transferase (MurA) that confers resistance to inactivation by the antibiotic fosfomycin. *Biochemistry* **1996**, *35*, 4923–4928. [CrossRef] [PubMed]

16. Herring, C.D.; Blattner, F.R. Conditional lethal amber mutations in essential *Escherichia coli* genes. *J. Bacteriol.* **2004**, *186*, 2673–2681. [CrossRef]

17. Falagas, M.E.; Vouloumanou, E.K.; Samonis, G.; Vardakas, K.Z. Fosfomycin. *Clin. Microbiol. Rev.* **2016**, *29*, 321–347. [CrossRef] [PubMed]

18. Lucas, A.E.; Ito, R.; Mustapha, M.M.; McElheny, C.L.; Mettus, R.T.; Bowler, S.L.; Kantz, S.F.; Pacey, M.P.; Pasculle, A.W.; Cooper, V.S.; et al. Frequency and mechanisms of spontaneous fosfomycin nonsusceptibility observed upon disk diffusion testing of *Escherichia coli*. *J. Clin. Microbiol.* **2017**, *56*, e01368-17. [CrossRef] [PubMed]

19. Tseng, S.P.; Wang, S.F.; Kuo, C.Y.; Huang, J.W.; Hung, W.C.; Ke, G.M.; Lu, P.L. Characterization of fosfomycin resistant extended-spectrum β-lactamase-producing *Escherichia coli* isolates from human and pig in Taiwan. *PLoS ONE* **2015**, *10*, e0135864. [CrossRef] [PubMed]

20. Wachino, J.; Yamane, K.; Suzuki, S.; Kimura, K.; Arakawa, Y. Prevalence of fosfomycin resistance among CTX-M-producing *Escherichia coli* clinical isolates in Japan and identification of novel plasmid-mediated fosfomycin-modifying enzymes. *Antimicrob. Agents Chemother.* **2010**, *54*, 3061–3064. [CrossRef] [PubMed]

21. Li, Y.; Zheng, B.; Li, Y.; Zhu, S.; Xue, F.; Liu, J. Antimicrobial susceptibility and molecular mechanisms of fosfomycin resistance in clinical *Escherichia coli* isolates in Mainland China. *PLoS ONE* **2015**, *10*, e0135269. [CrossRef]

22. Nilsson, A.I.; Berg, O.G.; Aspevall, O.; Kahlmeter, G.; Andersson, D.I. Biological costs and mechanisms of fosfomycin resistance in *Escherichia coli*. *Antimicrob. Agents Chemother.* **2003**, *47*, 2850–2858. [CrossRef] [PubMed]

23. Karageorgopoulos, D.E.; Wang, R.; Yu, X.H.; Falagas, M.E. Fosfomycin: Evaluation of the published evidence on the emergence of antimicrobial resistance in Gram-negative pathogens. *J. Antimicrob. Chemother.* **2012**, *67*, 255–268. [CrossRef] [PubMed]

24. Vervoort, J.; Xavier, B.B.; Stewardson, A.; Coenen, S.; Godycki-Cwirko, M.; Adriaenssens, N.; Kowalczyk, A.; Lammens, C.; Harbarth, S.; Goossens, H.; et al. An in vitro deletion in ribE encoding lumazine synthase contributes to nitrofurantoin resistance in *Escherichia coli*. *Antimicrob. Agents Chemother.* **2014**, *58*, 7225–7233. [CrossRef] [PubMed]

25. Zhang, X.; Zhang, Y.; Wang, F.; Wang, C.; Chen, L.; Liu, H.; Lu, H.; Wen, H.; Zhou, T. Unravelling mechanisms of nitrofurantoin resistance and epidemiological characteristics among *Escherichia coli* clinical isolates. *Int. J. Antimicrob. Agents.* **2018**, *52*, 226–232. [CrossRef]

26. Whiteway, J.; Koziarz, P.; Veall, J.; Sandhu, N.; Kumar, P.; Hoecher, B.; Lambert, I.B. Oxygen-insensitive nitroreductases: Analysis of the roles of nfsA and nfsB in development of resistance to 5-nitrofuran derivatives in *Escherichia coli*. *J. Bacteriol.* **1998**, *180*, 5529–5539. [CrossRef]

27. Ho, P.L.; Ng, K.Y.; Lo, W.U.; Law, P.Y.; Lai, E.L.; Wang, Y.; Chow, K.H. Plasmid-mediated oqxAB is an important mechanism for nitrofurantoin resistance in *Escherichia coli*. *Antimicrob. Agents Chemother.* **2015**, *60*, 537–543. [CrossRef]

28. Campos, J.; Mourão, J.; Marçal, S.; Machado, J.; Novais, C.; Peixe, L.; Antunes, P. Clinical *Salmonella Typhimurium* ST34 with metal tolerance genes and an IncHI2 plasmid carrying oqxAB-aac(6′)-Ib-cr from Europe. *J. Antimicrob. Chemother.* **2016**, *71*, 843–845. [CrossRef]

29. Dotto, G.; Giacomelli, M.; Grilli, G.; Ferrazzi, V.; Carattoli, A.; Fortini, D.; Piccirillo, A. High prevalence of oqxAB in *Escherichia coli* isolates from domestic and wild lagomorphs in Italy. *Microb. Drug Resist.* **2014**, *20*, 118–123. [CrossRef]

30. Rodríguez-Martínez, J.M.; Díaz de Alba, P.; Briales, A.; Machuca, J.; Lossa, M.; Fernández-Cuenca, F.; Rodríguez Baño, J.; Martínez-Martínez, L.; Pascual, Á. Contribution of OqxAB efflux pumps to quinolone resistance in extended-spectrum-β-lactamase-producing *Klebsiella pneumoniae*. *J. Antimicrob. Chemother.* **2013**, *68*, 68–73. [CrossRef]

 2020, 9, 375

Article

Genetic Characterization of Methicillin-Resistant *Staphylococcus aureus* Isolates from Human Bloodstream Infections: Detection of MLS$_B$ Resistance

Vanessa Silva [1,2,3,4], Sara Hermenegildo [1,2,3,4], Catarina Ferreira [5], Célia M. Manaia [5], Rosa Capita [6,7], Carlos Alonso-Calleja [6,7], Isabel Carvalho [1,2,3,4], José Eduardo Pereira [8], Luis Maltez [8], José L. Capelo [9,10], Gilberto Igrejas [2,3,4] and Patrícia Poeta [1,4,*]

[1] Microbiology and Antibiotic Resistance Team (MicroART), Department of Veterinary Sciences, University of Trás-os-Montes and Alto Douro (UTAD), 5001-801 Vila Real, Portugal; vanessasilva@utad.pt (V.S.); sara-1603@hotmail.com (S.H.); isabelcarvalho93@hotmail.com (I.C.)

[2] Department of Genetics and Biotechnology, University of Trás-os-Montes and Alto Douro (UTAD), 5001-801 Vila Real, Portugal; gigrejas@utad.pt

[3] Functional Genomics and Proteomics Unit, University of Trás-os-Montes and Alto Douro (UTAD), 5001-801 Vila Real, Portugal

[4] Associate Laboratory for Green Chemistry-LAQV, Chemistry Department, Faculty of Science and Technology, University Nova of Lisbon, 2829-516 Lisbon, Portugal

[5] CBQF—Centro de Biotecnologia e Química Fina-Laboratório Associado, Universidade Católica Portuguesa, Escola Superior de Biotecnologia, 4169-005 Porto, Portugal; aferreira@porto.ucp.pt (C.F.); cmanaia@porto.ucp.pt (C.M.M.)

[6] Department of Food Hygiene and Technology, Veterinary Faculty, University of León, 24071 León, Spain; rosa.capita@unileon.es (R.C.); carlos.alonso.calleja@unileon.es (C.A.-C.)

[7] Institute of Food Science and Technology, University of León, 24071 León, Spain

[8] Veterinary and Animal Science Research Center (CECAV), University of Trás-os-Montes and Alto Douro (UTAD), 5001-801 Vila Real, Portugal; jeduardo@utad.pt (J.E.P.); lmaltez@utad.pt (L.M.)

[9] BIOSCOPE Group, LAQV@REQUIMTE, Chemistry Department, Faculty of Science and Technology, NOVA University of Lisbon, 1349-008 Almada, Portugal; jlcm@fct.unl.pt

[10] Proteomass Scientific Society, 2825-466 Costa de Caparica, Portugal

* Correspondence: ppoeta@utad.pt; Tel.: +35-125-935-04-66; Fax: +35-125-935-06-29

Received: 15 June 2020; Accepted: 30 June 2020; Published: 3 July 2020

Abstract: In this study we aimed to characterize antimicrobial resistance in methicillin-resistant *Staphylococcus aureus* (MRSA) isolated from bloodstream infections as well as the associated genetic lineages of the isolates. Sixteen MRSA isolates were recovered from bacteremia samples from inpatients between 2016 and 2019. The antimicrobial susceptibility of these isolates was tested by the Kirby–Bauer disk diffusion method against 14 antimicrobial agents. To determine the macrolide–lincosamide–streptogramin B (MLS$_B$) resistance phenotype of the isolates, erythromycin-resistant isolates were assessed by double-disk diffusion (D-test). The resistance and virulence genes were screened by polymerase chain reaction (PCR). All isolates were characterized by multilocus sequence typing (MLST), *spa* typing, staphylococcal chromosomal cassette *mec* (SCC*mec*) typing, and accessory gene regulator (*agr*) typing. Isolates showed resistance to cefoxitin, penicillin, ciprofloxacin, erythromycin, fusidic acid, clindamycin, and aminoglycosides, confirmed by the presence of the *bla*Z, *erm*A, *erm*C, *mph*C, *msr*A/B, *aac*(6')-Ie-*aph*(2″)-Ia, and *ant*(4')-Ia genes. Three isolates were Panton–Valentine-leukocidin-positive. Most strains ($n = 12$) presented an inducible MLS$_B$ phenotype. The isolates were ascribed to eight *spa*-types (t747, t002, t020, t1084, t008, t10682, t18526, and t1370) and four MLSTs (ST22, ST5, ST105, and ST8). Overall, most ($n = 12$) MRSA isolates had a multidrug-resistance profile with inducible MLS$_B$ phenotypes and belonged to epidemic MRSA clones.

Keywords: MRSA; EMRSA-15; MLS$_B$; bacteremia; bloodstream infections

1. Introduction

Staphylococcus aureus is an opportunist human pathogen responsible for numerous types of infections, from skin infections, such as abscesses or infected wounds, to life-threatening conditions, such as endocarditis, osteomyelitis, or septicemia [1]. Some *S. aureus* strains can be quite virulent due to the combined action of several virulence factors, the most important being Panton–Valentine leukocidin (PVL) and toxic shock syndrome toxin, associated with immune evasion, tissue adhesion, and host cell injury [2]. *S. aureus* is known for its ability to acquire antibiotic resistance determinants. In fact, *S. aureus* has become an important cause of nosocomial infections, particularly methicillin-resistant *S. aureus* (MRSA), which is usually associated with a multidrug-resistance profile [3]. Consequently, MRSA infections are difficult to treat and are a leading cause of morbidity and mortality, especially among hospitalized patients and humans with weakened immune systems [4]. Due to the increase of MRSA strains, macrolides, lincosamides, and streptogramin B (MLS$_B$) were often used to treat MRSA infections, which led to a subsequent cross-resistance to these antibiotics [5]. Different mechanisms are responsible for the MLS$_B$ resistance, the most common being the target modification mediated by the *erm* (erythromycin ribosome methylase) gene [6]. In staphylococci, *erm*A and *erm*C are the main genes conferring the MLS$_B$ resistance phenotype, which can be constitutive or inducible [7]. Healthcare-associated MRSA rates vary considerably across countries in Europe, with a high prevalence in Southwest Europe and a lower prevalence in Northeast Europe [8]. The prevalence of MRSA in Portugal has remained one of the highest among the European countries in recent years—around 40% of *S. aureus* isolates from hospitalized individuals with infection in Portugal have been identified as MRSA. The predominant clonal complexes responsible for hospital infections in Portugal are CC22 and CC5, with the epidemic methicillin-resistant *Staphylococcus aureus* 15 (EMRSA-15) clone being the most prevalent [9].

S. aureus is considered one of the most important and common pathogens causing bloodstream infections and is the second leading cause of sepsis in industrialized countries [10]. Both hospital- and community-acquired MRSA bacteremia are associated with various clinical manifestations, such as metastatic infections, endocarditis, septic arthritis, osteomyelitis, and septic shock [11]. Community-acquired MRSA bacteremia has now surpassed hospital-acquired bacteremia worldwide, and it is frequently associated with other diseases, such as diabetes, ulcers, or chronic renal disease [12]. Despite the existence of an adequate treatment, MRSA is responsible for mortality rates of 20% to 40% in a period of 30 days [13]. Given the extreme severity of clinical complications from a generalized infection caused by *S. aureus* and its association with resistance to methicillin and most β-lactam antibiotics, it is extremely important to study the genetic characteristics of the most prevalent strains responsible for bacteremia in order to more effectively target the strategies for controlling these infections [14]. This study aimed to isolate and characterize the antimicrobial resistance and genetic lineages of MRSA strains isolated from bloodstream infections.

2. Results

A total of 16 MRSA isolates were obtained from 103 hospitalized patients with bacteremia over the 3-year study period, corresponding to a patient incidence of 15.5%. Table 1 shows the genotypical characterization of the MRSA strains. All isolates were resistant to cefoxitin and harbored the *mec*A gene. Eleven isolates belonged to SCC*mec* type IV and five to type II. The isolates were ascribed to eight *spa* types (t747, t002, t020, t1084, t008, t10682, t18526, and t1370). The 16 isolates were grouped into five different sequence types (STs), namely ST22 ($n = 9$), ST5 ($n = 2$), ST105 ($n = 2$), ST8 ($n = 2$), and ST5984 ($n = 1$). ST5984 was first described in this study and differs from ST105 by a one-point mutation on the *arc*C locus. The isolate categorized as ST5984 belonged to *spa*-type t1084, SCC*mec*, and *agr* type II and presented resistance to penicillin, erythromycin, ciprofloxacin, and fusidic acid.

Table 1. Antimicrobial resistance, virulence factors, and molecular characteristics of methicillin-resistant *Staphylococcus aureus* (MRSA) strains isolated from blood cultures.

Isolate	Antimicrobial Resistance		Virulence	Molecular Typing			
	Phenotype	Genotype [a]		MLST (CC)	*spa*	SCC*mec*	*agr*
VS2761	FOX, PEN, ERY, DA [1], CIP	*mec*A, *erm*C, *msr*(A/B)	*hl*A	22 (22)	t747	IV	I
VS2762	FOX, PEN, ERY, DA [2], CN, CIP	*mec*A, *blaZ*, *erm*A, *msr*(A/B), *aac*(6′)-Ie-*aph*(2″)-Ia	*hl*A	105 (5)	t002	II	II
VS2763	FOX, PEN, CIP	*mec*A, *blaZ*	*hl*A, *hl*B, *et*A	22 (22)	t747	IV	I
VS2764	FOX, PEN, ERY, DA [2], CIP	*mec*A, *blaZ*, *erm*C, *msr*(A/B), *mph*C	*hl*A, *et*A	22 (22)	t747	IV	I
VS2765	FOX, PEN, ERY, DA [2], CIP	*mec*A, *blaZ*, *erm*C, *msr*(A/B), *mph*C	*lukF/lukS*-PV, *hl*A, *hl*B, *et*A	22 (22)	t747	IV	I
VS2766	FOX, PEN, ERY, DA [1], CN, TOB, CIP	*mec*A, *blaZ*, *erm*C, *msr*(A/B), *mph*C, *aac*(6′)-Ie-*aph*(2″)-Ia, *ant*(4′)-Ia	*lukF/lukS*-PV, *hl*A, *hl*B, *et*A	22 (22)	t020	IV	I
VS2767	FOX, PEN, CIP	*mec*A, *blaZ*	*hl*A, *hl*B, *et*A	22 (22)	t747	IV	I
VS2768	FOX, PEN, ERY, DA [2], CIP	*mec*A, *blaZ*, *erm*C, *msr*(A/B), *mph*C	*lukF/lukS*-PV, *hl*A, *et*A	22 (22)	t747	IV	I
VS2769	FOX, PEN, ERY, DA [2], CIP	*mec*A, *blaZ*, *msr*(A/B), *mph*C	*hl*A, *hl*B	5 (5)	t002	II	II
VS2770	FOX, PEN, ERY, DA [2], CIP, FD	*mec*A, *blaZ*, *erm*A, *erm*C, *msr*(A/B), *mph*C	*hl*A, *hl*B	5984	t1084	II	II
VS2771	FOX, PEN, ERY, DA [2], CIP	*mec*A, *blaZ*, *erm*C, *msr*(A/B), *mph*C	*hl*A	8 (8)	t008	IV	I
VS2772	FOX, PEN, CIP	*mec*A, *blaZ*	*hl*B	5 (5)	t002	II	II
VS2773	FOX, PEN, ERY, DA [2], CIP	*mec*A, *blaZ*, *erm*A, *erm*C, *msr*(A/B), *mph*C	*hl*B	105 (5)	t10682	II	II
VS2774	FOX, PEN, ERY, DA [1], CIP	*mec*A, *blaZ*, *erm*A, *mph*C	*hl*B, *et*A	22 (22)	t18526	IV	I
VS2775	FOX, PEN, ERY, DA [2], CIP	*mec*A, *blaZ*, *erm*A, *erm*C, *msr*(A/B)	*hl*B, *et*A	22 (22)	t1370	IV	I
VS2776	FOX, PEN, FD	*mec*A, *blaZ*	*hl*B	8 (8)	t008	IV	I

[1] Constitutive MLS$_B$ (cMLS$_B$) phenotype; [2] Inducible MLS$_B$ (iMLS$_B$) phenotype; FOX: cefoxitin; PEN: penicillin; ERY: erythromycin; DA: clindamycin; CN: gentamicin; TOB: tobramycin; CIP: ciprofloxacin; FD: fusidic acid; MLST: multilocus sequence typing; ST: sequence type; CC: clonal complex; SCC*mec*: staphylococcal cassette chromosome *mec*; [a] *mec*A gene encodes the protein PBP2A; *blaZ* encodes the protein BlaZ; *erm* genes encode the rRNA adenine N-6-methyltransferase, *msr*(A/B) encodes the peptide methionine sulfoxide reductase; *mph*C encodes the macrolide 2′-phosphotransferase; *aac*(6′)-Ie-*aph*(2″)-Ia encodes the bifunctional enzyme AAC/APH; and *ant*(4′)-Ia encodes the aminoglycoside O-nucleotidyltransferase ANT(4′)-Ia

The majority of the isolates ($n = 9$) were typed as ST22 and SCC*mec* IV, also known as the EMRSA-15 clone. Six of these isolates were *spa*-type t747, and the other three were t020, t18526, and t1370. EMRSA-15 isolates were resistant to penicillin, and eight out of nine harbored the *blaZ* gene. Seven isolates showed resistance to erythromycin, and three were coresistant to clindamycin, showing a constitutive MLS$_B$ (cMLS$_B$) phenotype. Four erythromycin-resistant isolates did not show clindamycin resistance; however, they were positive upon D-testing and were considered inducible MLS$_B$ (iMLS$_B$) isolates. The observed MLS$_B$ phenotypes were mostly determined by identification of combinations of two or more genes: *erm*C + *msr*(A/B) ($n = 1$), *erm*A + *mph*C ($n = 1$), *erm*A + *erm*C + *msr*(A/B) ($n = 1$), and *erm*C + *msr*(A/B) + *mph*C ($n = 4$). Only one isolate was resistant to aminoglycosides, namely gentamicin and tobramycin, and harbored the resistance genes *aac*(6′)-Ie-*aph*(2″)-Ia and *ant*(4′)-Ia. Regarding the virulence factors, three isolates were PVL-positive, all isolates harbored the genes encoding hemolysins, and eight isolates carried the *eta* gene. All EMRSA-15 isolates belonged to *agr* type I.

Two isolates were typed as ST5-SCC*mec* II and one isolate as ST105-SCC*mec* II (New York/Japan and New York/Japan (related) clones, respectively). Both isolates were ascribed to *spa*-type t002. These isolates showed resistance to penicillin and ciprofloxacin and harbored the *blaZ* resistance gene. Two isolates showed an iMLS$_B$ phenotype and carried the gene combinations of *erm*A + *msr*(A/B) and *msr*(A/B) + *mph*C. One isolate carried the *aac*(6′)-Ie-*aph*(2″)-Ia gene encoding resistance to gentamicin. None of the isolates were positive for PVL; nevertheless, all isolates were positive for genes encoding hemolysins, and all were *agr* type II.

Finally, two isolates belonged to ST8 and SCC*mec* type IV (variant of the USA300 clone). Both isolates were typed as t008. One of the isolates showed a multidrug-resistant phenotype with resistance to penicillin, ciprofloxacin, and erythromycin and inducible resistance to clindamycin, harboring the respective resistance genes *blaZ*, *erm*C, *msr*(A/B), and *mph*C. The second isolate showed resistance to penicillin and fusidic acid. Both isolates belonged to *agr* type I and carried the genes encoding alpha- and beta-hemolysins.

3. Discussion

A total of 103 cases of bacteremia were identified at the local hospital between 2016 and 2019, of which 15.5% were caused by MRSA. All isolates were typically epidemic hospital-acquired MRSA (HA-MRSA) clones. MRSA bacteremia has been reported worldwide, and its frequency varies from one country to another. In 2018, the percentage of invasive MRSA in bacteremia in different parts of Europe varied from 0.0% to 43.0% with an average of 16.4% [15]. Notably, in Southern European countries, such as Italy, Romania, Greece, and Cyprus, these rates were higher than in other European countries. In Portugal, the percentage of invasive MRSA in bacteremia was 38.1%. MRSA bacteremia treatment is challenging, especially when dealing with multidrug-resistant strains. Indeed, in our study, 12 of the 16 isolates were considered multidrug-resistant since they presented resistance to antibiotics belonging to at least three distinct classes of antimicrobials.

EMRSA-15 is one of the most recurrent HA-MRSA clonal lineages in recent years [16,17]. This clone is known for its rapid spread and is responsible for causing several invasive infections, such as bacteremia [18]. EMRSA-15 carries SCC*mec* type IV, which is frequent in HA-MRSA clones and is smaller and has a lower fitness cost compared to SCC*mec* types II and III, increasing the clone's ability to spread worldwide [19]. In 2013, a study conducted by Faria et al. reported that EMRSA-15, followed by ST105-II, was the dominant clone among MRSA bloodstream infections in Portugal [20]. EMRSA-15 was also the predominant clone found in our study. Since 2001, this clone has been repeatedly isolated in hospitalized patients, communities, the environment, and animals in Portugal, replacing the resident HA-MRSA clones and becoming the main clone in this country [21]. Initially, the EMRSA-15 clone in Portugal was characterized by *spa*-types t747, t032, and t2357; however, as EMRSA-15 became the main clone, there was an increase in *spa* diversity [20]. Nevertheless, in this study, *spa*-type t747 was the most common. One of the EMRSA-15 isolates belonged to *spa*-type t020, which is a well-established type in Germany and is highly associated with EMRSA-15 [22,23]. *spa*-types t18526 and t1370 were also detected in this study, each in one isolate. The *spa*-type t18526 was reported for the first time in one of our previous studies conducted with samples from the same hospital with MRSA strains isolated from infected diabetic foot ulcers, in which the most prevalent clone and *spa*-type were also EMRSA-15 and t747 [9]. As for t1370, it was the predominant clone in an outbreak in a neonatal unit in the UK and in human patients in New Zealand and was always associated with EMRSA-15 [24,25]. The presence of PVL-encoding genes was only detected in EMRSA-15 isolates. Similar results were obtained by Goudarzi et al. when studying the molecular characteristics of MRSA strains from patients with bacteremia [26]. Although the PVL toxin is often associated with skin and soft-tissue infections, studies have shown an association between PVL and severe invasive infections [27]. PVL and SCC*mec* type IV presence are often used as markers of community-associated MRSA (CA-MRSA). This assumption may be inaccurate since it frequently includes the EMRSA-15 and USA800 clones, which are epidemiologically HA-MRSA [18]. The *eta* gene was found only among EMRSA-15 isolates, and consistent with other studies, all *eta* genes belonged to SCC*mec* type IV [26]. Furthermore, the *etb* gene was not detected in our study, which is in accordance with other studies that showed that invasive MRSA strains carried the *eta* gene but few or none carried the *etb* gene [26,28]. All EMRSA-15 isolates in our study belonged to *agr* type I, in agreement with other studies in which *agr* I, followed by *agr* II, was the most common type in MRSA bacteremia [29,30]. Furthermore, Ben Ayed et al. showed that *agr* I was associated with invasive infections, bacteremia in particular [31]. However, *agr* locus is strongly associated with bacterial genetic background, and therefore its prevalence may be driven by the genotypes circulating in each hospital. Goudarzi et al. reported that MRSA bacteremia isolates belonging to SCC*mec* IV and II isolates were distributed among *agr* type III [26]. In another study, *agr* II was the most common *agr* type in MRSA bacteremia strains, followed by *agr* I; however, the majority of strains belonged to CC5, which suggests that *agr* type may also be associated with clonal complex [32]. Indeed, Aschbacher et al. showed that bacteremia isolates belonging to CC22, CC5, and CC8 were *agr* types I, II, and I, respectively [30]. These results are in accordance with our study, in which all USA300 isolates belonged to *agr* type I, and all New York/Japan clones were *agr* type II. Three New York/Japan (or related) clones were detected in bacteremia isolates. This clone has been reported to be associated with bacteremia and is the most prevalent in France and South Korea [33,34]. The New York/Japan clone and ST105-MRSA-II were also found in bacteremia isolates in the study by Faria et al., who suggested that the ST105-MRSA-II clone could replace EMRSA-15 and be the next clonal wave of MRSA in Portuguese hospitals [20]. Nevertheless, this shift was not confirmed, since our studies and other recent studies in Portugal showed that there was no modification in the predominance of EMRSA-15 [9]. Furthermore, the New York/Japan clone seems to be significantly decreasing in prevalence in hospitals [35,36]. *spa*-type t002 is strongly linked with ST105-MRSA-II, since most New York/Japan isolates reported are type t002. Two ST8-MRSA-IV clones were also isolated in our study. These clones are a variant of the epidemic clone USA300, since this clone is linked to carriage of PVL and both of our isolates were PVL-negative. USA300 is sporadically isolated in MRSA infections in Portugal; this clone is frequently found in the United States (*spa*-type t008), where it is often responsible for bacteremia, and is found to a much lesser extent in Europe and the rest of the world [37].

Twelve of the sixteen isolates were resistant to erythromycin; however, only three showed coresistance to clindamycin and as such were categorized as cMLS$_B$. Those nine isolates were further characterized by D-test to the MLS$_B$ phenotype. All showed the inducible MLS$_B$ phenotype. Both cMLS$_B$ and iMLS$_B$ harbored several combinations of genes conferring resistance to macrolides and lincosamides; however, *erm*A, *erm*C, or both were present in all isolates. The *erm*A and *erm*C genes are the genes most commonly found in MLS$_B$-resistant *S. aureus*; nevertheless, MLS$_B$-resistant MRSA often carries combinations of two or more resistance genes [7,38]. In staphylococci showing an iMLS$_B$ phenotype, the methylase mRNA produced by bacteria is inactive, and the activation only occurs in the presence of a macrolide. In contrast, in the cMLS$_B$-resistance phenotype, active methylase mRNA is produced in the absence of a macrolide [38]. Therefore, identifying the MLS$_B$-resistance phenotype is essential since strains presenting an iMLS$_B$ phenotype may switch to a cMLS$_B$ phenotype under antibiotic pressure, which may lead to treatment failure [38].

4. Material and Methods

4.1. Bacterial Isolates

Blood samples were collected from 103 inpatients with bacteremia infection hospitalized at the Hospital Centre of Trás-os-Montes e Alto Douro E.P.E., Vila Real, Portugal, from 2016 to 2019. A small volume of blood culture was inoculated on an oxacillin-resistance-screening agar base (ORSAB) (OXOID) supplemented with 2 mg/L of oxacillin to isolate MRSA strains and incubated at 37 °C for 24 h. Four colonies from each plate were recovered and seeded onto Baird–Parker agar plates for further identification of possible *S. aureus*. MRSA strains were identified based on Gram staining, biochemical tests (catalase, DNase, and coagulase), and genotyping.

4.2. Antimicrobial Resistance Profile

MRSA strains were characterized according to their antibiotic resistance profiles using the Kirby–Bauer disk-diffusion method against 14 antimicrobial agents: cefoxitin (30 µg), chloramphenicol (30 µg), ciprofloxacin (5 µg), clindamycin (2 µg), erythromycin (15 µg), fusidic acid (10 µg), gentamicin (10 µg), kanamycin (30 µg), linezolid (10 µg), mupirocin (200 µg), penicillin (1U), tetracycline (30 µg), tobramycin (10 µg), and trimethoprim/sulfamethoxazole (1.25/23.75 µg). The tests were performed according to the European Committee on Antimicrobial Susceptibility Testing (EUCAST, 2018) guidelines except the test for kanamycin, which followed Clinical and Laboratory Standards Institute standards (CLSI, 2017). Isolates showing resistance to erythromycin were further characterized by double-disk diffusion (D-test) to determine the MLS$_B$ phenotype. Briefly, erythromycin and clindamycin disks were placed onto inoculated Muller–Hinton plates 15 mm apart from edge to edge. If the inhibition zone around the clindamycin disk showed a D-shape, the isolate was considered to have an inducible MLS$_B$ (iMLS$_B$) phenotype. Resistance to both erythromycin and clindamycin indicated a constitutive MLS$_B$ (cMLS$_B$) phenotype [39]. Quality control was performed with *S. aureus* strain ATCC 25923.

DNA was extracted from fresh cultures as previously described [9]. Briefly, two colonies of fresh cultures from each isolate were suspended in 45 µL of Milli-Q water. Five microliters of lysostaphin (1 mg/mL) was added, and the samples were incubated for 10 min at 37 °C. Then, 150 µL of Tris-HCl (0.1 M), 45 µL of Milli-Q water, and 5 µL of proteinase K (2 mg/mL) were added, and the samples were incubated at 67 °C for 10 min. Lastly, the samples were boiled for 5 min at 100 °C.

According to the phenotypic resistance of each isolate, the presence of the following antibiotic resistance genes was investigated by polymerase chain reaction (PCR): *blaZ*, *erm*(A), *erm*(B), *erm*(C), *erm*(T), *erm*(Y), *msr*(A/B), *mphC*, *linB*, *vgaB*, *vgaC*, *aac*(6′)-Ie-*aph*(2″)-Ia, *ant*(4′)-Ia, *fusB*, and *fusC* (Supplementary Table S1).

4.3. Characterization of Virulence Factors

The presence of the virulence genes encoding alpha- and beta-hemolysins (hla and hlb), exfoliative toxins (eta and etb), toxic shock syndrome toxin (tst), and Panton–Valentine leucocidin (PVL) (*lukF/lukS*-PV) was determined by PCR (Supplementary Table S1). Positive and negative controls used in all experiments belonged to the strain collection of University of Trás-os-Montes and Alto Douro.

4.4. Molecular Characterization

Multilocus sequence typing (MLST) and *spa* typing were performed for all isolates as previously described, supported by the public databases MLST and the Ridom SpaServer. According to the sequence type (ST), each isolate was grouped according to the corresponding clonal complex (CC).

All isolates were characterized by *agr* typing and staphylococcal chromosomal cassette *mec* (SCC*mec*) typing (I–V) using specific primers (Supplementary Table S1).

5. Conclusions

We found epidemic HA-MRSA clones, namely EMRSA-15, USA300, and New York/Japan, in samples recovered from bloodstream infections over a period of 3 years. Our results corroborate the relatively high prevalence of EMRSA-15 circulating in Portuguese hospitals. Most isolates were multidrug-resistant and presented an iMLSB or cMLSB phenotype, which may result in a therapeutic problem of inadequacy of antibiotic treatment and lead to high morbidity and mortality.

Supplementary Materials: The following are available online at http://www.mdpi.com/2079-6382/9/7/375/s1: Table S1: Primers used for molecular typing and detection of antimicrobial resistance genes in MRSA strains.

Author Contributions: Conceptualization, V.S. and P.P.; methodology, V.S., S.H., and C.F.; validation, C.M.M., R.C., C.A.-C., G.I., and P.P.; investigation, V.S., S.H., C.F., R.C., and C.A.-C.; resources, I.C.; writing—original draft preparation, V.S.; supervision, C.M.M., J.E.P., L.M., J.L.C., G.I., and P.P.; funding acquisition, R.C. and C.A.-C. All authors have read and agreed to the published version of the manuscript.

Funding: This research was funded by the Ministerio de Ciencia, Innovación y Universidades (Spain, Project RTI2018-098267-R-C33) and the Junta de Castilla y León (Consejería de Educación, Spain, Project LE164G18).

Acknowledgments: This work was funded by the R&D Project CAREBIO2—comparative assessment of antimicrobial resistance in environmental biofilms through proteomics—toward innovative theranostic biomarkers, with references NORTE-01-0145-FEDER-030101 and PTDC/SAU-INF/30101/2017, financed by the European Regional Development Fund (ERDF) through the Northern Regional Operational Program (NORTE 2020) and the Foundation for Science and Technology (FCT). This work was supported by the Associate Laboratory for Green Chemistry (LAQV), which is financed by national funds from FCT/MCTES (UIDB/50006/2020). Vanessa Silva and Isabel Carvalho are grateful to FCT (Fundação para a Ciência e a Tecnologia) for financial support through Ph.D. grants SFRH/BD/137947/2018 and SFRH/BD/133266/2017.

Conflicts of Interest: The authors declare no conflict of interest.

References

1. Mohanty, A.; Mohapatra, K.; Pal, B. Isolation and identification of staphylococcus aureus from skin and soft tissue infection in sepsis cases, Odisha. *J. Pure Appl. Microbiol.* **2018**, *12*, 419–424. [CrossRef]
2. Powers, M.E.; Wardenburg, J.B. Igniting the fire: *Staphylococcus aureus* virulence factors in the pathogenesis of sepsis. *PLoS Pathog.* **2014**, *10*, e1003871. [CrossRef] [PubMed]
3. Zha, G.-F.; Wang, S.-M.; Rakesh, K.P.; Bukhari, S.N.A.; Manukumar, H.M.; Vivek, H.K.; Mallesha, N.; Qin, H.-L. Discovery of novel arylethenesulfonyl fluorides as potential candidates against methicillin-resistant of Staphylococcus aureus (MRSA) for overcoming multidrug resistance of bacterial infections. *Eur. J. Med. Chem.* **2019**, *162*, 364–377. [CrossRef] [PubMed]
4. Kong, E.F.; Johnson, J.K.; Jabra-Rizk, M.A. Community-associated methicillin-resistant *Staphylococcus aureus*: An enemy amidst us. *PLoS Pathog.* **2016**, *12*, e1005837. [CrossRef]
5. Prabhu, K.; Rao, S.; Rao, V. Inducible clindamycin resistance in *Staphylococcus aureus* isolated from clinical samples. *J. Lab. Physicians* **2011**, *3*, 25–27. [CrossRef]
6. Coutinho, V.D.L.S.; Paiva, R.M.; Reiter, K.C.; de-Paris, F.; Barth, A.L.; Machado, A.B.M.P. Distribution of erm genes and low prevalence of inducible resistance to clindamycin among staphylococci isolates. *Braz. J. Infect. Dis.* **2010**, *14*, 564–568.
7. Khodabandeh, M.; Mohammadi, M.; Abdolsalehi, M.R.; Alvandimanesh, A.; Gholami, M.; Bibalan, M.H.; Pournajaf, A.; Kafshgari, R.; Rajabnia, R. Analysis of resistance to macrolide-lincosamide-streptogramin B among mecA-positive *Staphylococcus aureus* isolates. *Osong Public Health Res. Perspect.* **2019**, *10*, 25–31. [CrossRef]
8. Lakhundi, S.; Zhang, K. Methicillin-resistant *Staphylococcus aureus*: Molecular characterization, evolution, and epidemiology. *Clin. Microbiol. Rev.* **2018**, *31*, e00020-18. [CrossRef]

9. Silva, V.; Almeida, F.; Carvalho, J.A.; Castro, A.P.; Ferreira, E.; Manageiro, V.; Tejedor-Junco, M.T.; Caniça, M.; Igrejas, G.; Poeta, P. Emergence of community-acquired methicillin-resistant *Staphylococcus aureus* EMRSA-15 clone as the predominant cause of diabetic foot ulcer infections in Portugal. *Eur. J. Clin. Microbiol. Infect. Dis.* **2020**, *39*, 179–186. [CrossRef]

10. Dat, V.Q.; Vu, H.N.; Nguyen The, H.; Nguyen, H.T.; Hoang, L.B.; Vu Tien Viet, D.; Bui, C.L.; Van Nguyen, K.; Nguyen, T.V.; Trinh, D.T.; et al. Bacterial bloodstream infections in a tertiary infectious diseases hospital in Northern Vietnam: Aetiology, drug resistance, and treatment outcome. *BMC Infect. Dis.* **2017**, *17*, 493. [CrossRef]

11. Tong, S.Y.C.; Davis, J.S.; Eichenberger, E.; Holland, T.L.; Fowler, V.G., Jr. *Staphylococcus aureus* infections: Epidemiology, pathophysiology, clinical manifestations, and management. *Clin. Microbiol. Rev.* **2015**, *28*, 603–661. [CrossRef] [PubMed]

12. Hassoun, A.; Linden, P.K.; Friedman, B. Incidence, prevalence, and management of MRSA bacteremia across patient populations—A review of recent developments in MRSA management and treatment. *Crit. Care* **2017**, *21*, 211. [CrossRef] [PubMed]

13. Piechota, M.; Kot, B.; Frankowska-Maciejewska, A.; Gruzewska, A.; Woźniak-Kosek, A. Biofilm formation by Methicillin-resistant and Methicillin-sensitive *Staphylococcus aureus* strains from hospitalized patients in Poland. *Biomed. Res. Int.* **2018**, *2018*, 7. [CrossRef] [PubMed]

14. Jokinen, E.; Lindholm, L.; Huttunen, R.; Huhtala, H.; Vuento, R.; Vuopio, J.; Syrjänen, J. *Spa* type distribution in MRSA and MSSA bacteremias and association of *spa* clonal complexes with the clinical characteristics of bacteremia. *Eur. J. Clin. Microbiol. Infect. Dis.* **2018**, *37*, 937–943. [CrossRef]

15. Antimicrobial Consumption in the EU/EEA, Annual Epidemiological Report for 2018. Available online: https://www.ecdc.europa.eu/sites/default/files/documents/Antimicrobial-consumption-EU-EEA.pdf (accessed on 10 April 2020).

16. Lindsay, J.A. Hospital-associated MRSA and antibiotic resistance—What have we learned from genomics? *Int. J. Med. Microbiol.* **2013**, *303*, 318–323. [CrossRef]

17. Horváth, A.; Dobay, O.; Sahin-Tóth, J.; Juhász, E.; Pongrácz, J.; Iván, M.; Fazakas, E.; Kristóf, K. Characterisation of antibiotic resistance, virulence, clonality and mortality in MRSA and MSSA bloodstream infections at a tertiary-level hospital in Hungary: A 6-year retrospective study. *Ann. Clin. Microbiol. Antimicrob.* **2020**, *19*, 1–11. [CrossRef]

18. Bal, A.M.; Coombs, G.W.; Holden, M.T.G.; Lindsay, J.A.; Nimmo, G.R.; Tattevin, P.; Skov, R.L. Genomic insights into the emergence and spread of international clones of healthcare-, community-and livestock-associated meticillin-resistant *Staphylococcus aureus*: Blurring of the traditional definitions. *J. Glob. Antimicrob. Resist.* **2016**, *6*, 95–101. [CrossRef]

19. Knight, G.M.; Budd, E.L.; Lindsay, J.A. Large mobile genetic elements carrying resistance genes that do not confer a fitness burden in healthcare-associated meticillin-resistant *Staphylococcus aureus*. *Microbiology (UK)* **2013**, *159*, 1661–1672. [CrossRef]

20. Faria, N.A.; Miragaia, M.; de Lencastre, H.; The Multi Laboratory Project Collaborators. Massive dissemination of methicillin resistant *Staphylococcus aureus* in bloodstream infections in a high MRSA prevalence country: Establishment and diversification of EMRSA-15. *Microb. Drug Resist.* **2013**, *19*, 483–490. [CrossRef]

21. Aires-de-Sousa, M.; Correia, B.; de Lencastre, H.; Collaborators, M.P. Changing patterns in frequency of recovery of five methicillin-resistant *Staphylococcus aureus* clones in Portuguese hospitals: Surveillance over a 16-year period. *J. Clin. Microbiol.* **2008**, *46*, 2912–2917. [CrossRef]

22. Reinheimer, C.; Kempf, V.A.J.; Jozsa, K.; Wichelhaus, T.A.; Hogardt, M.; O'Rourke, F.; Brandt, C. Prevalence of multidrug-resistant organisms in refugee patients, medical tourists and domestic patients admitted to a German university hospital. *BMC Infect. Dis.* **2017**, *17*, 17. [CrossRef]

23. Chan, M.K.L.; Koo, S.H.; Quek, Q.; Pang, W.S.; Jiang, B.; Ng, L.S.Y.; Tan, S.H.; Tan, T.Y. Development of a real-time assay to determine the frequency of qac genes in methicillin resistant *Staphylococcus aureus*. *J. Microbiol. Methods* **2018**, *153*, 133–138. [CrossRef]

24. Williamson, D.A.; Roberts, S.A.; Ritchie, S.R.; Coombs, G.W.; Fraser, J.D.; Heffernan, H. Clinical and molecular epidemiology of methicillin-resistant *Staphylococcus aureus* in New Zealand: Rapid emergence of sequence type 5 (ST5)-SCCmec-IV as the dominant community-associated MRSA clone. *PLoS ONE* **2013**, *8*, e62020. [CrossRef]

25. Otter, J.A.; Klein, J.L.; Watts, T.L.; Kearns, A.M.; French, G.L. Identification and control of an outbreak of ciprofloxacin-susceptible EMRSA-15 on a neonatal unit. *J. Hosp. Infect.* **2007**, *67*, 232–239. [CrossRef]

26. Goudarzi, M.; Seyedjavadi, S.S.; Nasiri, M.J.; Goudarzi, H.; Sajadi Nia, R.; Dabiri, H. Molecular characteristics of methicillin-resistant *Staphylococcus aureus* (MRSA) strains isolated from patients with bacteremia based on MLST, SCCmec, spa, and agr locus types analysis. *Microb. Pathog.* **2017**, *104*, 328–335. [CrossRef]

27. Shallcross, L.J.; Fragaszy, E.; Johnson, A.M.; Hayward, A.C. The role of the Panton-Valentine leucocidin toxin in staphylococcal disease: A systematic review and meta-analysis. *Lancet. Infect. Dis.* **2013**, *13*, 43–54. [CrossRef]

28. Sila, J.; Sauer, P.; Kolar, M. Comparison of the prevalence of genes coding for enterotoxins, exfoliatins, Panton-Valentine leukocidin and TSST-1 between methicillin-resistant and methicillin-susceptible isolates of Staphylococcus Aureus at the University Hospital in Olomouc. *Biomed. Pap.* **2009**, *153*, 215–218. [CrossRef]

29. Abdulgader, S.M.; van Rijswijk, A.; Whitelaw, A.; Newton-Foot, M. The association between pathogen factors and clinical outcomes in patients with Staphylococcus aureus bacteraemia in a tertiary hospital, Cape Town. *Int. J. Infect. Dis.* **2020**, *91*, 111–118. [CrossRef]

30. Aschbacher, R.; Pagani, E.; Larcher, C.; Pichon, B.; Pike, R.; Ganner, M.; Hill, R.; Kearns, A.; Wootton, M.; Davies, L.; et al. Molecular epidemiology of methicillin-resistant *Staphylococcus aureus* from bacteraemia in northern Italy. *Infez. Med.* **2012**, *20*, 256–264.

31. Ben Ayed, S.; Boutiba-Ben Boubaker, I.; Samir, E.; Ben Redjeb, S. Prevalence of agr specificity groups among methicilin resistant Staphylococcus aureus circulating at Charles Nicolle hospital of Tunis. *Pathol. Biol.* **2006**, *54*, 435–438. [CrossRef]

32. He, W.; Chen, H.; Zhao, C.; Zhang, F.; Li, H.; Wang, Q.; Wang, X.; Wang, H. Population structure and characterisation of Staphylococcus aureus from bacteraemia at multiple hospitals in China: Association between antimicrobial resistance, toxin genes and genotypes. *Int. J. Antimicrob. Agents* **2013**, *42*, 211–219. [CrossRef]

33. Bouchiat, C.; Moreau, K.; Devillard, S.; Rasigade, J.-P.; Mosnier, A.; Geissmann, T.; Bes, M.; Tristan, A.; Lina, G.; Laurent, F.; et al. *Staphylococcus aureus* infective endocarditis versus bacteremia strains: Subtle genetic differences at stake. *Infect. Genet. Evol.* **2015**, *36*, 524–530. [CrossRef]

34. Kim, D.; Hong, J.S.; Yoon, E.-J.; Lee, H.; Kim, Y.A.; Shin, K.S.; Shin, J.H.; Uh, Y.; Shin, J.H.; Park, Y.S.; et al. Toxic shock syndrome toxin 1-producing methicillin-resistant *Staphylococcus aureus* of clonal complex 5, the New York/Japan epidemic clone, causing a high early-mortality rate in patients with bloodstream infections. *Antimicrob. Agents Chemother.* **2019**, *63*, e01362–e01419. [CrossRef]

35. Boswihi, S.S.; Udo, E.E.; Al-Sweih, N. Shifts in the Clonal Distribution of Methicillin-Resistant Staphylococcus aureus in Kuwait Hospitals: 1992–2010. *PLoS ONE* **2016**, *11*, e0162744. [CrossRef]

36. Lee, C.-Y.; Fang, Y.-P.; Chang, Y.-F.; Wu, T.-H.; Yang, Y.-Y.; Huang, Y.-C. Comparison of molecular epidemiology of bloodstream methicillin-resistant *Staphylococcus aureus* isolates between a new and an old hospital in central Taiwan. *Int. J. Infect. Dis.* **2019**, *79*, 162–168. [CrossRef]

37. David, M.Z. The Importance of *Staphylococcus aureus* genotypes in outcomes and complications of bacteremia. *Clin. Infect. Dis.* **2019**, *69*, 1878–1880. [CrossRef]

38. Mišić, M.; Čukić, J.; Vidanović, D.; Šekler, M.; Matić, S.; Vukašinović, M.; Baskić, D. Prevalence of genotypes that determine resistance of staphylococci to macrolides and lincosamides in Serbia. *Front. Public Health* **2017**, *5*, 200. [CrossRef]

39. Sasirekha, B.; Usha, M.S.; Amruta, J.A.; Ankit, S.; Brinda, N.; Divya, R. Incidence of constitutive and inducible clindamycin resistance among hospital-associated *Staphylococcus aureus*. *3 Biotech.* **2014**, *4*, 85–89. [CrossRef]

 *antibiotics*

Article

Antibiotic Resistance Characteristics of *Pseudomonas aeruginosa* Isolated from Keratitis in Australia and India

Mahjabeen Khan [1], Fiona Stapleton [1], Stephen Summers [2], Scott A. Rice [2,3] and Mark D. P. Willcox [1,*]

1   School of Optometry and Vision Science, UNSW, Sydney, NSW 2052, Australia; mahjabeen.khan@student.unsw.edu.au (M.K.); f.stapleton@unsw.edu.au (F.S.)
2   The Singapore Centre for Environment Life Sciences Engineering (SCELSE), The School of Biological Sciences, Nanyang Technological University, Singapore 639798, Singapore; ssummers@ntu.edu.sg (S.S.); RSCOTT@ntu.edu.sg (S.A.R.)
3   The ithree Institute, The University of Technology Sydney, Sydney, NSW 2007, Australia
*   Correspondence: m.willcox@unsw.edu.au; Tel.: +61-2-9385-4164

Received: 2 July 2020; Accepted: 9 September 2020; Published: 14 September 2020

Abstract: This study investigated genomic differences in Australian and Indian *Pseudomonas aeruginosa* isolates from keratitis (infection of the cornea). Overall, the Indian isolates were resistant to more antibiotics, with some of those isolates being multi-drug resistant. Acquired genes were related to resistance to fluoroquinolones, aminoglycosides, beta-lactams, macrolides, sulphonamides, and tetracycline and were more frequent in Indian (96%) than in Australian (35%) isolates ($p = 0.02$). Indian isolates had large numbers of gene variations (median 50,006, IQR = 26,967–50,600) compared to Australian isolates (median 26,317, IQR = 25,681–33,780). There were a larger number of mutations in the *mutL* and *uvrD* genes associated with the mismatch repair (MMR) system in Indian isolates, which may result in strains losing their efficacy for DNA repair. The number of gene variations were greater in isolates carrying MMR system genes or *exoU*. In the phylogenetic division, the number of core genes were similar in both groups, but Indian isolates had larger numbers of pan genes (median 6518, IQR = 6040–6935). Clones related to three different sequence types—ST308, ST316, and ST491—were found among Indian isolates. Only one clone, ST233, containing two strains was present in Australian isolates. The most striking differences between Australian and Indian isolates were carriage of *exoU* (that encodes a cytolytic phospholipase) in Indian isolates and *exoS* (that encodes for GTPase activator activity) in Australian isolates, large number of acquired resistance genes, greater changes to MMR genes, and a larger pan genome as well as increased overall genetic variation in the Indian isolates.

Keywords: antibiotic susceptibility; WGS; phylogenetic analysis; DNA mismatch repair system

1. Introduction

Pseudomonas aeruginosa is a ubiquitous bacterium which can cause opportunistic or nosocomial infections in immuno-compromised patients [1]. *P. aeruginosa* commonly causes corneal (keratitis) [2], respiratory, burn and wound infections, and infections related to medical or surgical devices including ventilator-associated pneumonia [3,4]. *P. aeruginosa* corneal infections are usually related to contact lens wear, but other risk factors for keratitis in non-contact lens wearers include ocular trauma, ocular surgery, and prior ocular surface disease [5–8].

The prevalence of multi-drug resistant (MDR) or extensively drug resistant strains of *P. aeruginosa* reduces treatment options, significantly increasing morbidity rates [9]. *P. aeruginosa* is naturally resistant to some antibiotics due to the possession of specific resistance genes such as *catB* that confers chloramphenicol resistance and an inducible *ampC* which encodes for a β-lactamase that hydrolyses cephalothin and ampicillin, conferring resistance to β-lactams [10]. Additionally, the regulation of efflux pumps also contributes towards an elevated resistance to antibiotics [11]. For example, expression of the efflux pump MexAB-OprM contributes towards intrinsic resistance to a broad spectrum of antibiotics [12], whereas the efflux pump MexXY-OprM is involved in the adaptive resistance to aminoglycosides [13]. Other resistance mechanisms in *P. aeruginosa* include the acquisition of transferrable resistance determinants, including those associated with transposons and integrons [14]. Antibiotic resistance of *P. aeruginosa* varies according to the region where the strains have been isolated [15,16] presumably due to the prescribing practices, availability of antibiotics, and perhaps their use in animal husbandry. Various epidemiological studies have identified MDR *P. aeruginosa* from different infections and these isolates have acquired different resistance characteristics. For example, aminoglycoside resistance [17] and ciprofloxacin persistence [18] are found in cystic fibrosis isolates of *P. aeruginosa*. Some of these MDR strains are clonal and such clonal strains are often the predominant global clinical MDR isolates [19] which spread resistance characteristics into the wider population which enables clonal lineages to expand with time.

ExoU has been associated with virulence of *P. aeruginosa* at the ocular surface. *ExoU* is a phospholipase that causes mammalian cell death [20] and *exoU* possession is common in strains isolated from ocular infections [21]. There is a correlation between carriage of *exoU* and elevated resistance to fluoroquinolones and aminoglycosides [22]. *ExoU* is carried by strains on a genomic island that also contains resistance genes for a range of antibiotics [23].

In addition to the acquisition of resistance genes, bacteria can develop resistance through mutation of genes so that antibiotic targets are modified. Mutation rates are elevated in strains that carry mutations in DNA mismatch repair (MMR) systems and hence such mutator strains will normally carry more mutations than non-mutator strains [24]. In *P. aeruginosa*, the MMR system is composed of *mutS*, *mutL*, and *UvrD* genes [25]. Strains of *P. aeruginosa* isolated from the lungs of cystic fibrosis patients have alterations in the DNA MMR system and this has been correlated with multiple antimicrobial resistance [23].

In Australia, there is a tight regulation of prescribing antibiotics, and antibiotics can only be obtained legally with a prescription from a qualified healthcare professional according to the Therapeutic Goods Act 1989. In India, on the other hand, whilst branded antibiotics exist, other forms such as counterfeit, substandard, and 'spurious' antibiotics have been reported [26], making surveillance and regulation difficult [27]. While the antibiotic consumption per person in Australia and India in 2010 was approximately similar, there was a more rapid increase between 2000 and 2010 in India [28]. These differences may affect antibiotic resistance development.

The aim of the current study was to compare the phenotypic resistance and genetic characteristics associated with resistance between strains isolated from Australia and India to better understand the underlying factors that may lead to an increased resistance in *P. aeruginosa* strains associated with ocular infection.

2. Results

2.1. Antibiotic Susceptibility

The minimum inhibitory concentrations (MICs) and minimum bactericidal concentrations (MBCs) of the *P. aeruginosa* isolates were determined (Table 1). Strains showing intermediate resistance (I) as well as full resistance to antibiotics were categorized as resistant (R) for subsequent analyses. Based on the Centers for Disease Control and Prevention's (CDC, Atlanta, GA, USA) definition of multi-drug resistance as "an isolate that is resistant to at least one antibiotic in three or more drug classes", isolates 198, 202, 216, 217, 218, 219, 220, and 221 were deemed to be multi-drug resistant. Australian isolates 223, 224, 225, 227, 233, and 235 were also resistant to three antibiotics but these antibiotics were not of different classes. Isolates 176, 193, and 206 were sensitive to all antibiotics, but all other isolates were resistant to at least one antibiotic. Overall, Indian isolates were more resistant to antibiotics compared to Australian isolates. Among Australian isolates ($n = 14$), resistance was 78% for imipenem, 57% for ceftazidime, 50% for ciprofloxacin, 21% for piperacillin, 14% for levofloxacin, 7% for tobramycin, and no isolates were resistant to gentamicin or polymyxin. In contrast, resistance in Indian isolates ($n = 12$) was 75% for ciprofloxacin, 58% for imipenem, 50% for levofloxacin, tobramycin, and ceftazidime, 41% for piperacillin, 40% for gentamicin, and 25% for polymyxin.

Table 1. Minimum inhibitory concentration (MIC) and minimum bactericidal concentration (MBC) of antibiotics to *Pseudomonas aeruginosa* keratitis isolates.

Strain Number	Fluoroquinolones *		Aminoglycosides		β-Lactams			Poly-Peptide
	2nd Generation	3rd Generation			Penicillin 4th Generation	Carba-Penem	Cephalosporin 3rd Generation	
	Cipro µg/mL ≤1, 2, ≥4 #	Levo µg/mL ≤2, 4, ≥8	Genta µg/mL ≤4, 8, ≥16	Tobra µg/mL ≤4, 8, ≥16	Pipera µg/mL ≤16	Imi µg/mL ≤2, 4, ≥8	Ceftaz µg/mL ≤8, 16, ≥32	PMB µg/mL ≤2, 4, ≥8
	MIC/MBC	MIC/MBC	MIC/MBC	MIC/MBC	MIC/MBC	MIC/MBC	MIC/MBC	MIC/MBC
123	1/1	1/1	0.25/0.5	4/4	8/16	4(I)/8	2/2	1280(R)/1280
126	0.5/1	0.5/1	0.5/1	0.25/0.5	8/16	8(R)/16	128(R)/256	1/1
127	1/2	0.25/1	2/4	32(R)/128	4/16	4(I)/8	128(R)/256	0.5/1
162	0.5/1	0.5/1	0.25/0.5	0.25/1	8/8	4(I)/4	2/4	0.25/0.5
169	2(I)/4	0.25/0.5	0.25/0.5	0.25/0.5	4/8	2/4	1/2	0.25/0.25
176	0.5/1	0.25/0.5	0.25/0.5	0.25/0.5	4/8	2/8	2/4	0.25/0.5
181	1/4	0.25/0.5	0.25/0.5	0.25/0.5	32(R)/64	4(I)/8	16(I)/32	0.5/1
182	1/2	0.25/0.5	0.25/0.5	0.25/0.5	4/8	8(R)/16	1/2	0.25/0.5
223	64(R)/128	1/2	0.5/1	0.5/1	160(R)/320	1/2	16(I)/32	2/4
224	16(R)/32	1/2	0.25/0.5	0.25/0.5	8/16	64(R)/128	16(I)/32	1/2
225	64(R)/128	16(R)/32	0.5/2	1/2	16/32	64(R)/128	8/16	0.25/0.5
227	64(R)/128	64(R)/128	0.5/1	0.25/1	16/32	16(R)/32	16(I)/32	0.25/0.5
233	8(R)/16	1/2	1/2	105/1	16/32	4(I)/8	160(I)/320	0.5/1
235	16(R)/32	0.5/1	2/4	0.5/1	64(R)/128	4(I)/8	64(I)/128	0.25
188	2(I)/4	1/2	0.5/1	32(R)/64	16/65	0.5/1	4/8	2/4
189	0.25/1	1/2	0.25/0.5	16(R)/32	4/8	2/1	8/16	2/4
193	1/1	0.25/1	0.25/0.25	0.25/0.5	4/8	2/4	2/2	0.5/1
198	1280(R)/2560	320(R)/1280	2560(R)/5120	16(R)/16	8/8	1/2	8/8	4(I)/4
202	640(R)/1280	320(R)/640	8(I)/32	320(R)/640	16/64	8(R)/32	8/32	0.25/0.25
206	1/1	0.5/0.5	1/1	0.25/0.5	8/8	2/4	2/4	0.25/0.5
216	64(R)/128	4 (I)/8	1/2	0.5/2	160(R)/320	16(R)/32	64(R)/128	64(R)/128
217	64(R)/128	32(R)/64	1/2	1/2	64(R)/128	8(R)/16	32(R)/64	0.25/1
218	8(R)/16	1/2	0.5/1	0.5/1	160(R)/320	8(R)/16	64(R)/128	1/4
219	≥5120(R)/≥5120	640(R)/1280	≥5120(R)/≥5120	1280(R)/2560	2560(R)/5120	40(R)/80	16(I)/32	0.25/1
220	2(I)/4	0.25/0.5	0.5/1	0.5/1	0.25/0.5	8(R)/16	160(R)/320	8(R)/16
221	2560(R)/5120	2560(R)/5120	2560(R)/5120	2560(R)/5120	64(R)/128	16(R)/32	32(R)/64	0.25/1

Data for Australian isolates (shaded in gray). Data for 123–182 is from a previously published study [29]. Strains 188–221 were Indian keratitis isolates. R = resistant, I = intermediate resistance. * Cipro = Ciprofloxacin, Levo = Levofloxacin, Genta = Gentamicin, Tobra = Tobramycin, Pipera = Piperacillin, Imi = Imipenem, Ceftaz = Ceftazidime, PMB = Polymyxin B; # = Antibiotic breakpoints for sensitive, intermediate, resistant classifications.

2.2. General Features of the Genomes

The isolates after de novo assembly consisted of different numbers of contigs ranging from 50 for isolate 169 to 1917 for isolate 216. The average number of coding sequences was 6162 ± 359.2 for the Australian isolates and 6544 ± 889 for the Indian isolates. Isolates had an average of 66.1% G + C content. The tRNA copy number for the isolates ranged from 57 to 86 (which may vary between studies that use different assembly methods). The general features of the isolates are provided in Supplementary Table S1.

2.3. Acquired Resistance Genes

P. aeruginosa isolates were examined for horizontally acquired antibiotic resistance genes (Table 2) using the Resfinder database. Altogether, 33 different acquired antibiotic resistance genes for various classes of antibiotics including aminoglycosides, fluoroquinolones, beta-lactams were found in these isolates (Table 2).

Table 2. Acquired resistance genes in *P. aeruginosa* isolates from India and Australia.

Genes	Australian Isolates														Indian Isolates											
	123	126	127	162	169	176	181	182	223	224	225	227	233	235	188	189	193	198	202	206	216	217	218	219	220	221
Aminoglycoside resistance genes																										
aph(3')-IIb	■	■	■	■	■	■	■	■	■	■	■	■	■	■	■	■	■	■	■	■	■	■	■	■	■	■
aph(6)-Id																		■	■			■			■	■
rmtD2																		■						■		■
rmtB																								■		■
aph(3')-VI																		■				■				
aph(3')-IIb																		■	■			■		■	■	■
aph(3'')-Ib							■											■						■		■
aac(6')-Ib3																		■	■			■		■	■	■
aac(3)-IId																		■	■							■
aadA1																								■		■
aac(6')-Ib-cr																		■	■					■		■
Fluoroquinolone resistance genes																										
crpP			■	■		■	■	■	■									■	■	■	■	■		■	■	■
qnrVC1																		■	■					■		■
Beta-lactamase resistance genes																										
blaPAO	■	■	■	■	■	■	■	■	■	■	■	■	■	■	■	■	■	■	■	■	■	■	■	■	■	■
blaLCR-1																		■								■
blaOXA-485										■											■			■		
blaOXA-486	■																					■	■			
blaOXA-488					■						■							■			■					■
blaOXA-396		■						■										■	■	■	■	■		■		■
blaOXA-395				■														■						■		■
blaOXA-50			■						■	■											■	■			■	
blaOXA-10																								■		
blaTEM-1B																								■		
blaVIM-2																										
blaPME-1																			■							
blaPAU-1																								■		
Sulphonamide, tetracycline, macrolide, fosfomycin, and chloramphenicol resistance genes																										
sul1																		■	■		■			■		■
tet(G)																		■	■					■		■
mph(E)																								■		
mph(A)																								■		
msr(E)*																								■		
fosA	■	■	■	■	■	■	■	■	■	■	■	■	■	■	■	■	■	■	■	■	■	■	■	■	■	■
catB	■	■	■	■	■	■	■	■	■	■	■	■	■	■	■	■	■	■	■	■	■	■	■	■	■	■

* *msr(E) encodes macrolide and lincosamide resistance.* Isolates shaded in grey indicate Australian strains. *Black color represents gene presence.*

An aminoglycoside resistance gene (*aph(3')-IIb*), a beta-lactam resistance gene (*blaPAO*), a fosfomycin resistance gene (*fosA*), and a chloramphenicol resistance gene (*catB7*) were common to all isolates. The Australian isolates (123–182) had acquired only eight resistance genes, while the Indian isolates (188–221) had acquired 26 different resistance genes (Table 2). Five Indian isolates (198, 202, 217, 219, and 221, with large pan genomes) acquired the largest number of resistance genes. Of these five isolates, the pairs 198/219 and 202/221 had the most similar resistance gene profiles and each member of the pair were of the same sequence type, ST308 and ST316 respectively. As acquired resistance genes may be carried on integrons, the genomes of the *P. aeruginosa* isolates were analyzed for integrons using Integron Finder version 1.5.1. *qnrCV1* was associated with a class 1 integron in isolates 202 and 221 and a Tn3 transposon in isolates 198 and 219.

Several types of non-synonymous variations were found in the core genome of the keratitis *P. aeruginosa* isolates when compared with the reference genome of PAO1 (Table 3). These non-synonymous mutations included single nucleotide polymorphisms (SNPs), multi-nucleotide polymorphisms (MNPs), deletions, insertions, and complex variations (where more than one change occurred at one specific location compared to the reference strain). The total variations in the isolates ranged from 76,080 in isolate 206 to 22,536 in isolate 181. There was a median of 26,317 (IQR = 25,681–33,780) variations in the genomes of Australian isolates and a median of 50,006 (IQR = 26,967–50,600) in the Indian isolates ($p = 0.09$). Based on the grouping of core genome phylogeny, isolates within group 2 (198, 202, 219, 220, 221, 233) had the most variations. Isolate 206, which had a unique sequence type and was placed in a separate group by pan genome analysis, had an exceptionally high number of variations (76,080) and SNPs (67,271).

Table 3. Frequency of different types of variation in the genes of *P. aeruginosa* isolates.

P. aeruginosa Isolates	Total Variants	Variant Complex	Variants Insertions	Variants Deletions	Variants MNP	Variant SNP
123	28,279	1593	187	163	398	25,938
126	26,258	1416	164	159	355	24,164
127	25,760	1362	163	176	391	23,668
162	50,999	3481	281	257	951	46,029
169	50,283	3359	269	245	922	45,488
176	26,065	1372	168	161	342	24,022
181	22,536	1063	162	133	283	20,895
182	25,684	1359	172	180	368	23,605
223	25,672	1358	167	176	402	23,568
224	26,376	1435	163	165	353	24,260
225	28,070	1566	167	156	385	25,796
227	28,000	1560	162	154	370	25,754
233	52,392	3590	285	263	956	47,298
235	24,919	1349	162	171	354	22,883
188	25,833	1435	164	154	351	23,729
189	25,910	1458	165	155	365	23,767
193	26,567	1445	180	147	389	24,406
198	50,631	3503	280	236	945	45,667
202	49,981	3461	257	236	902	45,125
206	76,180	6449	336	371	1653	67,271
216	28,166	1548	183	164	433	25,838
217	51,119	3575	290	226	944	46,084
218	29,161	1676	182	181	430	26,692
219	50,507	3484	273	237	925	45,588
220	50,180	3452	267	234	894	45,332
221	50,030	3477	260	237	906	45,150

SNP = single nucleotide polymorphism; MNP = multi-nucleotide polymorphism. Isolate numbers highlighted in gray are from Australia.

Non-synonymous mutations were assessed in resistance genes of the *P. aeruginosa* isolates (Supplementary Table S2). There were no large differences in the mutations in resistance genes of any of the isolates except the antibiotic efflux-related genes *opmH* and *rosC*. *opmH* had ≥9 mutations in all isolates except 127, 162,169, 202, 218, 220, and 221 (mostly isolates of group 2 of core and pan genome phylogenies except 127 and 218). *rosC* had 20 non-synonymous mutations including insertions/deletions in isolate 206, 11 in 233, and ≥5 mutations in isolates 162, 169, 176, 202, 216, 217, 219, 220, and 221 (mostly isolates of group 2 of core and pan genome phylogenies except 176, 216), but ≤3 mutations in isolates 123, 126, 123, 181, 182, 188, 189, 193, 198, and 218 (mostly isolates of group 1 of core and pan genome phylogenies except 198). Mutations in efflux genes encoding efflux pumps were also found, including *mexX*, *mexT*, *mexD*, *mexM*, and *mexY*, although there was no significant difference between two groups in the possession of mutations in these genes. All other mutations in the genes were random without any association to sequence type, phylogeny, or susceptibility to antibiotics.

2.4. Possession of exoU and Mutations in the DNA Mismatch Repair System

ExoU was present in the genomes of all isolates in group 2 (core and pan genome phylogenetic group) as well as isolates 123 and 127. All other isolates possessed *exoS* with the exception of isolate 126 which possessed both *exoU* and *exoS*. To address differences in the numbers of sequence variants between the isolates, the genes involved in the DNA mismatch repair (MMR) system *mutS* (that encodes a protein which binds to errors in DNA), *mutL* (that encodes a protein that works in synergy with MutS and activates UvrD), and *uvrD* (a DNA helicase active in DNA replication) were examined. The mutations in the MMR system included SNPs, indels, and complex variants. The number of mutations in *mutL* ranged from 1 to 2 and mutations in *mutS* (which ranged between 0 and 2) were found in seven isolates (Table 4). In *uvrD*, the number of mutations ranged between 0 and 5 (Table 4). *exoU* containing isolates possessed a median of two (IQR = 1–3) mutations in *mutL*, zero (IQR = 0–2) mutations in *mutS*, and four (IQR = 2–5) mutations in *uvrD*, whereas *exoS* containing isolates possessed a median of zero (IQR = 0–1) mutations in *mutL*, zero (IQR = 0–1) mutations in *mutS*, and two (IQR = 0–2) median mutations in *uvrD*. There were significant differences in the number of *mutL* ($p = 0.0021$) and *uvrD* ($p = 0.02$) mutations in *exoS* and *exoU* isolates but not with *mutS* ($p = 0.3$). Isolate 206, an *exoS* strain and an outlier in the core genome analysis, had one mutation in *mutL*. Details of mutations occurring in nucleotide and respective proteins are provided in Supplementary Table S3.

Table 4. Possession of *exoU* and *exoS* and number and type of non-synonymous mutations in the mismatch repair system genes in *P. aeruginosa* isolates.

P. aeruginosa Isolates	Type III Secretion System Genes	*mutL*	*mutS*	*uvrD*
123	*exoU*	1 SNP	0	1 complex
126	*exoU/exoS*	0	0	0
127	*exoU*	0	1 MNP	1 MNP, 1 complex
162	*exoU*	1 SNP	0	2 SNP, 1 MNP, 2 complexes
169	*exoU*	1 SNP	1 complex	2 SNP, 2 MNP, 1 complex
176	*exoS*	1 SNP	0	1 SNP
181	exoS	0	0	0
182	exoS	0	0	1 MNP 1complex
223	exoS	0	1 SNP	1 MNP, 1 complex
224	exoS	1 SNP	0	1 MNP, 1 complex
225	exoS	0	0	2 SNP, 2 MNP, 1 complex
227	exoS	0	0	2 SNP, 2 MNP, 1 complex
233	*exoU*	0	0	1 MNP, 1 complex
235	*exoS*	0	0	0
188	exoS	0	0	1 MNP, 1 complex
189	exoS	1 SNP	0	1 MNP, 1 complex
193	exoS	0	0	0
198	*exoU*	2 SNP	0	1 SNP, 3 complexes
202	*exoU*	1 SNP	1 complex	1 SNP, 2 MNP, 2 complexes
206	*exoS*	1 MNP	1 complex	0
216	*exoS*	0	0	0
217	*exoU*	2 SNP	1 complex	1 SNP, 2 MNP, 1 complex
218	*exoS*	0	0	0

Table 4. *Cont.*

P. aeruginosa Isolates	Type III Secretion System Genes	*mutL*	*mutS*	*uvrD*
219	*exoU*	2 SNP	0	1 SNP, 1 MNP, 2 complexes
220	*exoU*	1 SNP	1 complex	1 SNP, 2 MNP, 2 complexes
221	*exoU*	1 SNP	1 complex	1 SNP, 2 MNP, 2 complexes

SNP = single nucleotide polymorphism, MNP = multinucelotide polymorphism. Isolates shaded in grey indicate Australian strains.

2.5. Sequence Type Analysis and Phylogenetics

All Australian isolates were of different sequence types (ST), except 225 and 227 which belonged to ST233. Among the 12 Indian isolates, one isolate was designated as belonging to a new sequence type, two isolates (198 and 219) belonged to ST308, two others (188 and 189) belonged to ST491, and three isolates (202, 220, and 221) belonged to ST316 (Table 5).

Table 5. Sequence types of *P. aeruginosa* isolates.

P. aeruginosa Isolates	Sequence Types	Core Genes	Shell Genes	Pan/Total Genes
123	ST218	5496	508	6004
126	ST2726	5483	712	6195
127	ST845	5483	938	6421
162	ST298	5439	905	6344
169	ST1027	5456	694	6150
176	ST709	5547	1112	6659
181	ST244	5588	1047	6662
182	ST27	5486	1096	6582
223	ST17	5471	1232	6703
224	ST168	5483	607	6090
225 □	ST233	5515	1338	6853
227 □	ST233	5493	1304	6797
233	NEWST	5440	624	6064
235	ST262	5470	540	6010
188 *	ST491	5490	535	6025
189 *	ST491	5492	531	6023
193	ST760	5490	594	6084
198 †	ST308	5454	1428	6882
202 #	ST316	5425	1505	6930
206	NEWST	5331	1084	6415
216	ST1527	5480	1488	6968
217	ST1047	5448	1173	6621
218	ST3083	5513	488	6001
219 †	ST308	5451	1796	7247
220 #	ST316	5430	948	6378
221 #	ST316	5425	1511	6936
PA7	ST1196	3599	4586	8185
PA14	ST253	5436	790	6226

Gray shading denotes Australian isolates. *, †, #, □ indicates strains belong to the same sequence types (STs).

The number of core and total or pan (or total) genes were reported from the statistical summary of Roary v3.11.2. The core genomes of the isolates were aligned using PA7 (Accession number NC_009656.1), PA14 (Accession number NC_004863.1), and PAO1 (Accession number NC_002516.1) as reference strains. The eight published genomes of *P. aeruginosa* isolates from eye as well as strains from other sources were also included. The core genes of published isolates are provided in Supplementary Table S4. The isolates were sub-grouped based on the number of core genes; isolates with a similar number of core genes were closely aligned and isolates with the same sequence type were grouped together. The core genomes formed two groups in the phylogenetic tree (Figure 1). Isolates in group 1 tended to have a larger number of core genes than isolates in group 2. Isolate 206, PA57, and PA7 were outliers based on core genome phylogeny. The Australian and Indian isolates had a similar number of core genes, whereas the Indian isolates had a larger number of pan genes (10,889) due to the acquisition of shell genes (genes present in two or more strains) (Table 4).

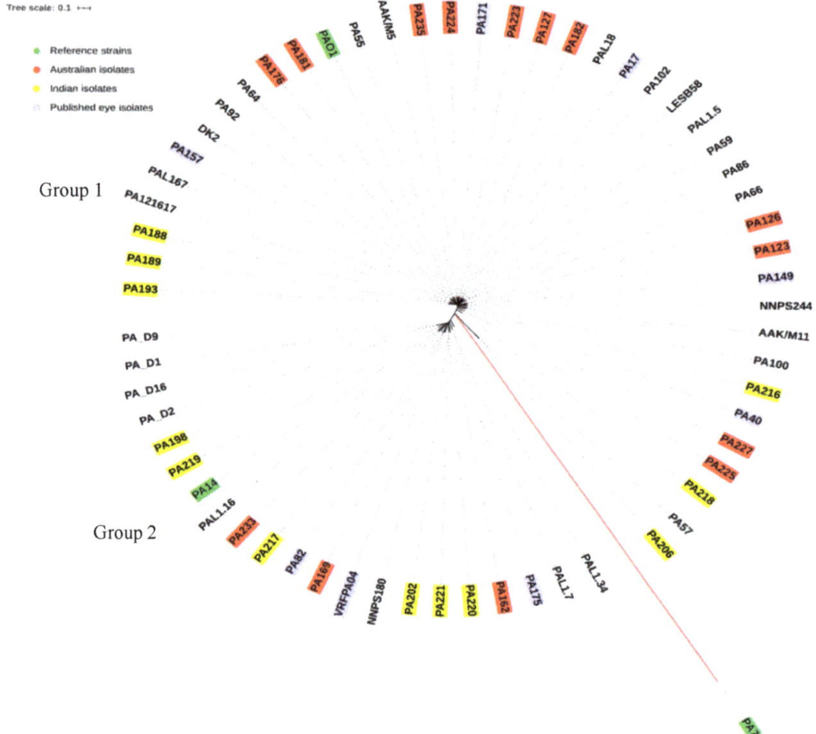

Figure 1. Core genome phylogeny of *P. aeruginosa* isolates using Parsnp. PAO1 was used as reference. PA7 and PA14 were also included.

The phylogenetic relationships of these *P. aeruginosa* isolates were assessed by aligning their pan genome against PAO1 as a reference. The output generated using Roary showing the gene presence or absence in all isolates is provided in Supplementary Figure S1. This again divided the *P. aeruginosa* isolates into two major groups. Six multi-drug resistant Indian isolates (198, 202, 217, 219, 220, 221) and the VRFPA04 isolate (isolated from the cornea) were clustered in one group, which also contained the two Australian isolates 162 and 169. The Indian isolate 216 was categorized in a separate sub-group due to the large number of shell genes and possession of *exoS*.

The second group (group 2 of the pan genome analysis) included most of the Australian (123, 126, 127, 162, 176, 181, 182, 223, 224, 225, 227, 235) and Indian (188, 189, 193, 216 218) isolates along with reference strain PAO1 (Figure 2). Overall, the multi-drug resistant Indian isolates had a large pan genome (total of 10,889 genes obtained from the statistical summary in Roary v3.11.2). The pan genome grouping of isolates was broadly based on the number of pan (or total) genes and possession of either *exoU* or *exoS* in each group, except two Australian isolates 123 and 127 which were in group 2 but possessed *exoU*. The other exception to this grouping pattern was for isolates 181 and 182 which had large pan genomes and were clustered into group 1 but carried *exoS*.

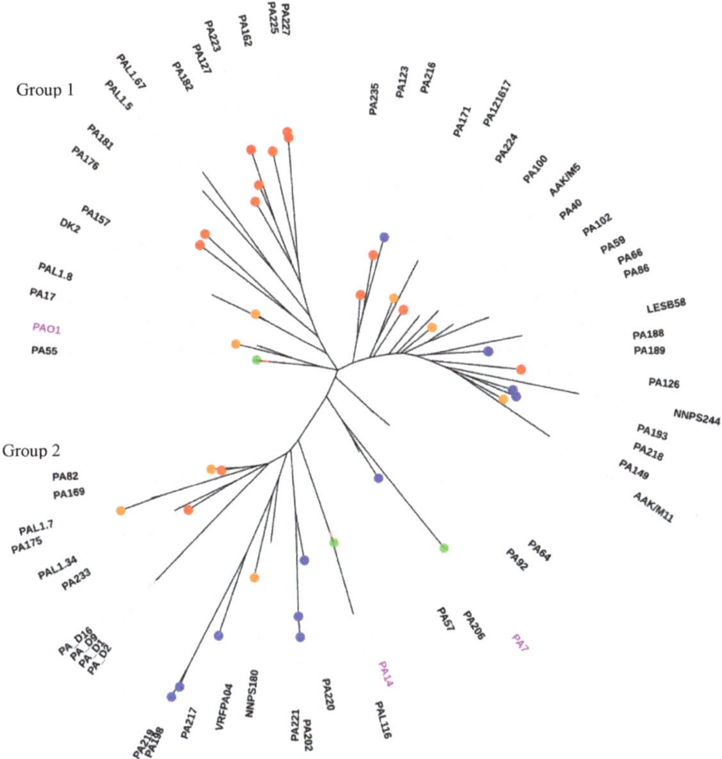

Figure 2. Pan genome phylogeny of *P. aeruginosa* isolates. Branches with no color representation indicate non-ocular isolates. Red color indicates Australian, blue color represents Indian, and orange color represents published eye isolates. Green color represents reference strains. Purple color represents reference strains.

The isolates of group 2 usually had a large number of pan genes and were *exoU*+. Isolates having similar numbers of pan genes were sub-grouped together. For example, isolate 193 (pan genes = 6084) and 218 (pan genes = 6001) were sub-grouped together. Isolate 218 had a similar number of pan genes to isolate 123 (pan genes = 6001), but isolate 218 possessed *exoS*, while 123 possessed *exoU*, and thus these were not grouped together. Isolates belonging to the same sequence type were also grouped together. The MDR isolates, the isolates with same STs, and isolates with large gene variations were clustered in one pan-group. The previously published isolates PA_D1, PA_D2, PA_D9, and PA_D16 with the same ST and those with large shell genes were grouped with the MDR isolates of the current study.

3. Discussion

This study investigated genomic differences in Australian and Indian *P. aeruginosa* isolates from keratitis. Phenotypically, more resistance was found in Indian isolates compared to Australian isolates as has been shown in previous studies [30,31]. Unregulated antibiotic use in India has been linked to increased antibiotic resistance [32]. Resistance to antibiotics is problematic even in the treatment of keratitis, where a topical application of antibiotics is used. Infection with antibiotic resistant strains results in prolonged infection [33], more severe keratitis [5], and an increase in the cost of treatment [34,35].

Indian *P. aeruginosa* strains harbored more resistance genes compared to Australian isolates, although *aph(3')-IIb*, *blaPAO1 (fosA)*, and *catB7* were found in all isolates, which was consistent with previous studies [31,36]. *qnrVC1* was found in four Indian isolates but no Australian isolates. This fluoroquinolone resistance gene has not been previously reported in *P. aeruginosa* ocular isolates [31], but it has been reported in burns isolates and has been identified as being carried on an integron [37]. Similarly, in the current study *qnrVC1* was carried on a class 1 integron in isolates 202 and 221, but integrated into a Tn3 transposon in isolates 198 and 219. This gene has also been isolated from the high risk ST773 clone of *P. aeruginosa* from urine in Hungary [38]. High risk clones are isolates with high mutational rates in resistance genes and those that have acquired a large number of resistance genes. As previously described, resistance to fluoroquinolones in keratitis *P. aeruginosa* isolates was also due to mutations in the quinolone resistance determining regions of *gyrA* and *parC* [15]. Possession of *qnrVC1* and mutations in *gyrA* and *parC* were associated with high levels of fluroquinolone resistance. The possession of large numbers of acquired resistance genes by Indian isolates likely contributed to the higher rates of resistance of these isolates. The Indian isolates 198, 202, 217, 219, 220, and 221 also had a high number of gene variations which is an independent mechanism of resistance.

The aminoglycoside resistance gene *aph(6)-Id* which encodes for streptomycin resistance was found in six Indian isolates, including the four that carried *qnrVC1*, but in no Australian isolates. Previously, *aph(6)-Id* was found in only one Indian ocular isolate from 1997 [31], but has been found in cystic fibrosis *P. aeruginosa* isolates [39] and has been associated with the transposon Tn5393 on a plasmid in one strain of *P. aeruginosa* [39]. As streptomycin is no longer used in clinical treatment [40], this resistance may not be clinically relevant but does suggest environmental selection for the persistence of this gene.

The total number of gene variants found in the Indian isolates 198, 202, 219, and 221 were greater than Australian isolates. However, there were a small number of SNPs found in the genes associated with resistance for these isolates. The Indian isolate 206 (NEWST) had a high number of SNPs in antibiotic resistance genes *mexC*, *mexD*, *mexM*, *mexX*, *mexS*, *opmE*, *mexP*, *mexK*, *oprJ*, *ampC*, *rosC*, and *mprF*. There was no difference in the mutations of other *mex* genes including *mexX*, *mexT*, *mexD*, *mexM*, and *mexY* between Australian and Indian isolates. Given that most isolates from both countries, whether they were sensitive or resistant, had a similar number of mutations in the resistance genes, it is likely that the resistance to antibiotics was related to the possession of acquired resistance genes rather than mutations in chromosomal genes.

In the Australian isolates, four out of the eight isolates (50%) carried *exoU*, while one isolate was both *exoU+/exoS+* and three (38%) were *exoS+*. In Indian isolates, 50% carried *exoU* and 50% carried *exoS*. A previous study has also shown an equal ratio of both genes [41] in keratitis isolates. The possession of the *exoU* genotype in *P. aeruginosa* ocular isolates has been related to elevated resistance to disinfectants [42], fluoroquinolones [43], and multiple antibiotics [41]. Furthermore, one study reported worst clinical outcomes and more resistance by *exoU* carrying isolates [43]. The isolates of this study showed similar findings because the *exoU+* isolates 198, 202, 217, 219, 220, 221, and 233 were also MDR.

The DNA mismatch repair system (MMR) in *P. aeruginosa* is based on the protein trimer MutS-MutL-UvrD and functions to correct errors and preserve the integrity of the genome [24,44,45]. The *mutH* component of MMR, which is important in other Gram negative bacteria, such as *E. coli* [46], has not been found previously in *P. aeruginosa* [25] and was not present in the isolates of the current study. Mutations in *mutS*, *mutL* and *uvrD* can reduce the ability of the bacterium to repair DNA lesions [46]. Strong mutator strains have defects in their MMR system and mutations in *mutS* predominate [47]. Mutations in the MMR can be a reason for the development of hypermutations in isolates. In cystic fibrosis, hypermutations were found to be a key factor in the development of MDR resistant *P. aeruginosa* strains [23]. Similar findings were found in this study where isolates 198, 202, 206, 219, and 221 had mutations in the MMR genes, and these isolates had an overall larger variation in their genomes. In the current study, isolates had more mutations in *mutL* and *uvrD*, suggesting the strains may not be strong mutators (which is usually associated with mutation in *mutS*), but nevertheless can undergo uncorrected genetic changes. Indeed, the *P. aeruginosa* isolates in the current study which had mutations in *mutL* and *uvrD* had greater numbers of SNPs, insertions and deletions, acquired genes, and had large pan genomes. Among these isolates, 198, 202, 219, and 221 possessed either the transposon Tn3 or class 1 integrons which carried the acquired genes. This also might be due to mutated MMR, as mutations in MMR genes increase the chances of horizontal gene transfer in mutator isolates [47]. The number of mutations in MMR was greater in *exoU* possessing isolates with large gene variations. *exoU* is carried on genomic islands [48,49] and these *exoU* carrying isolates had larger pan genomes with possession of mobile genetic elements. Therefore, the isolates with the mutated MMR systems may have a greater ability of strains to accumulate gene variations and the acquisition of *exoU*. Isolate 206, on the other hand, possessed *exoS* and was not MDR but possessed a large number of SNPs and a large pan genome with one mutation in each of *mutS* and *mutL*. Further in-depth studies are required to understand the influence of the MMR system on genomic changes in *P. aeruginosa*.

Analysis of the sequence types of the *P. aeruginosa* ocular isolates revealed the presence of three clones, two in the Indian and one in the Australian isolates. The isolates with the same STs had mostly the same phenotypic and genotypic features. The exception to this was isolate 220 that had acquired fewer resistance genes compared to the other two isolates 202 and 221 of ST316. Previously, five ocular *P. aeruginosa* strains from India isolated in 1997 were of sequence type ST308 [31]. The two isolates of ST308 in the current study, isolated in 2017 and 2018, had acquired more resistance genes compared to isolates from 1997 [31]. This indicates that the clonal isolates have continued to evolve over this time period, although the specific selection factors driving those changes are yet to be elucidated. None of the isolates were collected from the same patient. The majority of the isolates with the same STs grouped in the same phylogeny including previously published isolates (PA_D1, PA_D2, PA_D9, P_D16) with ST1971.

Core and pan genome phylogenies of the isolates produced two almost identical groups, which was in agreement with previously published studies [31,50]. Both phylogenies included isolates from either Australia or India, but those in group 2 tended to be the MDR Indian isolates and possessed higher numbers of antibiotic resistance genes. About 65% of all ocular isolates grouped together which indicated less diversity in the ocular *P. aeruginosa* isolates [31,51,52]. The grouping of MDR strains from this study with PA14 along a MDR ocular isolate VRFPA04 [36] in both core genome and pan genome analysis, and the grouping of the sensitive strains with PAO1 along the commonly studied cystic fibrosis isolates DK2 and LESB58, was similar to a previous study examining older isolates from India and Australia [31]. Isolate 206, which had the smallest number of core genes and was of a new sequence type, was an outlier in the core genome phylogeny similar to the taxonomic outlier PA7 [53]. However, isolate 206 was grouped together with other isolates in the pan genome because it had acquired a large number of genes. Acquired genes are part of the pan rather than the core genome [53] and the presence of larger pan genomes in MDR *P. aeruginosa* isolates points towards the acquisition of new genes [54]. Previously, a smaller core genome size of 4910 genes has been reported in ocular *P. aeruginosa* isolates [31]. However, the current study found a core genome size similar to *P. aeruginosa* from different sources, comprising 5316–5233 genes [55,56]. The core genome (which is almost 90% of total genome) refers to the conserved genes present in a species [57] which might differ in each individual strain within that species. Additionally, SNPs can be a result of poor sequencing quality and hence it is important to have a good sequencing depth at those positions to identify them as a mutation rather than sequencing error [58]. Grouping of all the isolates including ocular and non-ocular remained the same in both core and pan genome phylogeny.

4. Materials and Methods

4.1. P. aeruginosa Strains and Susceptibility Testing

Twenty-six *P. aeruginosa* keratitis isolates, eight isolated in Australia from 2004 to 2006, six from 2018 and 2019 (total 14 Australian isolates), and twelve isolated in India between 2017 and 2018, were included in this study. These isolates were selected from a larger collection of strains based on their antibiotic susceptibilities (those phenotypically resistant to multiple antibiotics, some resistant to one or multiple antibiotics, and some which were sensitive to all antibiotics). The susceptibilities of Australian strains (2004–2006) included in this study have been previously published [29]. Strains were selected after comparing their susceptibilities to antibiotics that are used to treat ocular infections. For genetic comparisons, the data of 34 *P. aeruginosa* isolates from eyes and other sources were also included. The general characteristics of these isolates are described in Supplementary Table S4. The genomes of these isolates were downloaded from the NCBI database and reannotated for this study using the same parameters as of the isolates of this study to avoid any bias in results.

The minimum inhibitory concentration (MIC) and minimum bactericidal concentration (MBC) of various antibiotics which are commonly used to treat *P. aeruginosa* keratitis [16] were assessed for the isolates using the broth microdilution method in 96-well plates following the Clinical and Laboratory Standard Institute guidelines [59]. The antibiotics tested were ciprofloxacin, levofloxacin, gentamicin, ceftazidime (Sigma-Aldrich, St. Louis, MO, USA), polymyxin B (Sigma-Aldrich, Vandtårnsvej, Søborg, Denmark), tobramycin, piperacillin (Cayman Chemical Company, Ann Arbor, MI, USA), and imipenem (LKT Laboratories Inc., St. Paul, MN, USA). The susceptibility results were interpreted using the EUCAST v9 [60] and CLSI [61] 2017 breakpoints.

4.2. Genomic Sequencing

DNeasy Blood and Tissue Kits (Qiagen, Hilden, Germany) were used for DNA extraction as per the manufacturer's recommendations. The Nextera XT DNA library preparation kit (Illumina, San Diego, CA, USA) was used to prepare paired-end libraries. All the libraries were multiplexed on one MiSeq run. FastQC version 0.117 (https://www.bioinformatics.babraham.ac.uk/projects/fastqc) was used to assess the quality of sequenced genomes using raw reads. Version 0.38 of Trimmomatic [61] was used for trimming the adapters from the reads following de novo assembly using Spades v3.13.0 [62]. Genomes were annotated using Prokka v1.12 [63].

Sequence types were investigated using PubMLST https://pubmlst.org/. Pan genomes of the *P. aeruginosa* isolates were analyzed using Roary v3.11.2 [64] using PAO1 as a reference, while core genome phylogeny was constructed using Harvest Suite Parsnp v1.2 [65] with strains PAO1, PA7, and PA14 used as reference strains. The output file 'genes_ presence_absence' was used to compare the *P. aeruginosa* isolates. Acquired resistance genes were identified using the online database Resfinder v3.1 (Centre for Genomic Epidemiology, DTU, Denmark) [66]. Integron Finder v1.5.1 was used to identify any integrons present in the isolates. Mutations in the genes were detected using Snippy V2 [67]. Isolates with same sequence types were compared for nucleotide similarities using the MUMmer online web tool (http://jspecies.ribohost.com/jspeciesws/#analyse).

Using the Pseudomonas genome database (http://www.pseudomonas.com) and comprehensive antibiotic resistance database (https://card.mcmaster.ca), 76 genes related to *P. aeruginosa* resistance were selected to investigate the presence of single nucleotide polymorphisms. All isolates were analyzed for the presence of the type III secretion system associated virulence factors *exoU* and *exoS* using the BlastN database.

4.3. Statistical Analysis

The statistical analysis was performed using GraphPad Prism v8. Medians were calculated with the 'descriptive statistics' option during analysis of variance (ANOVA). *P*-values less than 0.05 were considered as significant. Fischer's Exact test was used to find the difference between acquired genes.

To analyze the significant difference in the DNA mismatch repair genes between *exoU* and *exoS* isolates and gene variations in the isolates, the Mann–Whitney test was used.

5. Conclusions

Indian isolates and Australian isolates were clearly distinct in carrying a type III secretion system related to *exoU* and *exoS*. There was an association in the isolates for carrying acquired resistance genes with a large number of pan genes. Indian isolates were more resistant to antibiotics compared to Australian isolates. Additionally, isolates of *P. aeruginosa* from ocular infection had a large number of genetic variations (mutations) and a mutated mismatch repair system. However, the isolates collected from the same region or time will give a clearer idea of these differences.

Supplementary Materials: The following are available online at http://www.mdpi.com/2079-6382/9/9/600/s1, Table S1: Details of the *Pseudomonas aeruginosa* isolates used in the current study. Table S2: Gene variations of resistance genes in *Pseudomonas aeruginosa* isolates (the gray-shaded strains were isolated from Australia). Nucleotide accession: The nucleotide sequences are available in the GenBank under the Bio project accession number PRJNA590804. Table S3: Types of mutations in the mismatch repair system; Table S4: Genomics features of *P. aeruginosa* isolates; Figure S1: Pan-genome phylogenetic tree. The data on the right of the figure shows the presence and absence of genes. The tree was built using the genome of PAO1 as a reference.

Author Contributions: Conceptualization, M.K., F.S., and M.D.P.W.; methodology, M.K., F.S., S.S., S.A.R., and M.D.P.W.; writing—original draft preparation, M.K.; writing—review and editing, F.S., S.A.R., and M.D.P.W.; supervision, F.S. and M.D.P.W.; funding acquisition, F.S. All authors have read and agreed to the published version of the manuscript.

Funding: This research received no external funding.

Acknowledgments: The authors would like to acknowledge the Singapore Centre for Environmental Life Sciences Engineering (SCELSE), whose research is supported by the National Research Foundation Singapore, Ministry of Education, Nanyang Technological University, and National University of Singapore, under its Research Centre of Excellence Programme. We would also like to acknowledge Engineer Ahsan Ullah Khan, Independent Monitoring Unit Haripur KPK, Pakistan, for his help in the computational analysis and Doctor Nicole Carnt, School of Optometry and Vision Science UNSW, Sydney, for her help in the isolates collection. We are also thankful to UNSW high performance computing facility KATANA for providing us with the cluster time for the data analysis.

Conflicts of Interest: The authors declare no conflict of interest.

References

1. Richards, M.J.; Edwards, J.R.; Culver, D.H.; Gaynes, R.P. Nosocomial infections in medical intensive care units in the United States. National Nosocomial Infections Surveillance System. *Crit. Care Med.* **1999**, *27*, 887–892. [CrossRef] [PubMed]

2. Abjani, F.; Khan, N.A.; Jung, S.Y.; Siddiqui, R. Status of the effectiveness of contact lens disinfectants in Malaysia against keratitis-causing pathogens. *Exp. Parasitol.* **2017**, *183*, 187–193. [CrossRef] [PubMed]

3. Stapleton, F.; Carnt, N. Contact lens-related microbial keratitis: How have epidemiology and genetics helped us with pathogenesis and prophylaxis. *Eye (Lond.)* **2012**, *26*, 185–193. [CrossRef] [PubMed]

4. Trouillet, J.L.; Vuagnat, A.; Combes, A.; Kassis, N.; Chastre, J.; Gibert, C. *Pseudomonas aeruginosa* ventilator-associated pneumonia: Comparison of episodes due to piperacillin-resistant versus piperacillin-susceptible organisms. *Clin. Infect. Dis.* **2002**, *34*, 1047–1054. [CrossRef] [PubMed]

5. Green, M.; Apel, A.; Stapleton, F. Risk factors and causative organisms in microbial keratitis. *Cornea* **2008**, *27*, 22–27. [CrossRef]

6. Hooi, S.H.; Hooi, S.T. Culture-proven bacterial keratitis in a Malaysian general hospital. *Med. J. Malays.* **2005**, *60*, 614–623.

7. Parmar, P.; Salman, A.; Kalavathy, C.M.; Kaliamurthy, J.; Thomas, P.A.; Jesudasan, C.A. Microbial keratitis at extremes of age. *Cornea* **2006**, *25*, 153–158. [CrossRef]

8. Sharma, N.; Sinha, R.; Singhvi, A.; Tandon, R. *Pseudomonas* keratitis after laser in situ keratomileusis. *J. Cataract. Refract. Surg.* **2006**, *32*, 519–521. [CrossRef]

9. Chatterjee, S.; Agrawal, D. Multi-drug resistant *Pseudomonas aeruginosa* keratitis and its effective treatment with topical colistimethate. *Indian J. Ophthalmol.* **2016**, *64*, 153–157. [CrossRef]

10. Livermore, D.M. beta-Lactamases in laboratory and clinical resistance. *Clin. Microbiol. Rev.* **1995**, *8*, 557–584. [CrossRef]

11. Li, X.-Z.; Ma, D.; Livermore, D.M.; Nikaido, H. Role of efflux pump (s) in intrinsic resistance of *Pseudomonas aeruginosa*: Active efflux as a contributing factor to beta-lactam resistance. *Antimicrob. Agents Chemother.* **1994**, *38*, 1742–1752. [CrossRef] [PubMed]

12. Poole, K. *Pseudomonas aeruginosa*: Resistance to the max. *Front. Microbiol.* **2011**, *2*, 65. [CrossRef] [PubMed]

13. Hocquet, D.; Vogne, C.; El Garch, F.; Vejux, A.; Gotoh, N.; Lee, A.; Lomovskaya, O.; Plésiat, P. MexXY-OprM efflux pump is necessary for adaptive resistance of *Pseudomonas aeruginosa* to aminoglycosides. *Antimicrob. Agents Chemother.* **2003**, *47*, 1371–1375. [CrossRef] [PubMed]

14. Livermore, D.M. Multiple mechanisms of antimicrobial resistance in *Pseudomonas aeruginosa*: Our worst nightmare? *Clin. Infect. Dis.* **2002**, *34*, 634–640. [CrossRef] [PubMed]

15. Lomholt, J.A.; Kilian, M. Ciprofloxacin susceptibility of *Pseudomonas aeruginosa* isolates from keratitis. *Br. J. Ophthalmol.* **2003**, *87*, 1238–1240. [CrossRef]

16. Willcox, M.D. Review of resistance of ocular isolates of *Pseudomonas aeruginosa* and *staphylococci* from keratitis to ciprofloxacin, gentamicin and cephalosporins. *Clin. Exp. Optom.* **2011**, *94*, 161–168. [CrossRef]

17. Poonsuk, K.; Tribuddharat, C.; Chuanchuen, R. Aminoglycoside resistance mechanisms in *Pseudomonas aeruginosa* isolates from non-cystic fibrosis patients in Thailand. *Can. J. Microbiol.* **2013**, *59*, 51–56. [CrossRef]

18. Diver, J.M.; Schollaardt, T.; Rabin, H.R.; Thorson, C.; Bryan, L.E. Persistence mechanisms in *Pseudomonas aeruginosa* from cystic fibrosis patients undergoing ciprofloxacin therapy. *Antimicrob. Agents Chemother.* **1991**, *35*, 1538–1546. [CrossRef]

19. Guzvinec, M.; Izdebski, R.; Butic, I.; Jelic, M.; Abram, M.; Koscak, I.; Baraniak, A.; Hryniewicz, W.; Gniadkowski, M.; Andrasevic, A.T. Sequence types 235, 111, and 132 predominate among multidrug-resistant *Pseudomonas aeruginosa* clinical isolates in Croatia. *Antimicrob. Agents Chemother.* **2014**, *58*, 6277–6283. [CrossRef]

20. Phillips, R.M.; Six, D.A.; Dennis, E.A.; Ghosh, P. In vivo phospholipase activity of the *Pseudomonas aeruginosa* cytotoxin ExoU and protection of mammalian cells with phospholipase A2 inhibitors. *J. Biol. Chem.* **2003**, *278*, 41326–41332. [CrossRef]

21. Finck-Barbançon, V.; Goranson, J.; Zhu, L.; Sawa, T.; Wiener-Kronish, J.P.; Fleiszig, S.M.J.; Wu, C.; Mende-Mueller, L.; Frank, D.W. ExoU expression by *Pseudomonas aeruginosa* correlates with acute cytotoxicity and epithelial injury. *Mol. Microbiol.* **1997**, *25*, 547–557. [CrossRef] [PubMed]

22. Subedi, D.; Vijay, A.K.; Willcox, M. Overview of mechanisms of antibiotic resistance in *Pseudomonas aeruginosa*: An ocular perspective. *Clin. Exp. Optom.* **2018**, *101*, 162–171. [CrossRef]

23. Maciá, M.D.; Blanquer, D.; Togores, B.; Sauleda, J.; Pérez, J.L.; Oliver, A. Hypermutation is a key factor in development of multiple-antimicrobial resistance in *Pseudomonas aeruginosa* strains causing chronic lung infections. *Antimicrob. Agents Chemother.* **2005**, *49*, 3382–3386. [CrossRef] [PubMed]

24. Modrich, P. Mechanisms and biological effects of mismatch repair. *Annu. Rev. Genet.* **1991**, *25*, 229–253. [CrossRef]

25. Oliver, A.; Baquero, F.; Blazquez, J. The mismatch repair system (*mutS, mutL* and *uvrD* genes) in *Pseudomonas aeruginosa*: Molecular characterization of naturally occurring mutants. *Mol. Microbiol.* **2002**, *43*, 1641–1650. [CrossRef]

26. Alsan, M.; Schoemaker, L.; Eggleston, K.; Kammili, N.; Kolli, P.; Bhattacharya, J. Out-of-pocket health expenditures and antimicrobial resistance in low-income and middle-income countries: An economic analysis. *Lancet Infect. Dis.* **2015**, *15*, 1203–1210. [CrossRef]

27. Sánchez, M.; Sivaraman, S. News Media Reporting of Antimicrobial Resistance in Latin America and India. In *Antimicrobial Resistance in Developing Countries*; Sosa, A.d.J., Byarugaba, D.K., Amábile-Cuevas, C.F., Hsueh, P.-R., Kariuki, S., Okeke, I.N., Eds.; Springer: New York, NY, USA, 2010; pp. 525–537.

28. Van Boeckel, T.P.; Gandra, S.; Ashok, A.; Caudron, Q.; Grenfell, B.T.; Levin, S.A.; Laxminarayan, R. Global antibiotic consumption 2000 to 2010: An analysis of national pharmaceutical sales data. *Lancet Infect. Dis.* **2014**, *14*, 742–750. [CrossRef]

29. Khan, M.; Stapleton, F.; Willcox, M.D.P. Susceptibility of contact lens-related *Pseudomonas aeruginosa* keratitis isolates to multipurpose disinfecting solutions, disinfectants, and antibiotics. *Transl. Vis. Sci. Technol.* **2020**, *9*, 2. [CrossRef]

30. Choy, M.H.; Stapleton, F.; Willcox, M.D.P.; Zhu, H. Comparison of virulence factors in *Pseudomonas aeruginosa* strains isolated from contact lens- and non-contact lens-related keratitis. *J. Med. Microbiol.* **2008**, *57*, 1539–1546. [CrossRef] [PubMed]
31. Subedi, D.; Vijay, A.K.; Kohli, G.S.; Rice, S.A.; Willcox, M. Comparative genomics of clinical strains of *Pseudomonas aeruginosa* strains isolated from different geographic sites. *Sci. Rep.* **2018**, *8*, 15668. [CrossRef]
32. Porter, G.; Grills, N. Medication misuse in India: A major public health issue in India. *J. Public Health (Oxf.)* **2016**, *38*, e150–e157. [CrossRef] [PubMed]
33. Wilhelmus, K.R.; Abshire, R.L.; Schlech, B.A. Influence of fluoroquinolone susceptibility on the therapeutic response of fluoroquinolone-treated bacterial keratitis. *Arch. Ophthalmol.* **2003**, *121*, 1229–1233. [CrossRef] [PubMed]
34. Keay, L.; Edwards, K.; Naduvilath, T.; Taylor, H.R.; Snibson, G.R.; Forde, K.; Stapleton, F. Microbial keratitis predisposing factors and morbidity. *Ophthalmology* **2006**, *113*, 109–116. [CrossRef] [PubMed]
35. Keay, L.; Edwards, K.; Dart, J.; Stapleton, F. Grading contact lens-related microbial keratitis: Relevance to disease burden. *Optom. Vis. Sci.* **2008**, *85*, 531–537. [CrossRef]
36. Murugan, N.; Malathi, J.; Umashankar, V.; Madhavan, H.N. Unraveling genomic and phenotypic nature of multidrug-resistant (MDR) *Pseudomonas aeruginosa* VRFPA04 isolated from keratitis patient. *Microbiol. Res.* **2016**, *193*, 140–149.
37. Belotti, P.T.; Thabet, L.; Laffargue, A.; Andre, C.; Coulange-Mayonnove, L.; Arpin, C.; Messadi, A.; M'zali, F.; Quentin, C.; Dubois, V. Description of an original integron encompassing blaVIM-2, qnrVC1 and genes encoding bacterial group II intron proteins in *Pseudomonas aeruginosa*. *J. Antimicrob. Chemother.* **2015**, *70*, 2237–2240. [CrossRef]
38. Kocsis, B.; Toth, A.; Gulyas, D.; Ligeti, B.; Katona, K.; Rokusz, L.; Szabo, D. Acquired qnrVC1 and blaNDM-1 resistance markers in an international high-risk *Pseudomonas aeruginosa* ST773 clone. *J. Med. Microbiol.* **2019**, *68*, 336–338. [CrossRef]
39. Tauch, A.; Schluter, A.; Bischoff, N.; Goesmann, A.; Meyer, F.; Puhler, A. The 79,370-bp conjugative plasmid pB4 consists of an IncP-1beta backbone loaded with a chromate resistance transposon, the strA-strB streptomycin resistance gene pair, the oxacillinase gene bla(NPS-1), and a tripartite antibiotic efflux system of the resistance-nodulation-division family. *Mol. Genet. Genom.* **2003**, *268*, 570–584.
40. Sundin, G.W.; Bender, C.L. Dissemination of the strA-strB streptomycin-resistance genes among commensal and pathogenic bacteria from humans, animals, and plants. *Mol. Ecol.* **1996**, *5*, 133–143. [CrossRef]
41. Zhu, H.; Conibear, T.C.; Bandara, R.; Aliwarga, Y.; Stapleton, F.; Willcox, M.D. Type III secretion system-associated toxins, proteases, serotypes, and antibiotic resistance of *Pseudomonas aeruginosa* isolates associated with keratitis. *Curr. Eye Res.* **2006**, *31*, 297–306. [CrossRef]
42. Lakkis, C.; Fleiszig, S.M. Resistance of *Pseudomonas aeruginosa* isolates to hydrogel contact lens disinfection correlates with cytotoxic activity. *J. Clin. Microbiol.* **2001**, *39*, 1477–1486. [CrossRef] [PubMed]
43. Borkar, D.S.; Acharya, N.R.; Leong, C.; Lalitha, P.; Srinivasan, M.; Oldenburg, C.E.; Cevallos, V.; Lietman, T.M.; Evans, D.J.; Fleiszig, S.M. Cytotoxic clinical isolates of *Pseudomonas aeruginosa* identified during the Steroids for Corneal Ulcers Trial show elevated resistance to fluoroquinolones. *BMC Ophthalmol.* **2014**, *14*, 54. [CrossRef] [PubMed]
44. Tiraby, J.G.; Fox, M.S. Marker discrimination in transformation and mutation of *Pneumococcus*. *Proc. Natl. Acad. Sci. USA* **1973**, *70*, 3541–3545. [CrossRef] [PubMed]
45. Acharya, S.; Foster, P.L.; Brooks, P.; Fishel, R. The coordinated functions of the *E. coli* MutS and MutL proteins in mismatch repair. *Mol. Cell* **2003**, *12*, 233–246. [CrossRef]
46. Rayssiguier, C.; Thaler, D.S.; Radman, M. The barrier to recombination between *Escherichia coli* and *Salmonella typhimurium* is disrupted in mismatch-repair mutants. *Nature* **1989**, *342*, 396–401. [CrossRef]
47. Chopra, I.; O'Neill, A.J.; Miller, K. The role of mutators in the emergence of antibiotic-resistant bacteria. *Drug Resist. Updates* **2003**, *6*, 137–145. [CrossRef]
48. Sato, H.; Frank, D.W.; Hillard, C.J.; Feix, J.B.; Pankhaniya, R.R.; Moriyama, K.; Finck-Barbançon, V.; Buchaklian, A.; Lei, M.; Long, R.M.; et al. The mechanism of action of the *Pseudomonas aeruginosa*-encoded type III cytotoxin, ExoU. *EMBO J.* **2003**, *22*, 2959–2969. [CrossRef]
49. Kulasekara, B.R.; Kulasekara, H.D.; Wolfgang, M.C.; Stevens, L.; Frank, D.W.; Lory, S. Acquisition and evolution of the exoU locus in *Pseudomonas aeruginosa*. *J. Bacteriol.* **2006**, *188*, 4037–4050. [CrossRef]

50. Freschi, L.; Jeukens, J.; Kukavica-Ibrulj, I.; Boyle, B.; Dupont, M.J.; Laroche, J.; Larose, S.; Maaroufi, H.; Fothergill, J.L.; Moore, M.; et al. Clinical utilization of genomics data produced by the international *Pseudomonas aeruginosa* consortium. *Front. Microbiol.* **2015**, *6*, 1036. [CrossRef]

51. Winstanley, C.; Langille, M.G.; Fothergill, J.L.; Kukavica-Ibrulj, I.; Paradis-Bleau, C.; Sanschagrin, F.; Thomson, N.R.; Winsor, G.L.; Quail, M.A.; Lennard, N.; et al. Newly introduced genomic prophage islands are critical determinants of in vivo competitiveness in the Liverpool Epidemic Strain of *Pseudomonas aeruginosa*. *Genome Res.* **2009**, *19*, 12–23. [CrossRef]

52. Klockgether, J.; Cramer, N.; Wiehlmann, L.; Davenport, C.F.; Tummler, B. *Pseudomonas aeruginosa* genomic structure and diversity. *Front. Microbiol.* **2011**, *2*, 150. [CrossRef] [PubMed]

53. Roy, P.H.; Tetu, S.G.; Larouche, A.; Elbourne, L.; Tremblay, S.; Ren, Q.; Dodson, R.; Harkins, D.; Shay, R.; Watkins, K.; et al. Complete genome sequence of the multiresistant taxonomic outlier *Pseudomonas aeruginosa* PA7. *PLoS ONE* **2010**, *5*, e8842. [CrossRef] [PubMed]

54. Freschi, L.; Vincent, A.T.; Jeukens, J.; Emond-Rheault, J.G.; Kukavica-Ibrulj, I.; Dupont, M.J.; Charette, S.J.; Boyle, B.; Levesque, R.C. The *Pseudomonas aeruginosa* pan-genome provides new insights on its population structure, horizontal gene transfer, and pathogenicity. *Genome Biol. Evol.* **2019**, *11*, 109–120. [CrossRef] [PubMed]

55. Ozer, E.A.; Allen, J.P.; Hauser, A.R. Characterization of the core and accessory genomes of *Pseudomonas aeruginosa* using bioinformatic tools Spine and AGEnt. *BMC Genom.* **2014**, *15*, 737. [CrossRef]

56. Valot, B.; Guyeux, C.; Rolland, J.Y.; Mazouzi, K.; Bertrand, X.; Hocquet, D. What it takes to be a *Pseudomonas aeruginosa*? The core genome of the opportunistic pathogen updated. *PLoS ONE* **2015**, *10*, e0126468. [CrossRef] [PubMed]

57. Wolfgang, M.C.; Kulasekara, B.R.; Liang, X.; Boyd, D.; Wu, K.; Yang, Q.; Miyada, C.G.; Lory, S. Conservation of genome content and virulence determinants among clinical and environmental isolates of *Pseudomonas aeruginosa*. *Proc. Natl. Acad. Sci. USA* **2003**, *100*, 8484–8489. [CrossRef]

58. Smits, T.H.M. The importance of genome sequence quality to microbial comparative genomics. *BMC Genom.* **2019**, *20*, 662. [CrossRef]

59. Bolger, A.M.; Lohse, M.; Usadel, B. Trimmomatic: A flexible trimmer for Illumina sequence data. *Bioinformatics* **2014**, *30*, 2114–2120. [CrossRef]

60. EUCAST. *Breakpoint Tables for Interpretation of MICs and Zone Diameters*; Version 6.0; The European Committee on Antimicrobial Susceptibility Testing: Växjö, Sweden, 2019.

61. Patel, J.B.; Cockerill, F.R.; Alder, J.; Bradford, P.A.; Eliopoulos, G.M.; Hardy, D.J.; Hindler, J.A.; Jenkins, S.G.; Lewis, J.S.; Miller, L.A.; et al. *Performance Standards for Antimicrobial Susceptibility Testing*; Twenty-Fourth Informational Supplement; Clinical and Laboratory Standards Institute: Wayne, PA, USA, 2014.

62. Nurk, S.; Bankevich, A.; Antipov, D.; Gurevich, A.; Korobeynikov, A.; Lapidus, A.; Prjibelsky, A.; Pyshkin, A.; Sirotkin, A.; Sirotkin, Y.; et al. Assembling Genomes and Mini-metagenomes from Highly Chimeric Reads. In *Research in Computational Molecular Biology RECOMB 2013*; Deng, M., Jiang, R., Sun, F., Zhang, X., Eds.; Lecture Notes in Computer Science; Springer: Berlin/Heidelberg, Germany, 2013.

63. Seemann, T. Prokka: Rapid prokaryotic genome annotation. *Bioinformatics (Oxf. Engl.)* **2014**, *30*, 2068–2069. [CrossRef]

64. Page, A.J.; Cummins, C.A.; Hunt, M.; Wong, V.K.; Reuter, S.; Holden, M.T.; Fookes, M.; Falush, D.; Keane, J.A.; Parkhill, J. Roary: Rapid large-scale prokaryote pan genome analysis. *Bioinformatics* **2015**, *31*, 3691–3693. [CrossRef]

65. Treangen, T.J.; Ondov, B.D.; Koren, S.; Phillippy, A.M. The Harvest suite for rapid core-genome alignment and visualization of thousands of intraspecific microbial genomes. *Genome Biol.* **2014**, *15*, 524. [CrossRef] [PubMed]

66. Zankari, E.; Hasman, H.; Cosentino, S.; Vestergaard, M.; Rasmussen, S.; Lund, O.; Aarestrup, F.M.; Larsen, M.V. Identification of acquired antimicrobial resistance genes. *J. Antimicrob. Chemother.* **2012**, *67*, 2640–2644. [CrossRef] [PubMed]

67. Seeman, T. *Snippy: Fast Bacterial Variant Calling from NGS Reads*; GitHub Australia: Melbourne, Australia, 2015.

Article

Clonal Clusters, Molecular Resistance Mechanisms and Virulence Factors of Gram-Negative Bacteria Isolated from Chronic Wounds in Ghana

Denise Dekker [1,2,*], Frederik Pankok [3], Thorsten Thye [1], Stefan Taudien [3], Kwabena Oppong [4], Charity Wiafe Akenten [4], Maike Lamshöft [1,2], Anna Jaeger [1], Martin Kaase [3], Simone Scheithauer [3], Konstantin Tanida [5], Hagen Frickmann [5,6], Jürgen May [1,2,7] and Ulrike Loderstädt [3]

1   Department Infectious Disease Epidemiology, Bernhard Nocht Institute for Tropical Medicine Hamburg, 20359 Hamburg, Germany; thye@bnitm.de (T.T.); lamshoeft@bnitm.de (M.L.); anna.jaeger@bnitm.de (A.J.); may@bnitm.de (J.M.)
2   German Center for Infection Research (DZIF), Partner Site Hamburg-Lübeck-Borstel-Riems, 38124 Braunschweig, Germany
3   Institute for Infection Control and Infectious Diseases, University Medical Center Göttingen, 37075 Göttingen, Germany; frederik.pankok@med.uni-goettingen.de (F.P.); stefan.taudien@med.uni-goettingen.de (S.T.); martin.kaase@med.uni-goettingen.de (M.K.); simone.scheithauer@med.uni-goettingen.de (S.S.); ulrike.loderstaedt1@med.uni-goettingen.de (U.L.)
4   Kumasi Centre for Collaborative Research in Tropical Medicine (KCCR), South-End, Asuogya Road, Kumasi 039-5028, Ghana; oppong@kccr.de (K.O.); danquah@kccr.de (C.W.A.)
5   Department of Microbiology and Hospital Hygiene, Bundeswehr Hospital Hamburg, External Site at the Bernhard Nocht Institute for Tropical Medicine Hamburg, 20359 Hamburg, Germany; konstantintanida@bundeswehr.org (K.T.); frickmann@bnitm.de (H.F.)
6   Institute for Medical Microbiology, Virology and Hygiene, University Medicine Rostock, 18057 Rostock, Germany
7   Universitiy Medical Center Hamburg-Eppendorf (UKE), Tropical Medicine II, 20251 Hamburg, Germany
*   Correspondence: dekker@bnitm.de; Tel.: +49-40-4281-8535

Citation: Dekker, D.; Pankok, F.; Thye, T.; Taudien, S.; Oppong, K.; Akenten, C.W.; Lamshöft, M.; Jaeger, A.; Kaase, M.; Scheithauer, S.; et al. Clonal Clusters, Molecular Resistance Mechanisms and Virulence Factors of Gram-Negative Bacteria Isolated from Chronic Wounds in Ghana. *Antibiotics* **2021**, *10*, 339. https://doi.org/10.3390/antibiotics10030339

Academic Editor: Manuela Oliviera

Received: 24 February 2021
Accepted: 19 March 2021
Published: 22 March 2021

Publisher's Note: MDPI stays neutral with regard to jurisdictional claims in published maps and institutional affiliations.

Abstract: Wound infections are common medical problems in sub-Saharan Africa but data on the molecular epidemiology are rare. Within this study we assessed the clonal lineages, resistance genes and virulence factors of Gram-negative bacteria isolated from Ghanaian patients with chronic wounds. From a previous study, 49 *Pseudomonas aeruginosa*, 21 *Klebsiella pneumoniae* complex members and 12 *Escherichia coli* were subjected to whole genome sequencing. Sequence analysis indicated high clonal diversity with only nine *P. aeruginosa* clusters comprising two strains each and one *E. coli* cluster comprising three strains with high phylogenetic relationship suggesting nosocomial transmission. Acquired beta-lactamase genes were observed in some isolates next to a broad spectrum of additional genetic resistance determinants. Phenotypical expression of extended-spectrum beta-lactamase activity in the Enterobacterales was associated with $bla_{CTX-M-15}$ genes, which are frequent in Ghana. Frequently recorded virulence genes comprised genes related to invasion and iron-uptake in *E. coli*, genes related to adherence, iron-uptake, secretion systems and antiphagocytosis in *P. aeruginosa* and genes related to adherence, biofilm formation, immune evasion, iron-uptake and secretion systems in *K. pneumonia* complex. In summary, the study provides a piece in the puzzle of the molecular epidemiology of Gram-negative bacteria in chronic wounds in rural Ghana.

Keywords: wounds; Gram-negative bacteria; colonization; infection; clonal lineages; resistance genes; virulence factors

1. Introduction

The microbiology of chronic infected wounds, also on a molecular level, is poorly understood in sub-Saharan Africa (SSA) [1]. However, studies highlight the importance of antibiotic resistant Gram-negative bacteria [2–6].

From other parts in the world, in particular from industrialized countries, information on the microbiology and the role of biofilm-forming microorganisms causing such infections are well established [7–10].

In chronic wounds, *Pseudomonas aeruginosa* is amongst the most frequently isolated Gram-negative bacteria, associated with biofilm formation [11,12]. Tightly adhering biofilms pose a challenge in the diagnosis of *P. aeruginosa* using standard culturing methods [13].

In comparison, the role of Enterobacterales in chronic wounds has been much less characterized [14–17]. Studies have shown that geography seems to play a role in the estimation of their etiological relevance [18]. It was shown that skin colonization with Gram-negative bacteria is frequent in resource-limited (sub)tropical settings [19–21], in contrast to skin colonization of individuals from industrialized countries, where Gram-positive bacteria dominate [19]. Temperature and moisture have been discussed as likely reasons for the difference seen [22].

Isolation of potentially pathogenic bacteria from non-sterile sites like wounds does not necessarily indicate clinical relevance, which poses challenge to clinical interpretation.

In a recent study that focused on the overall bacterial composition of chronic wound infections in Ghana, from which the isolates for the present molecular analysis were taken, Enterobacterales and *Pseudomonas aeruginosa* constituted the majority of isolated bacterial strains [23]. A moderate proportion of ESBL-positive Enterobacterales suggests lower frequencies of antibiotic resistance [23] than what was recorded from other Ghanaian hospitals [5,24].

Within this study, we aim at characterizing clonal lineages, resistance-associated genetic elements and virulence genes of *P. aeruginosa*, the *Klebsiella pneumoniae* complex and *Escherichia coli*, which were recently isolated from chronic wounds of Ghanaian adult patients [23]. The molecular epidemiology of dominating clonal lineages and associated resistance genes will be assessed. Further, analysis of highly abundant virulence factors will be conducted.

2. Results

2.1. Clustering Based on Core Genome Multilocus Sequence Typing (cgMLST) Results

Of the 49 *P. aeruginosa* analyzed, a total of nine clusters comprising isolates without any recorded differences ($n = 2$) or with one or two alleles difference ($n = 7$) were found, suggesting closely related phylogeny (Figure 1). In addition to the clusters, 31 singletons with differences ranging from 80 to 3584 alleles were observed. MLST sequence types (ST) are indicated in Figure 1 and Tables A1 and A2. Cluster sequence types included the following: ST244, ST245, ST381, ST554, ST856, ST1485, ST2033, ST3227 and ST3590.

No clusters were identified among the 21 assessed *K. pneumonia* complex members, which were all singletons with differences ranging from 647 to 2244 alleles. *K. pneumoniae* complex sequence types are summarized in Figure 2. From the 12 *E. coli* isolates, three isolates in a cluster of close phylogenetic relationship were found ($1\times$ no allelic differences, 1×1 allele difference) (Figure 3). In addition to the cluster observed, nine singletons with differences ranging from 41 to 2365 alleles were recorded. The sequence type of the cluster was ST132 (Pasteur MLST scheme). Sequence types of all *E. coli* isolates are illustrated in Figure 3.

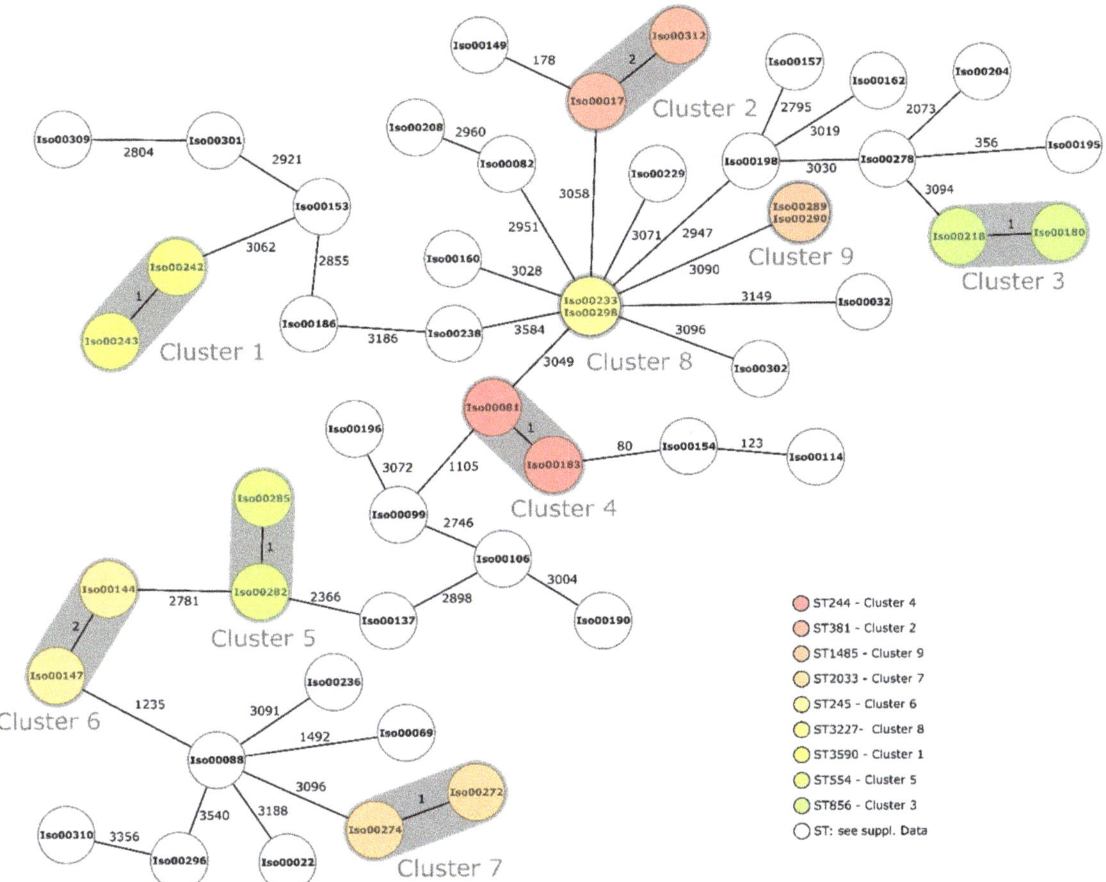

Figure 1. Minimum spanning tree of *P. aeruginosa* based on 3867 targets (core genome). Isolate numbers are found within the nodes, and numbers between nodes indicate the number of different alleles. Isolates within clusters are colored based on MLST sequence type (ST). The ST types of white nodes are indicated in Table A1.

2.2. Identified Molecular Resistance Mechanisms in Correlation to Previous Phenotypic Antibiotic Resistance

Table 1 summarizes acquired antimicrobial resistance determinants for *E. coli* and acquired genes mediating tolerance to disinfectants. Data for *P. aeruginosa* and *K. pneumoniae* are presented in Tables A1 and A2. Tables A3–A8 summarize the phenotypic resistance results as previously recorded [23].

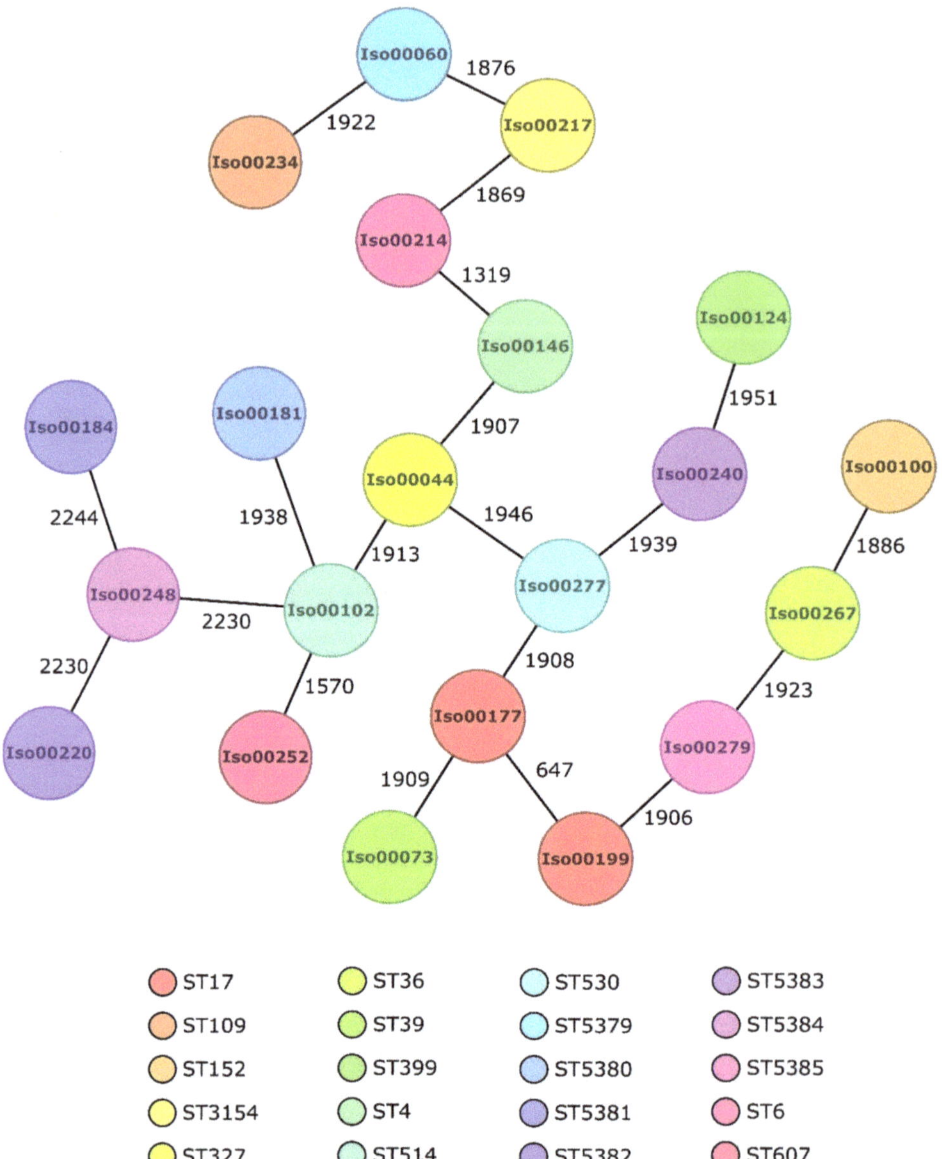

Figure 2. Minimum spanning tree of *K. pneumoniae* complex based on 2358 targets (core genome). Isolate numbers are found within the nodes, and the numbers between the nodes indicate the number of different alleles. Colors demonstrate the MLST sequence type of the isolates.

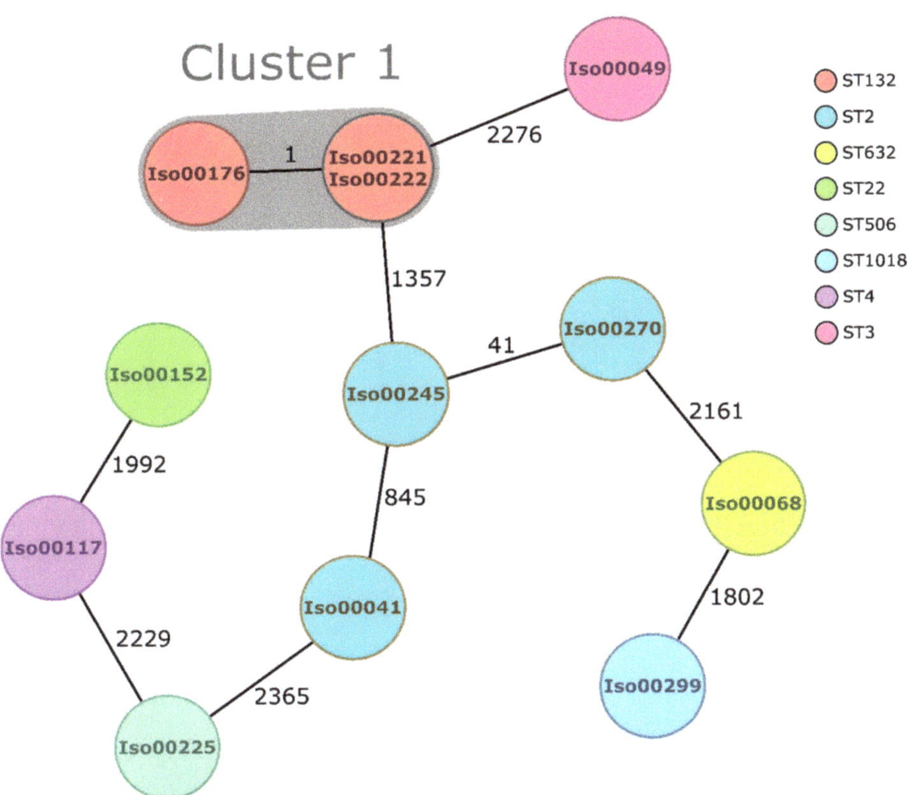

Figure 3. Minimum spanning tree of *E. coli* based on 2513 targets (core genome). Isolate numbers are found within the nodes, and the numbers between the nodes indicate the number of different alleles. Colors demonstrate the Pasteur sequence type of the isolates.

In the present study, phylogenetically identical or almost identical isolates also carried the same resistomes. All *E. coli* strains harbored acquired beta-lactamase genes with the majority coding for small spectrum beta-lactamases such as bla_{TEM-1} or bla_{OXA-1}. Only four strains carried the gene for an ESBL, in all cases $bla_{CTX-M-15}$. Among the *K. pneumoniae* complex strains, two belonged to the species *K. variicola*, one to the species *K. quasipneumoniae* and the remaining to the species *K. pneumoniae* sensu stricto as reflected by intrinsic bla_{LEN}, bla_{OKP} and $bla_{SHV-1\ like}$, respectively. Genes coding for ESBL ($bla_{CTX-M-15}$) were found solely in four out of 18 *K. pneumoniae* sensu stricto strains that also displayed resistance to oxyimino cephalosporins. In addition, several *K. pneumoniae* complex strains harbored bla_{TEM-1}, single strains also contained bla_{OXA-1} and bla_{SCO-1}.

With respect to *P. aeruginosa*, only one strain harbored acquired beta-lactamase genes (bla_{TEM-1} and bla_{SCO-1}). Increased minimum inhibitory concentrations (MICs) for carbapenems as observed in some *P. aeruginosa* strains were neither explained by matching acquired carbapenemase genes nor by full sequence analysis of the *oprD* gene. The associated amino acid sequences are shown in Figure A1. As indicated, the complete *oprD* gene was found in all 49 *P. aeruginosa* isolates; there was no evidence of protein truncation by premature stop of translation. The 49 isolates could be divided into 7 subgroups according to the protein sequence of the oprD protein, which differ in a total of 30 individual amino acid exchanges and in a single 12aa/10aa-stretch. Therefore, genotypic assessment could not identify the reason for the single carbapenem-resistant *P. aeuroginosa* isolate 088 (ST 1682).

Table 1. Analysis of antimicrobial resistance determinants, ordered by strain and MLST type, of the assessed *E. coli* isolates. ST = Sequence type.

Sample ID	ST-Type	Acquired Resistance Determinants Against										
		Beta lacatams	Sulfonamids	Trimethoprim	Makrodlids	Tetracyclins	Fluoroquinolones	Chloramphenicol	Aminoglycosides	Efflux pumps	Amino acid exchanges due to point mutations	Disinfectant resistance genes *
041	ST 2	*bla*OXA-1, *bla*TEM-1B, *bla*CTX-M-15	*sul1*	*dfrA17*	*mph(A)*	*tet(B)*	*aac(6′)-Ib-cr, aac(6′)-Ib-cr*	*catB3, catA1*	*aac(3)-IId, aac(6′)-Ib-cr, aadA5, aac(6′)-Ib-cr*	*mdf(A)*	*parE* p.S458A, *gyrA* p.S83L, *gyrA* p.D87N, *parC* p.S80I	*sitABCD, qacE*
049	ST 3	*bla*TEM-1B	*sul2, sul1,*	*dfrA12*	*mph(A)*	*tet(A)*			*aadA2, aph(3″)-Ib, aph(6)-Id*	*mdf(A)*		*sitABCD-like, qacE*
068	ST 632	*bla*TEM-1B	*sul3*	*dfrA12*		*tet(A)*		*cmlA1*	*aadA1, aadA2*	*mdf(A)-like*	*parE* p.S458A, *gyrA* p.S83L, *gyrA* p.D87N, *parC* p.S80I	
117	ST 4	*bla*TEM-1B	*sul1, sul2*	*dfrA7*		*tet(A)*		*catA1*	*aph(6)-Id, aph(3″)-Ib*	*mdf(A)-like*		*sitABCD-like, qacE*
152	ST 22	*bla*CARB-2, *bla*TEM-1B	*sul1*	*dfrA1*	*ere(B)*	*tet(B)*		*catA1*	*aadA1, aadA2b*	*mdf(A)-like*	*gyrA* p.S83L	*qacE, sitABCD*
176	ST 132	*bla*TEM-1B	*sul1*	*dfrA7*		*tet(A)*		*catA1*	*aph(3″)-Ib, aph(6)-Id*	*mdf(A)*		*qacE, sitABCD*
221	ST 132	*bla*TEM-1B	*sul1*	*dfrA7*		*tet(A)*		*catA1*	*aph(6)-Id, aph(3″)-Ib*	*mdf(A)*		*qacE, sitABCD*
222	ST 132	*bla*TEM-1B	*sul1, sul2*	*dfrA7*		*tet(A)*		*catA1*	*aph(3″)-Ib, aph(6)-Id*	*mdf(A)*		*qacE, sitABCD*
225	ST 506	*bla*TEM-1D, *bla*CTX-M-15	*sul1, sul2*	*dfrA17*	*mph(A)*	*tet(A)*		*catA1*	*aadA5, aph(6)-Id, aph(3″)-Ib*	*mdf(A)-like*	*gyrA* p.S83L, *parE* p.I529L	*sitABCD-like, qacE*
245	ST 2	*bla*TEM-1B	*sul1*	*dfrA12*	*mph(A)*	*tet(B)*	*qepA4 (neu)*	*catA1*	*aadA2, aac(3)-IId*	*mdf(A)*	*parE* p.S458A, *gyrA* p.S83L, *gyrA* p.D87N, *parC* p.S80I	*qacE*
270	ST 2	*bla*CTX-M-15				*tet(B)*		*catA1*		*mdf(A)*	*gyrA* p.S83L, *gyrA* p.D87N, *parE* p.S458A, *parC* p.S80I	
299	ST 1018	*bla*TEM-1B	*sul3*	*dfrA14*		*tet(A)*	*qnrS1*			*mdf(A)*		

* *sitABCD* = peroxides resistance, *qacE* = quaternary ammonium compounds resistance.

52

Other frequently detected resistance genes in *P. aeruginosa* were the fosfomycin resistance gene *fosA*, the chloramphenicol resistance gene *catB7*, the aminoglycoside resistance gene *aph(3′)-IIb* and the fluoroquinolone-resistance gene *crpP*. In the *Klebsiella pneumoniae* complex isolates, single amino acid exchanges and the fosmomycin resistance gene *fosA* were frequent. Various fluoroquinolone resistance genes and disinfectant tolerance mediating genes also quantitatively dominated. Finally, a broad spectrum of acquired genes causing resistance to the assessed classes of antimicrobial drugs and tolerance to disinfectants was observed in the *E. coli* strains.

2.3. Identified Molecular Virulence Mechanisms

Table 2 summarizes the analysis of virulence-related genes in *E. coli* (without genes mediating enteropathogenicity). Data for *P. aeruginosa* and *K. pneumoniae* are presented in Tables A9 and A10.

Table 2. Analysis of virulence determinants, ordered by strain and MLST type, of the assessed *E. coli* isolates. ST = Sequence type.

Sample ID	ST-Type	Pathogenicity Factor Groups					
		Adherence	Invasion	Toxin	Immune Evasion	Iron Uptake	Protease
041	ST 2	*fdeC*	*aslA, ompA*			*entA*-like, *entB, entC, entE, entF, entS, fepA, fepB, fepC, fepD, fepG,*	
049	ST 3		*aslA, kpsC, kpsD, kpsE, kpsF, kpsM, kpsU, kpsS*-like, *ompA*			chuS, chuU, chuV, chuW, chuY, *entA*-like, *entB, entC, entE, entF, entS, fepA, fepB, fepC, fepD, fepG*	
068	ST 632		*ompA*			*entA*-like, *entB, entC, entE, entF, entS, fepA, fepB, fepC, fepD, fepG*	
117	ST 4		*aslA, kpsC, kpsD, kpsE, kpsF, kpsM, kpsU; kpsS*-like, *ompA*	*hlyB, hlyC, hlyD,*	*tcpC*	chuA, chuS, chuT, chuU, chuV, chuW, chuX, chuY, *entA*-like, *entB, entC, entE, entF, entS, fepA, fepB, fepC, fepD, fepG, hlyA, iroN,*	*pic, sat, vat*
152	ST 22	*sfaB, sfaC, sfaD, sfaE, sfaF, sfaG, sfaH, sfaS, sfaX, sfaY*	*aslA, kpsC, kpsD, kpsE, kpsF, kpsM, kpsU; kpsS*-like, *ompA*	*cnf1; hlyA, hlyB, hlyC, hlyD,*	*tcpC*	chuA, chuS, chuT, chuU, chuV, chuW, chuX, chuY, entA-like, entB, entC, entE, entF, entS, fepA, fepB, fepC, fepD, fepG, iroN,	*vat*
176	ST 132		*aslA, kpsC, kpsD, kpsE, kpsF, kpsM, kpsU; kpsS*-like, *ompA*			*entA*-like, *entB, entC, entE, entF, entS, fepA, fepB, fepC, fepD, fepG,*	*sat*
221	ST 132		*aslA, kpsC, kpsD, kpsE, kpsM, kpsU; kpsS*-like, *ompA*			*entA*-like, *entB, entC, entE, entF, entS, fepA, fepB, fepC, fepD, fepG,*	*sat*

Table 2. *Cont.*

Sample ID	ST-Type	Pathogenicity Factor Groups					
		Adherence	Invasion	Toxin	Immune Evasion	Iron Uptake	Protease
222	ST 132		*aslA, kpsC, kpsD, kpsE, kpsF, kpsM, kpsU; kpsS*-like, *ompA*			*entA*-like, *entB, entC, entE, entF, entS, fepA, fepB, fepC, fepD, fepG,*	*sat*
225	ST 506		*aslA, kpsC, kpsD, kpsE, kpsF, kpsM, kpsU; kpsS*-like, *ompA*			*chuA, chuS, chuT, chuU, chuV, chuW, chuX, chuY, entA*-like, *entB, entC, entE, entF, entS, fepA, fepB, fepC, fepD, fepG,*	*sat*
245	ST 2		*aslA, ompA*			*entA*-like, *entB, entC, entE, entF, entS, fepB, fepC, fepD, fepG*	
270	ST 2		*aslA, ompA*			*entA*-like, *entB, entC, entE, entS, fepA, fepB, fepC, fepD, fepG*	
299	ST 1018		*ompA*			*entA*-like, *entB, entC, entE, entF, entS, fepA, fepB, fepD, fepG*	

The virulence-associated gene *exoU*, which has been described in association with the *P. aeruginosa* high-risk clone ST 135 [25], was recorded three times, associated with ST 135 (sample ID 296), ST 532 (sample ID 310) and ST 2483 (sample ID 22), respectively. Based on a Kleborate assessment, a positive virulence score was calculated for 7 out of 21 *K. pneumoniae* strains, comprising the known high-risk clones ST 17 (sample IDs 177, 199) and ST 152 (sample ID 100) [26], next to the clones ST 4 (sample ID 146), ST 6 (sample ID 214), ST 36 (sample ID 267) and ST 39 (sample ID 73), respectively. With focus on some important virulence associated genes in *Klebsiella* spp., *ybt* genes were detected in the abovementioned 7 samples, *iroE* was recorded in all 21 strains, while *clb* or *rpmA* genes were not detected.

Iron-uptake-related genes were numerous in all analyzed bacterial strains. For *P. aeruginosa* and *K. pneumoniae*, various secretion system-associated genes were found. Immune evasion-related genes were highly abundant in *K. pneumoniae* but not in *E. coli* isolates. Adherence-related genes were numerous in *P. aeruginosa* and in *K. pneumoniae* but not in *E. coli*.

Numerous invasion-associated genes were detected in *E. coli*, antiphagocytosis-associated genes were found in *P. aeruginosa*, and biofilm-associated genes in *K. pneumoniae*.

Less frequently detected were: toxin genes in *E. coli* and *K. pneumoniae*, protease genes in *E. coli* and *P. aeruginosa*, regulation genes in *P. aeruginosa* and *K. pneumoniae*, biosurfactant and pigment genes in *P. aeruginosa* and nutrition factor, efflux pumps and serum resistance genes in *K. pneumoniae*.

3. Discussion

Within this study, we aimed at filling information gaps on the molecular epidemiology of Gram-negative bacteria from chronic infected wounds in rural Ghana. Phylogenetic analyses based on core genome comparison indicated a high clonal diversity of the wound-associated isolates. Clonal clusters were restricted to nine *P. aeruginosa* clusters and one *E. coli* cluster, most likely indicating nosocomial transmission, which has most likely occurred in the wound dressing room that patients' visit on a weekly basis.

ST 135 and ST 244, which are among the worldwide top 10 *P. aeruginosa* high-risk clones [25], were found among the *P. aeruginosa* wound isolates. In detail, one ST 135 *Pseudomonas aeruginosa* isolate was detected, carrying the beta-lactamase-encoding genes bla_{TEM-1B} and bla_{SCO-1} and an *exoU* gene, next to five ST 244 without acquired beta-lactamases. Focusing on known pathogenic *K. pneumoniae* clones [26], two ST 17 strains, a clone reported to be associated with carbapenem-resistance, and one ST 152 strain, a clone known from the Caribbean as common carrier of multiple resistance genes, were detected. Strains carrying the *ybt* and *iro* genes were also identified as high-risk clones by the Kleborate software. From the observed *E. coli* ST types, none have been previously reported as being associated with pathogenic clones so far [27].

In line with the phenotypical antibiotic resistance results previously published [23], numerous acquired resistance determinants were detected in the bacterial strains under investigation. Focusing on the few observed clusters, comparable resistome compositions point towards recent nosocomial transmission. The gene $bla_{CTX-M-15}$ was identified as the determinant of the detected extended spectrum beta-lactamase (ESBL) expression in ESBL positive Enterobacterales [23]. This is in line with previous reports from both human and livestock-associated ESBL positive Enterobacterales in Ghana [28–34]. In *P. aeruginosa* and *K. pneumoniae*, bla_{SCO-1}, which has initially been described from an *Acinetobacter baumannii* isolate from Argentina [35], was observed. Beta-lactamases with high hydrolytic effects on carbapenems were lacking, the same applies to protein truncation by premature stop of translation of the *oprD* gene in *P. aeruginosa*. Accordingly, the genetic background of carbapenem resistance of a single *P. aeruginosa* strain could not be resolved, although downregulation of *oprD* expression due to mutations outside of the gene or *ampC* (class C betalactamase) overexpression could not be excluded as likely reasons.

Substance-specific genes and genes encoding efflux pumps mediating tolerance to disinfectants were observed in Enterobacterales. Therefore, further monitoring of the spread of disinfectant tolerance-associated genes and the effects of their abundance on disinfectant-based skin and wound decolonization strategies [36] seem advisable.

The importance of highly abundant virulence factors like iron-uptake- and secretion system-related genes in *P. aeruginosa* is comprehensively described in the literature [37,38]. Other genes reported in the literature like regulation-associated virulence genes, recently reported, were less frequently observed in our isolates [39,40]. However, due to lacking information on the individual etiological relevance of each isolate, any association with clinical effects remains speculative.

Further limitations of this study include a rather small sample size and the lack of a comparison strain collection containing isolates from other clinical specimens and environmental strains. Accordingly, the interpretation of the etiological relevance of individual strains remains challenging and is clearly beyond the scope of this work.

In summary, a broad spectrum of Gram-negative clones was isolated from the chronic wounds of the Ghanaian patients. Thereby, known high-risk clones [25–27] played only a minor role. Observed resistance patterns and mechanisms were in line with the spectrum expected from previous reports [23,28–34].

4. Materials and Methods

4.1. Sample Collection, Bacterial Culture and Antibiotic Susceptibility Testing

Single patient strains of *P. aeruginosa*, *E. coli* and *K. pneumoniae* complex were isolated from patients ≥ 15 years with an infected chronic wound at the Outpatient Department (OPD) of the Agogo Presbyterian Hospital, in the Asante Akim North District of rural Ghana. Patients typically visit the wound dressing room of the OPD on a weekly basis. Sampling was performed from January 2016 to November 2016. Sample collection and microbiological investigations were reported previously [23]. Antibiotic susceptibility was tested by the disk diffusion method and interpreted following the European Committee on Antimicrobial Susceptibility Testing (EUCAST) guidelines v.6.0 (http://www.eucast.org

(accessed on 15 January 2016)). Bacterial strains and antibiotic susceptibility were confirmed using the VITEK2 System. Those data have been published before [23].

4.2. DNA Isolation and Whole Genome Sequencing

Bacterial DNA was isolated using the MasterPure Complete DNA and RNA Purification Kit (LGC standards GmbH, Wesel, Germany) and sent for whole genome sequencing (WGS) to BGI Europe, Denmark, Copenhagen. A BGISEQ-500 device was used for sequencing, generating 2×150 bp paired-end reads with an aimed coverage of $100\times$. Original raw data were upload for public use to the short-read archive (SRA, NCBI) under the accession number PRJNA699140. Details on the strain-specific SRA accession numbers are provided in Table A11.

4.3. Whole Genome Sequencing and Data Analysis

All raw data passed quality control using FASTQC v.0.11.4 [41] and were used for further analysis. Taxonomic classification and contamination check of raw-reads was performed using KRAKEN2 v.2.0.8-beta [42]. Phylogenetic analysis based on core genome multi locus sequence typing (cgMLST) analysis was performed using the commercial software SeqSphere+ v. 7.2.0 (Ridom GmbH, Münster, Germany) [43]. The software pipeline included assessment of read data and adapter control using FASTQC followed by genome assembly using the internally provided assembler Velvet, applying default settings. The reference genomes NC_000913.3 (*E. coli*), NC_002516.2 (*P. aeruginosa*) and NC_01273.1 (*K. pneumoniae* species complex) were used for cgMLST analyses. Only samples with a ration of "good cgMLST targets" higher than 90% were included in the phylogenetic analysis. Novel cgMLST-based complex types (CT) were automatically assigned by the SeqSphere software. Unknown alleles and profiles of MLST genes were submitted to pubmlst.org or Institute Pasteur to establish novel sequence types (ST). Isolates were defined to be clonally identical with allele differences less than four. Moreover, raw data were assembled with SPAdes v3.13.11 [44] using the careful option. Scaffolds shorter than 500 bp or with a coverage smaller than ten were sorted out, using an in-house script. Abricate v.0.9.9 [45] was used to screen for resistance and virulence genes in SPAdes assembly files, using NCBI AMRFinderPlus [46] and VFDB [47] as reference databases (both updated 6 November 2020), respectively. Additionally, SPAdes assemblies were uploaded to ResFinder4.1 [48] to obtain WGS predicted phenotypes against different antimicrobials by using default settings (%ID > 90, minimum length > 60%) and to Kleborate to predict virulence genes in *Klebsiella* isolates.

4.4. Ethical Considerations

The Committee on Human Research, Publications and Ethics, School of Medical Science, Kwame Nkrumah University of Science and Technology in Kumasi, Ghana, approved this study (approval number CHRPE/AP/078/16).

5. Conclusions

In conclusion, this study provides a molecular insight into the epidemiology of Gram-negative bacteria isolated from chronic wound infections from patients in rural Ghana. Epidemiological data that focus on the distribution and spread of antimicrobial resistance determinants and associated virulence factors in resource-limited settings are scarce. Although the study is a small cross-sectional assessment, which cannot replace continuous surveillance programs, it might provide a glimpse of prevailing Gram-negative bacteria isolated from wound infections in this area of Ghana. Considering the ongoing need for resistance and virulence surveillance in tropical regions, larger future studies are desirable.

Author Contributions: U.L., D.D. and J.M. designed and coordinated this study. T.T., F.P. and S.T. performed bioinformatic analysis. M.L. supported the management of this study. A.J. managed the data collection. H.F., D.D. and U.L. wrote the first draft of this manuscript. K.O. conducted and supervised fieldwork. C.W.A. and K.T. conducted and supervised lab work. M.K. and S.S. supported the interpretation of the results, writing and editing the manuscript. All authors read and approved the final manuscript.

Funding: This study was funded by institutional funds of the Bernhard Nocht Institute for Tropical Medicine (BNITM).

Institutional Review Board Statement: The study was conducted according to guidelines of the Declaration of Helsinki. The Committee on Human Research, Publications and Ethics, School of Medical Science, Kwame Nkrumah University of Science and Technology in Kumasi, Ghana, approved this study (approval number CHRPE/AP/078/16).

Informed Consent Statement: Informed consent was obtained from all study participants.

Data Availability Statement: All relevant data have been provided in the paper and its Appendix A materials. Raw data are available applying the links as indicated in the methods chapter and can also be provided by the authors on reasonable request.

Acknowledgments: We thank all patients that participated in this study and the staff at the Agogo Presbyterian Hospital. Without their support, this research study would not have been possible. We thank the team of curators pubmlst.org and the Institute Pasteur MLST and whole genome MLST databases for curating the data and making them publicly available at http://bigsdb.pasteur.fr/ (accessed on 22 March 2021).

Conflicts of Interest: The authors declare no conflict of interest.

Appendix A

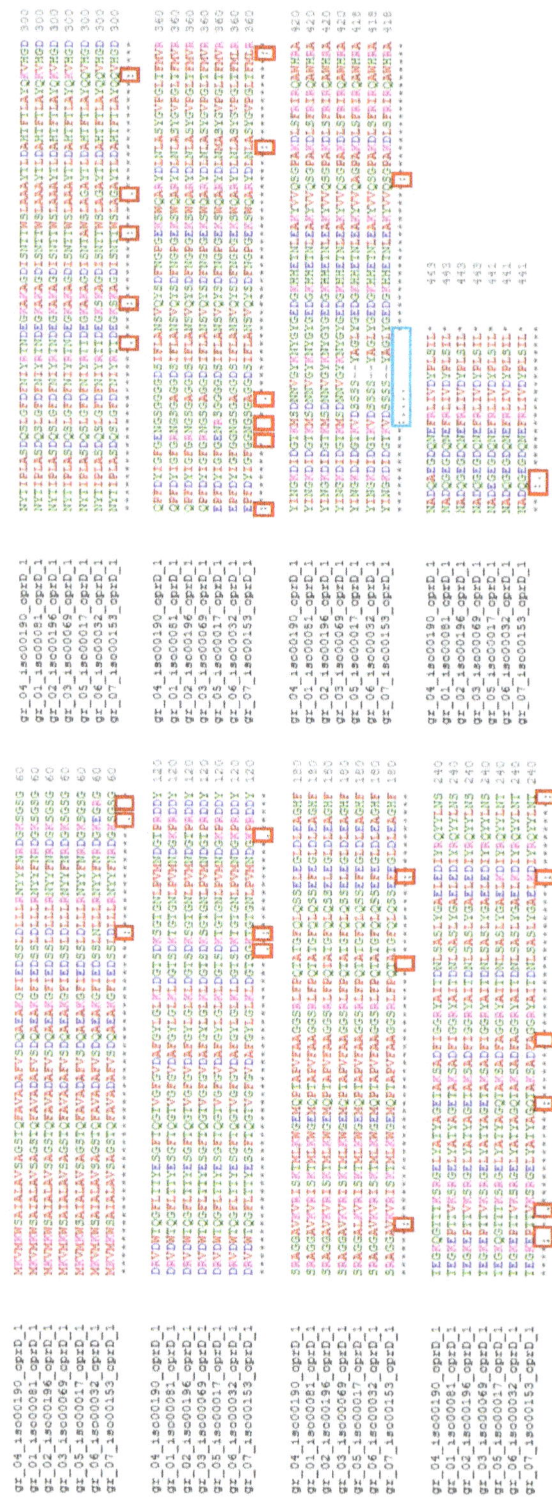

Red frames: 30 single amino acid differences between the groups were identified (16, 8 and 6 with strong, weak and no similarity, respectively *)

Blue frame: A 12 aa stretch is changed into a differing 10 aa stretch in groups 6 and 7

Figure A1. Clustal omega multiple alignment of oprD proteins—one example for the 7 detected subgroups.

*) similarities:

"*" indicates a site belonging to group exhibiting strong similarity.

":" indicates a site belonging to a group exhibiting weak similarity.

The criterion for distinguishing strong from weak similarity is as follows: Strong similarity corresponds to a PAM250 MATRIX score between amino acids of greater than 0.5, while weak similarity corresponds to a score of 0.5 or less.

Table A1. Analysis of antimicrobial resistance determinants, ordered by strain and MLST type, of the assessed *P. aeruginosa* isolates. ST = Sequence type.

Sample ID	ST-Type	Acquired Resistance Determinants Against												
		Beta Lacatams	Sulf-onamids	Fosfomy-cin	Trimetho-prim	Makro-lides	Tetracyc-linws	Fluoroqu-inolones	Chloram-phenicol	Rifam-picin	Amino-glycosides	Efflux Pumps	Amino Acid Exchanges Due to Point Mutations	Disinfectant Resistance Genes
017	ST 381			*fosA*					*catB7*		*aph(3′)-IIb*			
022	ST 2483			*fosA*					*catB7*		*aph(3′)-IIb*			
032	ST 3587		*sul1*	*fosA*	*dfrA15*		*tet*(G)		*catB7*		*aph(3′)-IIb*			
069	ST 360			*fosA*				*crpP*-like	*catB7*		*aph(3′)-IIb*			
081	ST 244			*fosA*					*catB7*		*aph(3′)-IIb*			
082	ST 514			*fosA*					*catB7*		*aph(3′)-IIb*			
088	ST 1682			*fosA*					*catB7*		*aph(3′)-IIb*			
099	ST 244			*fosA*					*catB7*		*aph(3′)-IIb*			
106	ST 1521			*fosA*					*catB7*		*aph(3′)-IIb*			
114	ST 244			*fosA*				*crpP*-like	*catB7*		*aph(3′)-IIb*-like			
137	ST 3014			*fosA*				*crpP*-like	*catB7*		*aph(3′)-IIb*-like			
144	ST 245			*fosA*				*crpP*-like	*catB7*		*aph(3′)-IIb*			
147	ST 245			*fosA*				*crpP*-like	*catB7*		*aph(3′)-IIb*			
149	ST 381			*fosA*				*crpP*-like	*catB7*		*aph(3′)-IIb*			
153	ST 704			*fosA*-like				*crpP*-like	*catB7*-like		*aph(3′)-IIb*-like			
154	ST 244	,		*fosA*				*crpP*-like	*catB7*		*aph(3′)-IIb*-like			
157	ST 2616			*fosA*					*catB7*-like		*aph(3′)-IIb*			
160	ST 170			*fosA*-like							*aph(3′)-IIb*			

Table A1. *Cont.*

Sample ID	ST-Type	Acquired Resistance Determinants Against												
		Beta Lacatams	Sulf-onamids	Fosfomy-cin	Trimetho-prim	Makro-lides	Tetracyc-linws	Fluoroqu-inolones	Chloram-phenicol	Rifam-picin	Amino-glycosides	Efflux Pumps	Amino Acid Exchanges Due to Point Mutations	Disinfectant Resistance Genes
162	ST 274			*fosA*				*crpP*-like	*catB7*		*aph(3′)-IIb*			
180	ST 856			*fosA*					*catB7*		*aph(3′)-IIb*			
183	ST 244			*fosA*					*catB7*		*aph(3′)-IIb*-like			
186	ST 3588			*fosA*-like					*catB7*-like		*aph(3′)-IIb*-like			
190	ST 871			*fosA*					*catB7*-like		*aph(3′)-IIb*			
195	ST 988			*fosA*				*crpP*-like	*catB7*-like		*aph(3′)-IIb*-like			
196	ST 2475			*fosA*				*crpP*-like	*catB7*		*aph(3′)-IIb*			
198	ST 2476			*fosA*				*crpP*-like	*catB7*		*aph(3′)-IIb*			
204	ST 639			*fosA*				*crpP*	*catB7*		*aph(3′)-IIb*-like			
208	ST 132			*fosA*				*crpP*-like	*catB7*		*aph(3′)-IIb*			
218	ST 856			*fosA*					*catB7*		*aph(3′)-IIb*			
229	ST 270			*fosA*				*crpP*-like	*catB7*		*aph(3′)-IIb*			
233	ST 3227			*fosA*					*catB7*		*aph(3′)-IIb*			
236	ST 266			*fosA*					*catB7*		*aph(3′)-IIb*			
238	ST 3589			*fosA*-like				*crpP*-like	*catB7*-like		*aph(3′)-IIb*-like			
242	ST 3590			*fosA*-like										
243	ST 3590			*fosA*-like					*catB7*-like		*aph(3′)-IIb*-like			

Table A1. *Cont.*

Sample ID	ST-Type	Acquired Resistance Determinants Against												
		Beta Lacatams	Sulf-onamids	Fosfomy-cin	Trimetho-prim	Makro-lides	Tetracyc-linws	Fluoroqu-inolones	Chloram-phenicol	Rifam-picin	Amino-glycosides	Efflux Pumps	Amino Acid Exchanges Due to Point Mutations	Disinfectant Resistance Genes
272	ST 2033			*fosA*					*catB7*-like		*aph(3′)-IIb*-like			
274	ST 2033			*fosA*					*catB7*-like		*aph(3′)-IIb*			
278	ST 988			*fosA*				*crpP*-like	*catB7*-like					
282	ST 554			*fosA*				*crpP*-like	*catB7*		*aph(3′)-IIb*			
285	ST 554			*fosA*					*catB7*		*aph(3′)-IIb*			
289	ST 1485			*fosA*					*catB7*		*aph(3′)-IIb*			
290	ST 1485			*fosA*					*catB7*		*aph(3′)-IIb*			
296	ST 235	*bla*$_{TEM-1B}$, *bla*$_{SCO-1}$	*sul1*	*fosA*			*tet(G)*		*catB7*-like		*aph(3′)-IIb*-like, *aac(3)-IIa*			
298	ST 3227			*fosA*					*catB7*		*aph(3′)-IIb*			
301	ST 3593			*fosA*-like					*catB7*-like		*aph(3′)-IIb*-like			
302	ST 1755			*fosA*					*catB7*		*aph(3′)-IIb*			
309	ST 3592			*fosA like*				*crpP*-like	*catB7*-like		*aph(3′)-IIb*-like			
310	ST 532		*sul1*	*fosA*					*catB7*-like		*aph(3″)-Ib*, *aph(6)-Id*, *aph(3′)-IIb*			
312	ST 381			*fosA*					*catB7*		*aph(3′)-IIb*			

Acquired resistance genes for macrolides, rifampicin, resistance-associated point mutations, genes for efflux pumps or genes mediating tolerance against disinfectants were not detected.

Table A2. Analysis of antimicrobial resistance determinants, ordered by strain and MLST type, of the assessed *K. pneumoniae* isolates. ST = Sequence type.

Sample ID	ST-Type	Acquired Resistance Determinants Against											Amino Acid Exchanges Due to Point Mutations	Disinfectant Resistance Genes *
		Beta Lacatams	Sulfo-namids	Fosfomy-cin	Trimetho-prim	Macro-lides	Tetracyc-lines	Fluoro-quinolones	Chloram-phenicol	Rifam-picin	Amino-glycosides	Efflux Pumps		
044	ST 327			*fosA*				*oqxB, oqxA*					*ompK37* p.I70M, *ompK37* p.I128M,*ompK37* p.I128M, *ompK36* p.L59V, *ompK36* p.L191S, *ompK36* p.F207W, *ompK36* p.A217S, *ompK36* p.N218H, *ompK36* p.D224E, *ompK36* p.L228V, *ompK36* p.E232R, *ompK36* p.T254S, *acrR* p.P161R, *acrR* p.G164A, *acrR* p.F172S, *acrR* p.R173G, *acrR* p.L195V, *acrR* p.F197I, *acrR* p.K201M	*oqxB, oqxA*
060	ST 5379	*bla*TEM-1C	*sul1, sul2*	*fosA*	*dfrA12*	*mph(A)*		*oqxA, oqxB, qnrS1*	*catA2*-like		*aph(6)-Id, aph(3″)-Ib, aph(3′)-Ia, aadA2, aac(3)-IIa*		*acrR* p.P161R, *acrR* p.G164A, *acrR* p.F172S, *acrR* p.R173G, *acrR* p.L195V, *acrR* p.F197I, *acrR* p.K201M, *ompK36* p.N49S, *ompK36* p.L59V, *ompK36* p.T184P, *ompK37* p.I70M, *ompK37* p.I128M	*oqxA, qacE, oqxB*

Table A2. *Cont.*

Sample ID	ST-Type	Acquired Resistance Determinants Against												
		Beta Lacatams	Sulfo-namids	Fosfomy-cin	Trimetho-prim	Macro-lides	Tetracyc-lines	Fluoro-quinolones	Chloram-phenicol	Rifam-picin	Amino-glycosides	Efflux Pumps	Amino Acid Exchanges Due to Point Mutations	Disinfectant Resistance Genes *
073	ST 39	*bla*TEM-1B , *bla*CTX-M-15	*sul1,*	*fosA*	*dfrA27*	*erm(B), mph(A)*	*tet(D)*	*oqxB, oqxA, aac(6′)-Ib-cr, qnrB2, aac(6′)-Ib-cr*	*catA2-like*	*ARR-3*	*aac(6′)-Ib-cr, aadA16, aac(3)-IIa, aac(6′)-Ib-cr, aph(3″)-Ib, aph(6)-Id*		*acrR* p.P161R, *acrR* p.G164A, *acrR* p.F172S, *acrR* p.R173G, *acrR* p.L195V, *acrR* p.F197I, *acrR* p.K201M, *ompK37* p.I70M, *ompK37* p.I128M, *ompK37* p.N230G, *ompK36* p.N49S, *ompK36* p.L59V, *ompK36* p.L191S, *ompK36* p.F207W, *ompK36* p.A217S, *ompK36* p.N218H, *ompK36* p.D224E, *ompK36* p.L228V, *ompK36* p.E232R, *ompK36* p.T254S	*oqxB, oqxA, qacE*
100	ST 152	*bla*CTX-M-15 , *bla*OXA-1 , *bla*TEM-1B	*sul2, sul1*	*fosA*	*dfrA1, dfrA27*	*mph(A)*	*tet(D)*	*aac(6′)-Ib-cr, oqxB, qnrB6, oqxA, aac(6′)-Ib-cr*	*catB3, catA1, catB3*	*ARR-3*	*aac(3)-IIa, aph(6)-Id, aph(3″)-Ib, aadA1, aadA16, aph(3′)-Ia, aac(6′)-Ib-cr, aac(6′)-Ib-cr*		*ompK36* p.N49S, *ompK36* p.L59V, *ompK36* p.G189T, *ompK36* p.F198Y, *ompK36* p.F207Y, *ompK36* p.A217S, *ompK36* p.T222L, *ompK36* p.D223G, *ompK36* p.E232R, *ompK36* p.N304E, *acrR* p.P161R, *acrR* p.G164A, *acrR* p.F172S, *acrR* p.R173G, *acrR* p.L195V, *acrR* p.F197I, *acrR* p.K201M, *ompK37* p.I70M, *ompK37* p.I128M, *ompK37* p.N230G	*oqxB, oqxA*

Table A2. *Cont.*

Sample ID	ST-Type	Acquired Resistance Determinants Against											Amino Acid Exchanges Due to Point Mutations	Disinfectant Resistance Genes *
		Beta Lacatams	Sulfo-namids	Fosfomy-cin	Trimetho-prim	Macro-lides	Tetracyc-lines	Fluoro-quinolones	Chloram-phenicol	Rifam-picin	Amino-glycosides	Efflux Pumps		
102	ST 514			*fosA*			*tet(C)*	*oqxB, oqxA*	*catA1*				*ompK36* p.N49S, *ompK36* p.L59V, *ompK36* p.L191S, *ompK36* p.F207W, *ompK36* p.A217S, *ompK36* p.N218H, *ompK36* p.D224E, *ompK36* p.L228V, *ompK36* p.E232R, *ompK36* p.T254S, *ompK37* p.I70M, *ompK37* p.I128M, *ompK37* p.N230G, *acrR* p.P161R, *acrR* p.G164A, *acrR* p.F172S, acrR p.R173G, *acrR* p.L195V, *acrR* p.F197I, *acrR* p.K201M	*oqxB, oqxA*
124	ST 399			*fosA*				*oqxA, oqxB*	*catA1*				*ompK36* p.N49S, *ompK36* p.L59V, *ompK36* p.G189T, *ompK36* p.F198Y, *ompK36* p.F207Y, *ompK36* p.A217S, *ompK36* p.T222L, *ompK36* p.D223G, *ompK36* p.E232R, *acrR* p.P161R, *acrR* p.G164A, *acrR* p.F172S, acrR p.R173G, *acrR* p.F197I, *acrR* p.K201M, *ompK37* p.I70M, *ompK37* p.I128M	*oqxA, oqxB*
146	ST 4		*sul2*	*fosA*			*tet(D)*	*oqxA, oqxB*	*catA2*-like					*oqxA, oqxB*

Table A2. *Cont.*

Sample ID	ST-Type	Acquired Resistance Determinants Against												
		Beta Lacatams	Sulfo-namids	Fosfomy-cin	Trimetho-prim	Macro-lides	Tetracyc-lines	Fluoro-quinolones	Chloram-phenicol	Rifam-picin	Amino-glycosides	Efflux Pumps	Amino Acid Exchanges Due to Point Mutations	Disinfectant Resistance Genes *
177	ST 17		*sul1, sul2*	*fosA*	*dfrA15*		*tet(A)*	*oqxA, oqxB*-like	*catA1*		*aadA1, aph(3″)-Ib, aph(6)-Id*		*ompK37* p.I70M, *ompK37* p.I128M, *acrR* p.P161R, *acrR* p.G164A, *acrR* p.F172S, *acrR* p.R173G, *acrR* p.L195V, *acrR* p.F197I, *acrR* p.K201M, *ompK36* p.N49S, *ompK36* p.L59V,*ompK36* p.L191S, *ompK36* p.F207W, *ompK36* p.A217S, *ompK36* p.N218H, *ompK36* p.D224E, *ompK36* p.L228V, *ompK36* p.E232R*ompK36* p.T254S	*qacE, oqxB*-like, *oqxA*
181	ST 5380			*fosA*				*oqxA, oqxB*					*ompK36* p.N49S, *ompK36* p.L59V, *ompK36* p.L191S, *ompK36* p.F207W, *ompK36* p.A217S, *ompK36* p.N218H, *ompK36* p.D224E, *ompK36* p.L228V, *ompK36* p.E232R, *ompK36* p.T254S,*ompK37* p.I70M, *ompK37* p.I128M, *acrR* p.P161R, *acrR* p.G164A, *acrR* p.F172S, *acrR* p.R173G, *acrR* p.L195V, *acrR* p.F197I, *acrR* p.K201M	*oqxA, oqxB*

Table A2. *Cont.*

Sample ID	ST-Type	Acquired Resistance Determinants Against												
		Beta Lacatams	Sulfo-namids	Fosfomy-cin	Trimetho-prim	Macro-lides	Tetracyc-lines	Fluoro-quinolones	Chloram-phenicol	Rifam-picin	Amino-glycosides	Efflux Pumps	Amino Acid Exchanges Due to Point Mutations	Disinfectant Resistance Genes *
184	ST 5381			*fosA*				*oqxA*-like, *oqxB*-like					*ompK37* p.I70M, *ompK37* p.I128M, *ompK36* p.N49S, *ompK36* p.L59V, *ompK36* p.L191Q, *ompK36* p.F198Y, *ompK36* p.A217S, *ompK36* p.N218H, *ompK36* p.Q227N, *ompK36* p.L229V, *ompK36* p.N304E, *acrR* p.P161R, *acrR* p.G164A, *acrR* p.F172S, *acrR* p.R173G, *acrR* p.L195V, *acrR* p.F197I, *acrR* p.K201M	*oqxA*-like, *oqxB*-like
199	ST 17	*bla*CTX-M-15, *bla*TEM-1B	*sul2, sul1*	*fosA*-like	*dfrA16*			*oqxA, oqxB*			*aadA2b, aac(3)-IIa*		*acrR* p.P161R, *acrR* p.G164A, *acrR* p.F172S, *acrR* p.R173G, *acrR* p.L195V, *acrR* p.F197I, *acrR* p.K201M, *ompK37* p.I70M, *ompK37* p.I128M, *ompK36* p.N49S, *ompK36* p.L59V, *ompK36* p.T86V, *ompK36* p.S89T, *ompK36* p.D91K, *ompK36* p.A93S, *ompK36* p.L191Q, *ompK36* p.F207W, *ompK36* p.A217S, *ompK36* p.N218H, *ompK36* p.Q227N, *ompK36* p.L229V, *ompK36* p.E232R, *ompK36* p.H235D, *ompK36* p.T254S	*oqxA, oqxB, qacE*

Table A2. *Cont.*

Sample ID	ST-Type	Acquired Resistance Determinants Against												
		Beta Lacatams	Sulfo-namids	Fosfomy-cin	Trimetho-prim	Macro-lides	Tetracyc-lines	Fluoro-quinolones	Chloram-phenicol	Rifam-picin	Amino-glycosides	Efflux Pumps	Amino Acid Exchanges Due to Point Mutations	Disinfectant Resistance Genes *
214	ST 6		*sul1*	*fosA*-like	*dfrA14*			*oqxB*-like, *oqxA*	*catA1*		*aph(3′)-Ia*		*ompK37* p.I70M, *ompK37* p.I128M, *ompK36* p.N49S, *ompK36* p.L59V, *ompK36* p.G189T, *ompK36* p.F198Y, *ompK36* p.F207Y, *ompK36* p.A217S, *ompK36* p.T222L, *ompK36* p.D223G, *ompK36* p.E232R, *acrR* p.P161R, *acrR* p.G164A, *acrR* p.F172S, *acrR* p.R173G, *acrR* p.L195V, *acrR* p.F197I, *acrR* p.K201M	*oqxB*- like, *oqxA*
217	ST 3154	*bla*SCO-1 , *bla*TEM-1B	*sul1*, *sul2*	*fosA*	*dfrA12*, *dfrA14*		*tet(A)*	*oqxA*, *oqxB*-like	*catA2*-like		*aph(6)-Id*, *aph(3″)-Ib*, *aac(3)-IIa*, *aadA2*		*acrR* p.P161R, *acrR* p.G164A, *acrR* p.F172S, *acrR* p.R173G, *acrR* p.L195V, *acrR* p.F197I, *acrR* p.K201M, *ompK37* p.I70M, *ompK37* p.I128M, *ompK36* p.N49S, *ompK36* p.L59V, *ompK36* p.G189T, *ompK36* p.F198Y, *ompK36* p.F207Y, *ompK36* p.A217S, *ompK36* p.T222L, *ompK36* p.D223G, *ompK36* p.E232R, *ompK36* p.N304E	*oqxA*, *qacE*, *oqxB*-like

Table A2. *Cont.*

Sample ID	ST-Type													Acquired Resistance Determinants Against	
		Beta Lacatams	Sulfo-namids	Fosfomy-cin	Trimetho-prim	Macro-lides	Tetracyc-lines	Fluoro-quinolones	Chloram-phenicol	Rifam-picin	Amino-glycosides	Efflux Pumps		Amino Acid Exchanges Due to Point Mutations	Disinfectant Resistance Genes *
220	ST 5382			*fosA*-like				*oqxB*-like, *oqxA*-like	*catA1*					*ompK37* p.I70M, *ompK37* p.I128M, *ompK36* p.N49S, *ompK36* p.L59V, *ompK36* p.L191Q, *ompK36* p.A217S, *ompK36* p.N218H, *ompK36* p.Q227N, *ompK36* p.L229V, *ompK36* p.N304E, *acrR* p.P161R, *acrR* p.F172S, *acrR* p.R173G, *acrR* p.L195V, *acrR* p.F197I, *acrR* p.K201M	*oqxB*-like, *oqxA*-like
234	ST 109			*fosA*				*oqxA*, *oqxB*-like						*ompK36* p.N49S, *ompK36* p.L59V, *ompK36* p.L191S, *ompK36* p.F207W, *ompK36* p.A217S, *ompK36* p.N218H, *ompK36* p.D224E, *ompK36* p.L228V, *ompK36* p.E232R, *ompK36* p.T254S, *acrR* p.P161R, *acrR* p.G164A, *acrR* p.F172S,*acrR* p.R173G, *acrR* p.L195V, *acrR* p.F197I, *acrR* p.K201M, *ompK37* p.I70M, *ompK37* p.I128M	*oqxA*, *oqxB*-like

Table A2. *Cont.*

Sample ID	ST-Type	Acquired Resistance Determinants Against												
		Beta Lacatams	Sulfo-namids	Fosfomy-cin	Trimetho-prim	Macro-lides	Tetracyc-lines	Fluoro-quinolones	Chloram-phenicol	Rifam-picin	Amino-glycosides	Efflux Pumps	Amino Acid Exchanges Due to Point Mutations	Disinfectant Resistance Genes *
240	ST 5383			*fosA*-like			*tet(D)*	*oqxA, oqxB*-like					*ompK36* p.N49S, *ompK36* p.L59V, *ompK36* p.L191S, *ompK36* p.F207W, *ompK36* p.A217S, *ompK36* p.N218H, *ompK36* p.D224E, *ompK36* p.L228V, *ompK36* p.E232R, *ompK37* p.I70M, *ompK37* p.I128M, *acrR* p.P161R, *acrR* p.G164A, *acrR* p.F172S, *acrR* p.R173G, *acrR* p.L195V, *acrR* p.F197I,*acrR* p.K201M	*oqxA, oqxB*-like
248	ST 5384			*fosA*-like			*tet(A)*	*oqxB*-like, *oqxA*-like	*catA1*				*ompK36* p.N49S, *ompK36* p.L59V, *ompK36* p.L191S, *ompK36* p.F198Y, *ompK36* p.F207W, *ompK36* p.A217S, *ompK36* p.N218H, *ompK36* p.D224E, *ompK36* p.L228V, *ompK36* p.E232R, *ompK37* p.I70M, *ompK37* p.I128M, *acrR* p.P161R, *acrR* p.G164A, *acrR* p.F172S, *acrR* p.R173G, *acrR* p.L195V, *acrR* p.F197I, *acrR* p.K201M	*oqxB*-like, *oqxA*-like

Table A2. *Cont.*

Sample ID	ST-Type	Acquired Resistance Determinants Against												
		Beta Lacatams	Sulfo-namids	Fosfomy-cin	Trimetho-prim	Macro-lides	Tetracyc-lines	Fluoro-quinolones	Chloram-phenicol	Rifam-picin	Amino-glycosides	Efflux Pumps	Amino Acid Exchanges Due to Point Mutations	Disinfectant Resistance Genes *
252	ST 607	*bla*_{TEM-1B}	*sul2, sul1*	*fosA*-like	*dfrA7*		*tet(A)*	*oqxB*-like, *oqxA*	*catA1*		*aph(3″)-Ib, aph(6)-Id*		*ompK37* p.I70M, *ompK37* p.I128M, *ompK37* p.N230G, *ompK36* p.N49S, *ompK36* p.L59V, *ompK36* p.L191S, *ompK36* p.F207W, *ompK36* p.A217S, *ompK36* p.N218H, ompK36 p.D224E, *ompK36* p.L228V, *ompK36* p.E232R, *ompK36* p.T254S, *acrR* p.P161R,*acrR* p.G164A, *acrR* p.F172S, *acrR* p.R173G, *acrR* p.L195V, *acrR* p.F197I, *acrR* p.K201M	*oqxB*-like, *oqxA, qacE*
267	ST 36	*bla*_{CTX-M-15}, *bla*_{TEM-1B}	*sul2, sul1*	*fosA*	*dfrA27*		*tet(D)*	*aac(6′)-Ib-cr, oqxA, oqxB*	*catA2*-like	ARR-3	*aph(6)-Id, aph(3″)-Ib, aac(6′)-Ib-cr, aadA16, aac(3)-IIa, aph(6)-Id*		*ompK36* p.N49S, *ompK36* p.L59V, *ompK36* p.T184P, *ompK37* p.I70M, *ompK37* p.I128M, *ompK37* p.N230G, *acrR* p.P161R, *acrR* p.G164A, *acrR* p.F172S, *acrR* p.R173G, *acrR* p.L195V, *acrR* p.F197I, *acrR* p.K201M	*oqxA, qacE, oqxB*

Table A2. *Cont.*

Sample ID	ST-Type	Acquired Resistance Determinants Against												
		Beta Lacatams	Sulfo-namids	Fosfomy-cin	Trimetho-prim	Macro-lides	Tetracyc-lines	Fluoro-quinolones	Chloram-phenicol	Rifam-picin	Amino-glycosides	Efflux Pumps	Amino Acid Exchanges Due to Point Mutations	Disinfectant Resistance Genes *
277	ST 530	*bla*_{TEM-35}	*sul2*	*fosA*-like	*dfrA14*		*tet(D)*	*oqxA*, *oqxB*-like			*aph(3″)-Ib*, *aph(6)-Id*		*acrR* p.P161R, *acrR* p.G164A, *acrR* p.F172S, *acrR* p.R173G, *acrR* p.L195V, *acrR* p.F197I, *acrR* p.K201M, *ompK36* p.N49S, *ompK36* p.L59V, *ompK36* p.L191S, *ompK36* p.F207W, *ompK36* p.A217S, *ompK36* p.N218H, *ompK36* p.D224E, *ompK36* p.L228V, *ompK36* p.E232R,*ompK36* p.T254S, *ompK37* p.I70M, *ompK37* p.I128M	*oqxA*, *oqxB*-like
279	ST 5385			*fosA*				*oqxA, oqxB*					*acrR* p.P161R, *acrR* p.G164A, *acrR* p.F172S, *acrR* p.R173G, *acrR* p.L195V, *acrR* p.F197I, *acrR* p.K201M, *ompK36* p.N49S, *ompK36* p.L59V, *ompK36* p.T184P, *ompK37* p.I70M, *ompK37* p.I128M	*oqxA, oqxB*

* *qacE* = quaternary ammonium compounds resistance and *oqxB* and *oqxA* = efflux pumps mediating resistance against disinfectants.

Table A3. Phenotypic resistance the *P. aeruginosa* strains. Data are missing for strains 198, 218 and 312, due to loss during subcultivation. MIC = minimum inhibitory concentration. N.a. = value missing due to loss of strain or failed reaction.

Sample ID	Piperacillin		Piperacillin/Tazobactam		Ceftrazidime		Cefepime		Imipenem		Meropenem		Gentamicin	
	MIC	Interpretation	MIC	Interpretation	MIC	Interpretation	MIC	Interpretation	MIC	Interpretation	MIC	Interpretation	MIC	Interpretation
17	$\leq$4	S	$\leq$4	S	$\leq$1	S	$\leq$1	S	$\leq$0.25	S	$\leq$0.25	S	$\leq$1	S
22	$\leq$4	S	$\leq$4	S	$\leq$1	S	$\leq$1	S	$\leq$0.25	S	$\leq$0.25	S	$\leq$1	S
32	$\leq$4	S	8	S	2	S	2	S	1	S	$\leq$0.25	S	$\leq$1	S
69	$\leq$4	S	$\leq$4	S	4	S	2	S	1	S	$\leq$0.25	S	$\leq$1	S
81	$\leq$4	S	8	S	2	S	$\leq$1	S	1	S	1	S	$\leq$1	S
82	16	S	8	S	4	S	2	S	2	S	$\leq$0.25	S	$\leq$1	S
88	$\geq$128	R	$\geq$128	R	$\geq$64	R	32	R	$\geq$16	R	4	I	$\leq$1	S
99	8	S	8	S	4	S	2	S	2	S	1	S	$\leq$1	S
106	$\leq$4	S	8	S	2	S	2	S	2	S	1	S	$\leq$1	S
114	$\leq$4	S	8	S	2	S	$\leq$1	S	2	S	1	S	$\leq$1	S
137	16	S	8	S	4	S	2	S	2	S	2	S	$\leq$1	S
144	16	S	8	S	4	S	2	S	2	S	1	S	$\leq$1	S
147	8	S	$\leq$4	S	4	S	8	S	2	S	0.5	S	4	S
149	8	S	8	S	4	S	2	S	$\leq$0.25	S	$\leq$0.25	S	$\leq$1	S
153	$\leq$4	S	8	S	2	S	$\leq$1	S	1	S	$\leq$0.25	S	$\leq$1	S
154	64	R	$\leq$4	S	$\leq$1	S	$\leq$1	S	2	S	0.5	S	$\leq$1	S
157	16	S	8	S	4	S	4	S	2	S	$\leq$0.25	S	2	S
160	$\geq$128	R	32	R	16	R	32	R	8	I	8	I	8	R
162	64	R	32	R	8	S	8	S	2	S	1	S	2	S
180	16	S	8	S	4	S	2	S	2	S	$\leq$0.25	S	$\leq$1	S
183	8	S	8	S	4	S	2	S	2	S	0.5	S	$\leq$1	S
186	16	n.a.	n.a.	S	4	S	2	S	2	S	$\leq$0.25	S	$\leq$1	S
190	16	S	8	S	4	S	2	S	2	S	0.5	S	$\leq$1	S
195	8	S	8	S	4	S	$\leq$1	S	2	S	$\leq$0.25	S	$\leq$1	S
196	$\leq$4	S	$\leq$4	S	2	S	$\leq$1	S	2	S	0.5	S	$\leq$1	S
198	n.a.	n.a.	n.a.	n.a.	n.a.	n.a.	n.a.	n.a.	n.a.	n.a.	n.a.	n.a.	n.a.	n.a.

Table A3. *Cont.*

Sample ID	Piperacillin		Piperacillin/Tazobactam		Ceftrazidime		Cefepime		Imipenem		Meropenem		Gentamicin	
	MIC	Interpretation	MIC	Interpretation	MIC	Interpretation	MIC	Interpretation	MIC	Interpretation	MIC	Interpretation	MIC	Interpretation
204	8	S	8	S	2	S	$\leq$1	S	2	S	$\leq$0.25	S	$\leq$1	S
208	8	S	8	S	4	S	2	S	2	S	$\leq$0.25	S	$\leq$1	S
218	n.a.	n.a.	n.a.	n.a.	n.a.	n.a.	n.a.	n.a.	n.a.	n.a.	n.a.	n.a.	n.a.	n.a.
229	8	S	8	S	4	S	2	S	1	S	$\leq$0.25	S	$\leq$1	S
233	$\geq$128	R	$\geq$128	R	32	R	8	S	8	I	4	I	$\leq$1	S
236	16	S	16	S	4	S	2	S	2	S	0.5	S	2	S
238	8	S	16	S	4	S	2	S	2	S	$\leq$0.25	S	$\leq$1	S
242	8	S	8	S	4	S	2	S	1	S	$\leq$0.25	S	$\leq$1	S
243	16	S	8	S	4	S	2	S	1	S	$\leq$0.25	S	$\leq$1	S
272	64	R	64	R	8	S	4	S	2	S	1	S	$\leq$1	S
274	16	S	8	S	4	S	2	S	2	S	0.5	S	$\leq$1	S
278	$\leq$4	S	$\leq$4	S	2	S	$\leq$1	S	2	S	$\leq$0.25	S	$\leq$1	S
282	$\leq$4	S	8	S	$\leq$1	S	$\leq$1	S	2	S	1	S	$\leq$1	S
285	$\leq$4	S	$\leq$4	S	$\leq$1	S	$\leq$1	S	2	S	1	S	$\leq$1	S
289	$\leq$4	S	8	S	$\leq$1	S	$\leq$1	S	2	S	1	S	$\leq$1	S
290	8	S	8	S	$\leq$1	S	$\leq$1	S	2	S	1	S	$\leq$1	S
296	$\geq$128	R	64	R	4	S	8	S	1	S	1	S	$\geq$16	R
298	$\geq$128	R	$\geq$128	R	$\geq$64	R	8	S	8	I	4	I	$\leq$1	S
301	16	S	8	S	4	S	2	S	2	S	0.5	S	$\leq$1	S
302	8	S	8	S	4	S	$\leq$1	S	2	S	0.5	S	$\leq$1	S
309	16	S	8	S	4	S	4	S	2	S	1	S	$\leq$1	S
310	32	R	16	S	4	S	4	S	2	S	1	S	$\leq$1	S
312	n.a.	n.a.	n.a.	n.a.	n.a.	n.a.	n.a.	n.a.	n.a.	n.a.	n.a.	n.a.	n.a.	n.a.

Table A4. Phenotypic resistance of *P. aeruginosa* strains. Data are missing for strains 198, 218 and 312 due to loss during subcultivation. MIC = minimum inhibitory concentration. N.a. = value missing due to loss of strain or failed reaction.

Sample ID	Ciprofloxacin		Moxifloxacin		Aztreonam		Amikacin		Tobramycin		Fosfomycin		Colistin	
	MIC	Interpretation	MIC	Interpretation	MIC	Interpretation	MIC	Interpretation	MIC	Interpretation	MIC	Interpretation	MIC	Interpretation
17	$\leq$0.25	S	1	R	4	I	$\leq$2	S	$\leq$1	S	128	R	$\leq$0.5	S
22	$\leq$0.25	S	1	R	2	I	$\leq$2	S	$\leq$1	S	128	R	1	S
32	$\leq$0.25	S	2	R	4	I	$\leq$2	S	$\leq$1	S	128	R	$\leq$0.5	S
69	$\leq$0.25	S	0.5	R	4	I	$\leq$2	S	$\leq$1	S	128	R	$\leq$0.5	S
81	$\leq$0.25	S	0.5	R	4	I	$\leq$2	S	$\leq$1	S	$\geq$256	R	$\leq$0.5	S
82	$\leq$0.25	S	1	R	16	I	$\leq$2	S	$\leq$1	S	$\leq$16	R	$\leq$0.5	S
88	2	R	$\geq$8	R	32	R	$\leq$2	S	$\leq$1	S	128	R	$\leq$0.5	S
99	$\leq$0.25	S	2	R	16	I	$\leq$2	S	$\leq$1	S	128	R	$\leq$0.5	S
106	2	R	1	R	8	I	$\leq$2	S	$\leq$1	S	$\geq$256	R	$\leq$0.5	S
114	$\leq$0.25	S	0.5	R	4	I	$\leq$2	S	$\leq$1	S	128	R	$\leq$0.5	S
137	$\leq$0.25	S	1	R	16	I	$\leq$2	S	$\leq$1	S	$\geq$256	R	$\leq$0.5	S
144	$\leq$0.25	S	1	R	16	I	$\leq$2	S	$\leq$1	S	$\geq$256	R	$\leq$0.5	S
147	$\leq$0.25	S	2	R	4	I	8	S	$\leq$1	S	$\geq$256	R	$\leq$0.5	S
149	$\leq$0.25	S	1	R	16	I	$\leq$2	S	$\leq$1	S	$\geq$256	R	$\leq$0.5	S
153	$\leq$0.25	S	2	R	4	I	$\leq$2	S	$\leq$1	S	$\leq$16	R	$\leq$0.5	S
154	$\leq$0.25	S	0.5	R	4	I	$\leq$2	S	$\leq$1	S	$\geq$256	R	$\leq$0.5	S
157	$\leq$0.25	S	1	R	16	I	$\leq$2	S	$\leq$1	S	$\geq$256	R	$\leq$0.5	S
160	1	R	$\geq$8	R	$\geq$64	R	16	I	$\leq$1	S	64	R	$\leq$0.5	S
162	0.5	S	2	R	32	R	4	S	$\leq$1	S	128	R	$\leq$0.5	S
180	$\leq$0.25	S	1	R	16	I	$\leq$2	S	$\leq$1	S	$\geq$256	R	$\leq$0.5	S
183	$\leq$0.25	S	0.5	R	4	I	$\leq$2	S	$\leq$1	S	$\geq$256	R	$\leq$0.5	S
186	$\leq$0.25	S	1	R	16	I	$\leq$2	S	$\leq$1	S	32	R	$\leq$0.5	S
190	$\leq$0.25	S	1	R	16	I	4	S	$\leq$1	S	32	R	$\leq$0.5	S
195	$\leq$0.25	S	0.5	R	4	I	$\leq$2	S	$\leq$1	S	64	R	$\leq$0.5	S
196	$\leq$0.25	S	0.5	R	4	I	$\leq$2	S	$\leq$1	S	64	R	$\leq$0.5	S
198	n.a.	n.a.	n.a.	n.a.	n.a.	n.a.	n.a.	n.a.	n.a.	n.a.	n.a.	n.a.	n.a.	n.a.

Table A4. *Cont.*

Sample ID	Ciprofloxacin		Moxifloxacin		Aztreonam		Amikacin		Tobramycin		Fosfomycin		Colistin	
	MIC	Interpretation	MIC	Interpretation	MIC	Interpretation	MIC	Interpretation	MIC	Interpretation	MIC	Interpretation	MIC	Interpretation
204	$\leq$0.25	S	1	R	2	I	$\leq$2	S	$\leq$1	S	32	R	2	S
208	$\leq$0.25	S	1	R	8	I	$\leq$2	S	$\leq$1	S	128	R	$\leq$0.5	S
218	n.a.	n.a.	n.a.	n.a.	n.a.	n.a.	n.a.	n.a.	n.a.	n.a.	n.a.	n.a.	n.a.	n.a.
229	$\leq$0.25	S	1	R	4	I	$\leq$2	S	$\leq$1	S	128	R	$\leq$0.5	S
233	$\leq$0.25	S	2	R	16	I	$\leq$2	S	$\leq$1	S	$\leq$16	R	$\leq$0.5	S
236	$\leq$0.25	S	1	R	16	I	8	S	$\leq$1	S	64	R	$\leq$0.5	S
238	$\leq$0.25	S	2	R	8	I	$\leq$2	S	$\leq$1	S	$\leq$16	R	$\leq$0.5	S
242	$\leq$0.25	S	1	R	8	I	$\leq$2	S	$\leq$1	S	128	R	$\leq$0.5	S
243	$\leq$0.25	S	2	R	16	I	$\leq$2	S	$\leq$1	S	64	R	$\leq$0.5	S
272	$\leq$0.25	S	2	R	32	R	$\leq$2	S	$\leq$1	S	$\geq$256	R	$\leq$0.5	S
274	$\leq$0.25	S	2	R	16	I	$\leq$2	S	$\leq$1	S	$\geq$256	R	$\leq$0.5	S
278	$\leq$0.25	S	0.5	R	4	I	$\leq$2	S	$\leq$1	S	128	R	$\leq$0.5	S
282	$\leq$0.25	S	1	R	2	I	$\leq$2	S	$\leq$1	S	128	R	$\leq$0.5	S
285	$\leq$0.25	S	1	R	2	I	$\leq$2	S	$\leq$1	S	128	R	$\leq$0.5	S
289	$\leq$0.25	S	0.5	R	4	I	$\leq$2	S	$\leq$1	S	128	R	$\leq$0.5	S
290	$\leq$0.25	S	0.5	R	4	I	$\leq$2	S	$\leq$1	S	128	R	$\leq$0.5	S
296	$\geq$4	R	$\geq$8	R	32	R	$\leq$2	S	$\geq$16	R	64	R	$\leq$0.5	S
298	$\leq$0.25	S	2	R	16	I	$\leq$2	S	$\leq$1	S	$\leq$16	R	$\leq$0.5	S
301	$\leq$0.25	S	2	R	16	I	$\leq$2	S	$\leq$1	S	$\leq$16	R	$\leq$0.5	S
302	$\leq$0.25	S	1	R	4	I	$\leq$2	S	$\leq$1	S	128	R	$\leq$0.5	S
309	$\leq$0.25	S	1	R	4	I	$\leq$2	S	$\leq$1	S	$\leq$16	R	$\leq$0.5	S
310	$\leq$0.25	S	2	R	16	I	$\leq$2	S	$\leq$1	S	128	R	$\leq$0.5	S
312	n.a.	n.a.	n.a.	n.a.	n.a.	n.a.	n.a.	n.a.	n.a.	n.a.	n.a.	n.a.	n.a.	n.a.

Table A5. Phenotypic resistance of the *Klebsiella* strains. MIC = minimum inhibitory concentration. ESBL = signal in phenotypic testing for extended-spectrum beta-lactamases.

Sample ID	ESBL	Ampicillin		Ampicillin/ Sulbactam		Piperacillin/ Tazobactam		Cefuroxime		Cefuroxime Axetil		Cefpodoxime		Cefotaxime		Ceftrazidime	
		MIC	Inter pretation	MIC	Inter pretation	MIC	Inter pretation	MIC	Inter pretation	MIC	Inter pretation	MIC	Inter pretation	MIC	Inter pretation	MIC	Inter pretation
44	negative	≥32	R	16	R	≤4	S	4	I	4	I	≤0.25	S	≤1	S	≤1	S
60	negative	≥32	R	16	R	≤4	S	≤1	I	≤1	S	≤0.25	S	≤1	S	≤1	S
73	positive	≥32	R	≥32	R	≥128	R	≥64	R	≥64	R	≥8	R	≥64	R	8	R
100	positive	≥32	R	≥32	R	≥128	R	≥64	R	≥64	R	≥8	R	≥64	R	16	R
102	negative	≥32	R	≤2	I	8	S	8	I	8	I	≤0.25	S	≤1	S	≤1	S
124	negative	≥32	R	≤2	I	≤4	S	2	I	2	I	≤0.25	S	≤1	S	≤1	S
146	negative	≥32	R	≤2	I	8	S	2	I	2	I	≤0.25	S	≤1	S	≤1	S
177	positive	≥32	R	≥32	R	≥128	R	2	I	2	S	≤0.25	S	≤1	S	≤1	I
181	negative	≥32	R	≤2	I	≤4	S	2	I	2	I	≤0.25	S	≤1	S	≤1	S
184	negative	16	R	≤2	I	≤4	S	2	I	2	S	≤0.25	S	≤1	S	≤1	S
199	positive	≥32	R	≥32	R	8	R	≥64	R	≥64	R	≥8	R	≥64	R	16	R
214	negative	≥32	R	≤2	I	≤4	S	≤1	I	≤1	S	≤0.25	S	≤1	S	≤1	S
217	negative	≥32	R	≥32	R	≥128	R	4	I	4	I	≤0.25	S	≤1	S	≤1	S
220	negative	≥32	R	≤2	I	≤4	S	4	I	4	I	≤0.25	S	≤1	S	≤1	S
234	negative	≥32	R	≤2	I	≤4	S	2	I	2	I	≤0.25	S	≤1	S	≤1	S
240	negative	≥32	R	≤2	I	≤4	S	≤1	I	≤1	S	≤0.25	S	≤1	S	≤1	S
248	negative	≥32	R	≤2	I	≤4	S	2	I	2	I	≤0.25	S	≤1	S	≤1	S
252	negative	≥32	R	16	R	≤4	S	2	I	2	I	≤0.25	S	≤1	S	≤1	S
267	positive	≥32	R	≥32	R	32	R	≥64	R	≥64	R	≥8	R	≥64	R	16	R
277	negative	≥32	R	≥32	R	64	R	2	I	2	I	≤0.25	S	≤1	S	≤1	S
279	negative	≥32	R	≤2	I	≤4	S	2	I	2	I	≤0.25	S	≤1	S	≤1	S

Table A6. Phenotypic resistance of the *Klebsiella* strains. MIC = minimum inhibitory concentration. ESBL = signal in phenotypic testing for extended-spectrum beta-lactamases.

Sample ID	ESBL	Ertapenem		Imipenem		Meropenem		Gentamicin		Ciprofloxacin		Moxifloxacin		Tigecycline		Trimethoprim/ Sulfamethoxazole	
		MIC	Inter pretation	MIC	Inter pretation	MIC	Inter pretation	MIC	Inter pretation	MIC	Inter pretation	MIC	Inter pretation	MIC	Inter pretation	MIC	Inter pretation
44	negative	$\leq$0.5	S	$\leq$0.25	S	$\leq$0.25	S	$\leq$1	S	$\leq$0.25	S	$\leq$0.25	S	$\leq$0.5	S	$\leq$20	S
60	negative	$\leq$0.5	S	$\leq$0.25	S	$\leq$0.25	S	$\geq$16	R	1	R	2	R	$\leq$0.5	S	$\geq$320	R
73	positive	$\leq$0.5	S	$\leq$0.25	S	$\leq$0.25	S	$\geq$16	R	1	R	2	R	$\leq$0.5	S	$\geq$320	R
100	positive	$\leq$0.5	S	$\leq$0.25	S	$\leq$0.25	S	$\geq$16	R	$\geq$4	R	$\geq$8	R	$\leq$0.5	S	$\geq$320	R
102	negative	$\leq$0.5	S	$\leq$0.25	S	$\leq$0.25	S	$\leq$1	S	$\leq$0.25	S	0.5	R	4	R	$\leq$20	S
124	negative	$\leq$0.5	S	$\leq$0.25	S	$\leq$0.25	S	$\leq$1	S	$\leq$0.25	S	$\leq$0.25	S	$\leq$0.5	S	$\leq$20	S
146	negative	$\leq$0.5	S	$\leq$0.25	S	$\leq$0.25	S	$\leq$1	S	$\leq$0.25	S	$\leq$0.25	S	1	S	$\leq$20	S
177	negative	$\leq$0.5	S	$\leq$0.25	S	$\leq$0.25	S	$\leq$1	S	$\leq$0.25	S	$\leq$0.25	S	2	I	$\geq$320	R
181	negative	$\leq$0.5	S	$\leq$0.25	S	$\leq$0.25	S	$\leq$1	S	$\leq$0.25	S	$\leq$0.25	S	$\leq$0.5	S	$\leq$20	S
184	negative	$\leq$0.5	S	$\leq$0.25	S	$\leq$0.25	S	$\leq$1	S	$\leq$0.25	S	$\leq$0.25	S	$\leq$0.5	S	$\leq$20	S
199	positive	$\leq$0.5	S	$\leq$0.25	S	$\leq$0.25	S	$\geq$16	R	$\leq$0.25	S	$\leq$0.25	S	$\leq$0.5	S	$\geq$320	R
214	negative	$\leq$0.5	S	$\leq$0.25	S	$\leq$0.25	S	$\leq$1	S	$\leq$0.25	S	$\leq$0.25	S	$\leq$0.5	S	$\geq$320	R
217	negative	$\leq$0.5	S	$\leq$0.25	S	$\leq$0.25	S	$\geq$16	R	$\leq$0.25	S	$\leq$0.25	S	2	I	$\geq$320	R
220	negative	$\leq$0.5	S	$\leq$0.25	S	$\leq$0.25	S	$\leq$1	S	$\leq$0.25	S	0.5	R	1	S	$\leq$20	S
234	negative	$\leq$0.5	S	$\leq$0.25	S	$\leq$0.25	S	$\leq$1	S	$\leq$0.25	S	0.5	R	1	S	$\leq$20	S
240	negative	$\leq$0.5	S	$\leq$0.25	S	$\leq$0.25	S	$\leq$1	S	$\leq$0.25	S	$\leq$0.25	S	$\leq$0.5	S	$\leq$20	S
248	negative	$\leq$0.5	S	$\leq$0.25	S	$\leq$0.25	S	$\leq$1	S	$\leq$0.25	S	$\leq$0.25	S	2	I	$\leq$20	S
252	negative	$\leq$0.5	S	$\leq$0.25	S	$\leq$0.25	S	$\leq$1	S	$\leq$0.25	S	$\leq$0.25	S	1	S	$\geq$320	R
267	positive	$\leq$0.5	S	$\leq$0.25	S	$\leq$0.25	S	$\geq$16	R	$\leq$0.25	S	$\leq$0.25	S	$\leq$0.5	S	$\geq$320	R
277	negative	$\leq$0.5	S	$\leq$0.25	S	$\leq$0.25	S	$\leq$1	S	$\leq$0.25	S	$\leq$0.25	S	1	S	$\geq$320	R
279	negative	$\leq$0.5	S	$\leq$0.25	S	$\leq$0.25	S	$\leq$1	S	$\leq$0.25	S	$\leq$0.25	S	$\leq$0.5	S	$\leq$20	S

Table A7. Phenotypic resistance of *Escherichia coli* strains. MIC = minimum inhibitory concentration. ESBL = signal in phenotypic testing for extended-spectrum beta-lactamases.

Sample ID	ESBL	Ampicillin		Ampicillin/ Sulbactam		Piperacillin/ Tazobactam		Cefuroxime		Cefuroxime Axetil		Cefpodoxime		Cefotaxime		Ceftrazidime	
		MIC	Inter pretation	MIC	Inter pretation	MIC	Inter pretation	MIC	Inter pretation	MIC	Inter pretation	MIC	Inter pretation	MIC	Inter pretation	MIC	Inter pretation
41	positive	≥32	R	≥32	R	64	R	≥64	R	≥64	R	≥8	R	≥64	R	16	R
49	negative	≥32	R	16	R	≤4	S	4	I	4	S	≤0.25	S	≤1	S	≤1	S
68	negative	≥32	R	16	R	≤4	S	≤1	I	≤1	S	≤0.25	S	≤1	S	≤1	S
117	negative	≥32	R	≥32	R	64	R	4	I	4	S	≤0.25	S	≤1	S	≤1	S
152	negative	≥32	R	≥32	R	≥128	R	4	I	4	S	0.5	S	≤1	S	≤1	S
176	negative	≥32	R	≥32	R	≤4	I	2	I	2	S	≤0.25	S	≤1	S	≤1	S
221	negative	≥32	R	≥32	R	≤4	I	2	I	2	S	≤0.25	S	≤1	S	≤1	S
222	negative	≥32	R	≥32	R	≤4	I	4	I	4	S	≤0.25	S	≤1	S	≤1	S
225	positive	≥32	R	≥32	R	≤4	R	≥64	R	≥64	R	≥8	R	≥64	R	16	R
245	positive	≥32	R	≥32	R	16	I	16	R	16	R	1	S	2	I	≤1	S
270	positive	≥32	R	16	R	≤4	R	≥64	R	≥64	R	≥8	R	≥64	R	≥64	R
299	negative	≥32	R	≤2	I	≤4	S	4	I	4	S	≤0.25	S	≤1	S	≤1	S

Table A8. Phenotypic resistance of *Escherichia coli* strains. MIC = minimum inhibitory concentration. ESBL = signal in phenotypic testing for extended-spectrum beta-lactamases.

Sample ID	ESBL	Ertapenem		Imipenem		Meropenem		Gentamicin		Ciprofloxacin		Moxifloxacin		Tigecycline		Trimethoprim/ Sulfamethoxazole	
		MIC	Inter pretation	MIC	Inter pretation	MIC	Inter pretation	MIC	Inter pretation	MIC	Inter pretation	MIC	Inter pretation	MIC	Inter pretation	MIC	Inter pretation
41	positive	≤0.5	S	≤0.25	S	≤0.25	S	≥16	R	≥4	R	≥8	R	≤0.5	S	≥320	R
49	negative	≤0.5	S	≤0.25	S	≤0.25	S	≤1	S	≤0.25	S	≤0.25	S	≤0.5	S	≥320	R
68	negative	≤0.5	S	≤0.25	S	≤0.25	S	≤1	S	≥4	R	≥8	R	≤0.5	S	≥320	R
117	negative	≤0.5	S	≤0.25	S	≤0.25	S	≤1	S	≤0.25	S	≤0.25	S	≤0.5	S	≥320	R
152	negative	≤0.5	S	≤0.25	S	≤0.25	S	2	S	1	R	2	R	≤0.5	S	≥320	R
176	negative	≤0.5	S	≤0.25	S	≤0.25	S	≤1	S	≤0.25	S	≤0.25	S	≤0.5	S	≥320	R
221	negative	≤0.5	S	≤0.25	S	≤0.25	S	≤1	S	≤0.25	S	≤0.25	S	≤0.5	S	≥320	R
222	negative	≤0.5	S	≤0.25	S	≤0.25	S	≤1	S	≤0.25	S	≤0.25	S	≤0.5	S	≥320	R
225	positive	≤0.5	S	≤0.25	S	≤0.25	S	≤1	S	0.5	I	1	R	≤0.5	S	≥320	R
245	positive	≤0.5	S	0.5	S	≤0.25	S	≥16	R	≥4	R	≥8	R	≤0.5	S	≥320	R
270	positive	≤0.5	S	≤0.25	S	≤0.25	S	≤1	S	≥4	R	≥8	R	≤0.5	S	≤20	S
299	negative	≤0.5	S	≤0.25	S	≤0.25	S	≤1	S	0.5	I	2	R	≤0.5	S	≥320	R

Table A9. Analysis of virulence determinants, ordered by strain and MLST type, of the assessed *P. aeruginosa* isolates. ST = Sequence type.

Sample ID	ST-Type	Pathogenicity Factor Groups								
		Adherence	Anti-Phagocytosis	Biosurfactant	Iron Uptake	Pigment	Protease	Toxin	Regulation	Secretion System
017	ST 381	*waaA, waaC, waaF, waaG, waaP, wzy, wzz, chpA, chpB, chpC, chpD, chpE, fimV, pilB, pilD, pilF, pilG, pilH, pilI, pilK, pilM, pilN, pilO, pilP, pilQ, pilR, pilS, pilT, pilU, pilC like, xcpA/pilD*	*alg44, alg8, algA, algB, algC, algD, algE, algF, algG, algP/algR3 algI, algJ, algK, algL, algQ, algR, algU, algW, algX, algZ, mucA, mucB, mucC*	*rhlA, rhlB*	*fptA, fpvA, pchA, pchB, pchC, pchD, pchE, pchF, pchG, pchH, pchI, pchR, pvdA, pvdD, pvdE*	*phzM, phzS*	*aprA, lasA*	*toxA, plcH*	*lasI, rhlI*	*xcpP, xcpQ, xcpR, xcpS, xcpT, xcpU, xcpV, xcpW, xcpX, xcpY, xcpZ*
022	ST 2483	*waaA, waaC, waaF, waaG, waaP, chpA, chpB, chpC, chpD, chpE, fimT, fimU, fimV, pilB, pilD, pilE, pilF, pilG, pilH, pilI, pilK, pilM, pilN, pilO, pilP, pilQ, pilR, pilS, pilT, pilU, pilV, pilW, pilX pilY1, pilY2, pilC, xcpA/pilD*	*alg44, alg8, algA, algB, algC, algD, algE, algF, algG, algI, algJ, algK, algL, algQ, algR, algU, algW, algX, algZ, mucA, mucB, mucC*	*rhlA, rhlB*	*fptA, pchA, pchB, pchC, pchD, pchE, pchF, pchG, pchH, pchI, pchR, pvdA*	*phzM, phzS*	*aprA, lasA*	*toxA, plcH*	*lasI, rhlI*	*xcpP, xcpQ, xcpR, xcpS, xcpT, xcpU, xcpV, xcpW, xcpX, xcpY, xcpZ*

Table A9. *Cont.*

Sample ID	ST-Type	Pathogenicity Factor Groups								
		Adherence	Anti-Phagocytosis	Biosurfactant	Iron Uptake	Pigment	Protease	Toxin	Regulation	Secretion System
032	ST 3587	*waaA, waaC, waaF, waaG, waaP, chpA, chpB, chpC, chpD, chpE, fimT, fimU, fimV, pilB, pilD, pilE, pilF, pilG, pilH, pilI, pilK, pilM, pilN, pilO, pilP, pilQ, pilR, pilS, pilT, pilU, pilV, pilW, pilX pilY1, pilY2, pilC like, xcpA/pilD*	*alg44, alg8, algA, algB, algC, algD, algE, algF, algG, algI, algJ, algK, algL, algQ, algR, algU, algW, algX, algZ, mucA, mucB, mucC*	*rhlA, rhlB*	*fptA, pchA, pchB, pchC, pchD, pchE, pchF, pchG, pchH, pchI, pchR, pvdA*	*phzM, phzS*	*aprA, lasA*	*toxA, plcH*	*lasI, rhlI*	*xcpP, xcpQ, xcpR, xcpS, xcpT, xcpU, xcpV, xcpW, xcpX, xcpY, xcpZ*
069	ST 360	*waaA, waaC, waaF, waaG, waaP, wzy, wzz, chpA, chpB, chpC, chpD, chpE, fimT, fimU, fimV, pilB, pilD, pilE, pilF, pilG, pilH, pilI, pilK, pilM, pilN, pilO, pilP, pilQ, pilR, pilS, pilT, pilU, pilV, pilW, pilX pilY1, pilY2, pilC like, xcpA/pilD*	*alg44, alg8, algA, algB, algC, algD, algE, algF, algG, algI, algJ, algK, algL, algQ, algR, algU, algW, algX, algZ, mucA, mucB, mucC*	*rhlA, rhlB*	*fptA, pchA, pchB, pchC, pchD, pchE, pchF, pchG, pchH, pchI, pchR, pvdA*	*phzM, phzS*	*aprA, lasA*	*toxA, plcH*	*lasI, rhlI*	*xcpP, xcpQ, xcpR, xcpS, xcpT, xcpU, xcpV, xcpW, xcpX, xcpY, xcpZ*

Table A9. *Cont.*

Sample ID	ST-Type	Pathogenicity Factor Groups								
		Adherence	Anti-Phagocytosis	Biosurfactant	Iron Uptake	Pigment	Protease	Toxin	Regulation	Secretion System
081	ST 244	*waaA, waaC, waaF, waaG, waaP, wzy, wzz, chpA, chpB, chpC, chpD, chpE, fimT, fimU, fimV, pilA like, pilB, pilD, pilE, pilF, pilG, pilH, pilI, pilK, pilM, pilN, pilO, pilP, pilQ, pilR, pilS, pilT, pilU, pilV, pilW, pilX pilY1, pilY2, pilC, xcpA/pilD*	*alg44, alg8, algA, algB, algC, algD, algE, algF, algG, algI, algJ, algK, algL, algQ, algR, algU, algW, algX, algZ, mucA, mucB, mucC*	*rhlA, rhlB*	*fptA, fpvA, pchA, pchB, pchC, pchD, pchE, pchF, pchG, pchH, pchI, pchR, pvdA, pvdE*	*phzM, phzS*	*aprA, lasA*	*toxA, plcH*	*lasI, rhlI*	*xcpP, xcpQ, xcpR, xcpS, xcpT, xcpU, xcpV, xcpW like, xcpX, xcpY, xcpZ*
082	ST 514	*waaA, waaC, waaF, waaG, waaP, chpA, chpB, chpC, chpD, chpE, fimV, pilB, pilD, pilF, pilG, pilH, pilI, pilK, pilM, pilN, pilO, pilP, pilQ, pilR, pilS, pilT, pilU, pilC, xcpA/pilD*	*alg44, alg8, algA, algB, algC, algD, algE, algF, algG, algI, algJ, algK, algL, algP/algR3, algQ, algR, algU, algW, algX, algZ, mucA, mucB, mucC*	*rhlA, rhlB*	*fptA, pchA, pchB, pchC, pchD, pchE, pchF, pchG, pchH, pchI, pchR, pvdA*	*phzM, phzS*	*aprA, lasA*	*toxA, plcH*	*lasI, rhlI*	*xcpP, xcpQ, xcpR, xcpS, xcpT, xcpU, xcpV, xcpW, xcpX, xcpY, xcpZ*
088	ST 1682	*waaA, waaC, waaF, waaG, waaP, wzy, wzz, chpA, chpB, chpC, chpD, chpE, fimT, fimU, fimV, pilB, pilD, pilE, pilF, pilG, pilH, pilI, pilK, pilM, pilN, pilO, pilP, pilQ, pilR, pilS, pilT, pilU, pilV, pilW, pilX pilY1, pilY2, xcpA/pilD*	*alg44, alg8, algA, algB, algC, algD, algE, algF, algG, algI, algJ, algK, algL, algQ, algR, algU, algW, algX, algZ, mucA, mucB, mucC*	*rhlA, rhlB*	*fptA, pchA, pchB, pchC, pchD, pchE, pchF, pchG, pchH, pchI, pchR, pvdA*	*phzM, phzS*	*aprA, lasA*	*toxA, plcH*	*lasI, rhlI*	*xcpP, xcpQ, xcpR, xcpS, xcpT, xcpU, xcpV, xcpW, xcpX, xcpY, xcpZ*

Table A9. *Cont.*

Sample ID	ST-Type	Pathogenicity Factor Groups								
		Adherence	Anti-Phagocytosis	Biosurfactant	Iron Uptake	Pigment	Protease	Toxin	Regulation	Secretion System
099	ST 244	*waaA, waaC, waaF, waaG, waaP, wzy, wzz, chpA, chpB, chpC, chpD, chpE, fimT, fimU, fimV, pilA, pilB, pilD, pilE, pilF, pilG, pilH, pilI, pilK, pilM, pilN, pilO, pilP, pilQ, pilR, pilS, pilT, pilU, pilV, pilW, pilX pilY1, pilY2, pilC, xcpA/pilD*	*alg44, alg8, algA, algB, algC, algD, algE, algF, algG, algI, algJ, algK, algL, algP/algR3, algQ, algR, algU, algW, algX, algZ, mucA, mucB, mucC*	*rhlA, rhlB*	*fptA, fpvA, pchA, pchB, pchC, pchD, pchE, pchF, pchG, pchH, pchI, pchR, pvdA, pvdE*	*phzM, phzS*	*aprA, lasA*	*toxA, plcH*	*lasI, rhlI*	*xcpP, xcpQ, xcpR, xcpS, xcpT, xcpU, xcpV, xcpW, xcpX, xcpY, xcpZ*
106	ST 1521	*waaA, waaC, waaF, waaG, waaP, wzy, wzz, chpA, chpB, chpC, chpD, chpE, fimT, fimU, fimV, pilA, pilB, pilD, pilE, pilF, pilG, pilH, pilI, pilK, pilM, pilN, pilO, pilP, pilQ, pilR, pilS, pilT, pilU, pilV, pilW, pilX pilY1, pilY2, pilC, xcpA/pilD*	*alg44, alg8, algA, algB, algC, algD, algE, algF, algG, algI, algJ, algK, algL, algQ, algR, algU, algW, algX, algZ, mucA, mucB, mucC*	*rhlA, rhlB*	*fptA, fpvA, pchA, pchB, pchC, pchD, pchE, pchF, pchG, pchH, pchI, pchR, pvdA, pvdD, pvdE*	*phzM, phzS*	*aprA, lasA*	*toxA, plcH*	*lasI, rhlI*	*xcpP, xcpQ, xcpR, xcpS, xcpT, xcpU, xcpV, xcpW, xcpX, xcpY, xcpZ*

Table A9. *Cont.*

Sample ID	ST-Type	Pathogenicity Factor Groups								
		Adherence	Anti-Phagocytosis	Biosurfactant	Iron Uptake	Pigment	Protease	Toxin	Regulation	Secretion System
114	ST 244	*waaA, waaC, waaF, waaG, waaP, wzy, wzz, chpA, chpB, chpC, chpD, chpE, fimT, fimU, fimV, pilA like, pilB, pilD, pilE, pilF, pilG, pilH, pilI, pilK, pilM, pilN, pilO, pilP, pilQ, pilR, pilS, pilT, pilU, pilV, pilW, pilX pilY1, pilY2, pilC, xcpA/pilD*	*alg44, alg8, algA, algB, algC, algD, algE, algF, algG, algI, algJ, algK, algL, algQ, algR, algU, algW, algX, algZ, mucA, mucB, mucC*	*rhlA, rhlB*	*fptA, fpvA, pchA, pchB, pchC, pchD, pchE, pchF, pchG, pchH, pchI, pchR, pvdA, pvdE*	*phzM, phzS*	*aprA, lasA*	*toxA, plcH*	*lasI, rhlI*	*xcpP, xcpQ, xcpR, xcpS, xcpT, xcpU, xcpV, xcpW like, xcpX, xcpY, xcpZ*
137	ST 3014	*waaA, waaC, waaF, waaG, waaP, wzy, wzz, chpA, chpB, chpC, chpD, chpE, fimT, fimU, fimV, pilA like, pilB, pilD, pilE, pilF, pilG, pilH, pilI, pilK, pilM, pilN, pilO, pilP, pilQ, pilR, pilS, pilT, pilU, pilV, pilW, pilX pilY1, pilY2, pilC, xcpA/pilD*	*alg44, alg8, algA, algB, algC, algD, algE, algF, algG, algI, algJ, algK, algL, algP/algR3, algQ, algR, algU, algW, algX, algZ, mucA, mucB, mucC*	*rhlA, rhlB*	*fptA, fpvA, pchA, pchB, pchC, pchD, pchE, pchF, pchG, pchH, pchI, pchR, pvdA, pvdE*	*phzM, phzS*	*aprA, lasA*	*toxA, plcH*	*lasI, rhlI*	*xcpP, xcpQ, xcpR, xcpS, xcpT, xcpU, xcpV, xcpW, xcpX, xcpY, xcpZ*

Table A9. *Cont.*

Sample ID	ST-Type	Pathogenicity Factor Groups								
		Adherence	Anti-Phagocytosis	Biosurfactant	Iron Uptake	Pigment	Protease	Toxin	Regulation	Secretion System
144	ST 245	*waaA, waaC, waaF, waaG, waaP, wzy, wzz, chpA, chpB, chpC, chpD, chpE, fimT, fimU, fimV, pilA, pilB, pilD, pilE, pilF, pilG, pilH, pilI, pilK, pilM, pilN, pilO, pilP, pilQ, pilR, pilS, pilT, pilU, pilV, pilW, pilX pilY1, pilY2, pilC like, xcpA/pilD*	*alg44, alg8, algA, algB, algC, algD, algE, algF, algG, algI, algJ, algK, algL, algQ, algR, algU, algW, algX, algZ, mucA, mucB, mucC*	*rhlA, rhlB*	*fptA, pchA, pchB, pchC, pchD, pchE, pchF, pchG, pchH, pchI, pchR, pvdA*	*phzM, phzS*	*aprA, lasA*	*toxA, plcH*	*lasI, rhlI*	*xcpP, xcpQ, xcpR, xcpS, xcpT, xcpU, xcpV, xcpW, xcpX, xcpY, xcpZ*
147	ST 245	*waaA, waaC, waaF, waaG, waaP, wzy, wzz, chpA, chpB, chpC, chpD, chpE, fimT, fimU, fimV, pilA, pilB, pilD, pilE, pilF, pilG, pilH, pilI, pilK, pilM, pilN, pilO, pilP, pilQ, pilR, pilS, pilT, pilU, pilV, pilW, pilX pilY1, pilY2, pilC like, xcpA/pilD*	*alg44, alg8, algA, algB, algC, algD, algE, algF, algG, algI, algJ, algK, algL, algQ, algR, algU, algW, algX, algZ, mucA, mucB, mucC*	*rhlA, rhlB*	*fptA, pchA, pchB, pchC, pchD, pchE, pchF, pchG, pchH, pchI, pchR, pvdA*	*phzM, phzS*	*aprA, lasA*	*toxA, plcH*	*lasI, rhlI*	*xcpP, xcpQ, xcpR, xcpS, xcpT, xcpU, xcpV, xcpW, xcpX, xcpY, xcpZ*

Table A9. *Cont.*

Sample ID	ST-Type	Pathogenicity Factor Groups								
		Adherence	Anti-Phagocytosis	Biosurfactant	Iron Uptake	Pigment	Protease	Toxin	Regulation	Secretion System
149	ST 381	*waaA, waaC, waaF, waaG, waaP, wzy, wzz, chpA, chpB, chpC, chpD, chpE, fimV, pilA, pilB, pilD, pilF, pilG, pilH, pilI, pilK, pilM, pilN, pilO, pilP, pilQ, pilR, pilS, pilT, pilU, pilC like, xcpA/pilD*	*alg44, alg8, algA, algB, algC, algD, algE, algF, algG, algI, algJ, algK, algL, algP/algR3, algQ, algR, algU, algW, algX, algZ, mucA, mucB, mucC*	*rhlA, rhlB*	*fptA, fpvA, pchA, pchB, pchC, pchD, pchE, pchF, pchG, pchH, pchI, pchR, pvdA, pvdE*	*phzM, phzS*	*aprA, lasA*	*toxA, plcH*	*lasI, rhlI*	*xcpP, xcpQ, xcpR, xcpS, xcpT, xcpU, xcpV, xcpW, xcpX, xcpY, xcpZ*
153	ST 704	*waaA, waaC, waaF, waaG, waaP, wzy, wzz, chpA, chpB, chpC, chpD, chpE, fimT, fimU, fimV, pilA, pilB, pilD, pilE, pilF, pilG, pilH, pilI, pilK, pilM, pilN, pilO, pilP, pilQ, pilR, pilS, pilT, pilU, pilV, pilW, pilX pilY1, pilY2, pilC like, xcpA/pilD*	*alg44, alg8, algA, algB, algC, algD, algE, algF, algG, algI, algJ, algK, algL, algQ, algR, algU, algW, algX, algZ, mucA, mucB, mucC*	*rhlA, rhlB*	*fptA, pchA, pchB, pchC, pchD, pchE, pchF, pchG, pchH, pchI, pchR, pvdA*	*phzM, phzS*	*aprA, lasA*	*plcH*	*lasI*	*xcpP, xcpQ, xcpR, xcpS, xcpT, xcpU, xcpV, xcpW, xcpX, xcpY, xcpZ*
154	ST 244	*waaA, waaC, waaF, waaG, waaP, wzy, wzz, chpA, chpB, chpC, chpD, chpE, fimT, fimU, fimV, pilA like, pilB, pilD, pilE, pilF, pilG, pilH, pilI, pilK, pilM, pilN, pilO, pilP, pilQ, pilR, pilS, pilT, pilU, pilV, pilW, pilX pilY1, pilY2, pilC, xcpA/pilD*	*alg44, alg8, algA, algB, algC, algD, algE, algF, algG, algI, algJ, algK, algL, algP/algR3, algQ, algR, algU, algW, algX, algZ, mucA, mucB, mucC*	*rhlA, rhlB*	*fptA, fpvA, pchA, pchB, pchC, pchD, pchE, pchF, pchG, pchH, pchI, pchR, pvdA, pvdE*	*phzM, phzS*	*aprA, lasA*	*toxA, plcH*	*lasI, rhlI*	*xcpP, xcpQ, xcpR, xcpS, xcpT, xcpU, xcpV, xcpW like, xcpX, xcpY, xcpZ*

Table A9. *Cont.*

Sample ID	ST-Type	Pathogenicity Factor Groups								
		Adherence	Anti-Phagocytosis	Biosurfactant	Iron Uptake	Pigment	Protease	Toxin	Regulation	Secretion System
157	ST 2616	*waaA, waaC, waaF, waaG, waaP, chpA, chpB, chpC, chpD, chpE, fimV, pilA, pilB, pilD, pilF, pilG, pilH, pilI, pilK, pilM, pilN, pilO, pilP, pilQ, pilR, pilS, pilT, pilU, pilC like, xcpA/pilD*	*alg44, alg8, algA, algB, algC, algD, algE, algF, algG, algI, algJ, algK, algL, algQ, algR, algU, algW, algX, algZ, mucA, mucB, mucC*	*rhlA, rhlB*	*fptA, fpvA, pchA, pchB, pchC, pchD, pchE, pchF, pchG, pchH, pchI, pchR, pvdA, pvdE*	*phzM, phzS*	*aprA, lasA*	*toxA, plcH*	*lasI, rhlI*	*xcpP, xcpQ, xcpR, xcpS, xcpT, xcpU, xcpV, xcpW, xcpX, xcpY, xcpZ*
160	ST 170	*waaA, waaC, waaF, waaG, waaP, chpA, chpB, chpC, chpD, chpE, fimV, pilA, pilB, pilD, pilF, pilG, pilH, pilI, pilK, pilM, pilN, pilO, pilP, pilQ, pilR, pilS, pilT, pilU, pilC like, xcpA/pilD*	*alg44, alg8, algA, algB, algC, algD, algE, algF, algG, algI, algJ, algK, algL, algP/algR3, algQ, algR, algU, algW, algX, algZ, mucA, mucB, mucC*	*rhlA, rhlB*	*fptA, pchA, pchB, pchC, pchD, pchE, pchF, pchG, pchH, pchI, pchR, pvdA*	*phzM, phzS*	*aprA, lasA*	*toxA, plcH*	*lasI, rhlI*	*xcpP, xcpQ, xcpR, xcpS, xcpT, xcpU, xcpV, xcpW, xcpX, xcpY, xcpZ*
162	ST 274	*waaA, waaC, waaF, waaG, waaP, chpA, chpB, chpC, chpD, chpE, fimV, pilA, pilB, pilD, pilF, pilG, pilH, pilI, pilK, pilM, pilN, pilO, pilP, pilQ, pilR, pilS, pilT, pilU, pilC like, xcpA/pilD*	*alg44, alg8, algA, algB, algC, algD, algE, algF, algG, algI, algJ, algK, algL, algP/algR3, algQ, algR, algU, algW, algX, algZ, mucA, mucB, mucC*	*rhlA, rhlB*	*fptA, pchA, pchB, pchC, pchD, pchE, pchF, pchG, pchH, pchI, pchR, pvdA*	*phzM, phzS*	*aprA, lasA*	*toxA, plcH*	*lasI, rhlI*	*xcpP, xcpQ, xcpR, xcpS, xcpT, xcpU, xcpV, xcpW, xcpX, xcpY, xcpZ*

Table A9. *Cont.*

Sample ID	ST-Type	Pathogenicity Factor Groups									
		Adherence	Anti-Phagocytosis	Biosurfactant	Iron Uptake	Pigment	Protease	Toxin	Regulation	Secretion System
180	ST 856	*waaA, waaC, waaF, waaG, waaP, chpA, chpB, chpC, chpD, chpE, fimT, fimU, fimV, pilA, pilB, pilD, pilE, pilF, pilG, pilH, pilI, pilK, pilM, pilN, pilO, pilP, pilQ, pilR, pilS, pilT, pilU, pilV, pilW, pilX pilY1, pilY2, pilC like, xcpA/pilD*	*alg44, alg8, algA, algB, algC, algD, algE, algF, algG, algI, algJ, algK, algL, algP/algR3, algQ, algR, algU, algW, algX, algZ, mucA, mucB, mucC*	*rhlA, rhlB*	*fptA, pchA, pchB, pchC, pchD, pchE, pchF, pchG, pchH, pchI, pchR, pvdA*	*phzM, phzS*	*aprA, lasA*	*toxA, plcH*	*lasI, rhlI*	*xcpP, xcpQ, xcpR, xcpS, xcpT, xcpU, xcpV, xcpW, xcpX, xcpY, xcpZ*
183	ST 244	*waaA, waaC, waaF, waaG, waaP, wzy, wzz, chpA, chpB, chpC, chpD, chpE, fimT, fimU, fimV, pilA like, pilB, pilD, pilE, pilF, pilG, pilH, pilI, pilK, pilM, pilN, pilO, pilP, pilQ, pilR, pilS, pilT, pilU, pilV, pilW, pilX pilY1, pilY2, pilC, xcpA/pilD*	*alg44, alg8, algA, algB, algC, algD, algE, algF, algG, algI, algJ, algK, algL, algQ, algR, algU, algW, algX, algZ, mucA, mucB, mucC*	*rhlA, rhlB*	*fptA, fpvA, pchA, pchB, pchC, pchD, pchE, pchF, pchG, pchH, pchI, pchR, pvdA, pvdE*	*phzM, phzS*	*aprA, lasA*	*toxA, plcH*	*lasI, rhlI*	*xcpP, xcpQ, xcpR, xcpS, xcpT, xcpU, xcpV, xcpW like, xcpX, xcpY, xcpZ*

Table A9. *Cont.*

Sample ID	ST-Type	Pathogenicity Factor Groups								
		Adherence	Anti-Phagocytosis	Biosurfactant	Iron Uptake	Pigment	Protease	Toxin	Regulation	Secretion System
186	ST 3588	*waaA, waaC, waaF, waaG, waaP, chpA, chpB, chpC, chpD, chpE, fimT, fimU, fimV, pilA like, pilB, pilD, pilE, pilF, pilG, pilH, pilI, pilK, pilM, pilN, pilO, pilP, pilQ, pilR, pilS, pilT, pilU, pilV, pilW, pilX pilY1, pilY2, pilC, xcpA/pilD*	*alg44, alg8, algA, algB, algC, algD, algE, algF, algG, algI, algJ, algK, algL, algQ, algR, algU, algW, algX, algZ, mucA, mucB, mucC*	*rhlA, rhlB*	*fptA, pchA, pchB, pchC, pchD, pchE, pchF, pchG, pchH, pchI, pchR, pvdA*	*phzM, phzS*	*aprA, lasA*	*plcH*	*lasI, rhlI*	*xcpP, xcpQ, xcpR, xcpS, xcpT, xcpU, xcpV, xcpW, xcpX, xcpY, xcpZ*
190	ST 871	*waaA, waaC, waaF, waaG, waaP, chpA, chpB, chpC, chpD, chpE, fimT, fimU, fimV, pilA, pilB, pilD, pilE, pilF, pilG, pilH, pilI, pilK, pilM, pilN, pilO, pilP, pilQ, pilR, pilS, pilT, pilU, pilV, pilW, pilX pilY1, pilY2, pilC like, xcpA/pilD*	*alg44, alg8, algA, algB, algC, algD, algE, algF, algG, algI, algJ, algK, algL, algP/algR3, algQ, algR, algU, algW, algX, algZ, mucA, mucB, mucC*	*rhlA, rhlB*	*fptA, pchA, pchB, pchC, pchD, pchE, pchF, pchG, pchH, pchI, pchR, pvdA*	*phzM, phzS*	*aprA, lasA*	*toxA, plcH*	*lasI, rhlI*	*xcpP, xcpQ, xcpR, xcpS, xcpT, xcpU, xcpV, xcpW, xcpX, xcpY, xcpZ*
195	ST 988	*waaA, waaC, waaF, waaG, waaP, chpA, chpB, chpC, chpD, chpE, fimT, fimU, fimV, pilA, pilB, pilD, pilE, pilF, pilG, pilH, pilI, pilK, pilM, pilN, pilO, pilP, pilQ, pilR, pilS, pilT, pilU, pilV, pilW, pilX pilY1, pilY2, pilC like, xcpA/pilD*	*alg44, alg8, algA, algB, algC, algD, algE, algF, algG, algI, algJ, algK, algL, algP/algR3, algQ, algR, algU, algW, algX, algZ, mucA, mucB, mucC*	*rhlA, rhlB*	*fptA, pchA, pchB, pchC, pchD, pchE, pchF, pchG, pchH, pchI, pchR, pvdA*	*phzM, phzS*	*aprA, lasA*	*toxA, plcH*	*lasI, rhlI*	*xcpP, xcpQ, xcpR, xcpS, xcpT, xcpU, xcpV, xcpW, xcpX, xcpY, xcpZ*

Table A9. *Cont.*

Sample ID	ST-Type	Pathogenicity Factor Groups									
		Adherence	Anti-Phagocytosis	Biosurfactant	Iron Uptake	Pigment	Protease	Toxin	Regulation	Secretion System
196	ST 2475	*waaA, waaC, waaF, waaG, waaP, wzy, wzz, chpA, chpB, chpC, chpD, chpE, fimT, fimU, fimV, pilA, pilB, pilD, pilE, pilF, pilG, pilH, pilI, pilK, pilM, pilN, pilO, pilP, pilQ, pilR, pilS, pilT, pilU, pilV, pilW, pilX pilY1, pilY2, pilC, xcpA/pilD*	*alg44, alg8, algA, algB, algC, algD, algE, algF, algG, algI, algJ, algK, algL, algP/algR3, algQ, algR, algU, algW, algX, algZ, mucA, mucB, mucC*	*rhlA, rhlB*	*fptA, fpvA pchA, pchB, pchC, pchD, pchE, pchF, pchG, pchH, pchI, pchR, pvdA, pvdE*	*phzM, phzS*	*aprA, lasA*	*toxA, plcH*	*lasI, rhlI*	*xcpP, xcpQ, xcpR, xcpS, xcpT, xcpU, xcpV, xcpW, xcpX, xcpY, xcpZ*
198	ST 2476	*waaA, waaC, waaF, waaG, waaP, chpA, chpB, chpC, chpD, chpE, fimV, pilB, pilD, pilF, pilG, pilH, pilI, pilK, pilM, pilN, pilO, pilP, pilQ, pilR, pilS, pilT, pilU, pilC like, xcpA/pilD*	*alg44, alg8, algA, algB, algC, algD, algE, algF, algG, algI, algJ, algK, algL, algP/algR3, algQ, algR, algU, algW, algX, algZ, mucA, mucB, mucC*	*rhlA, rhlB*	*fptA, pchA, pchB, pchC, pchD, pchE, pchF, pchG, pchH, pchI, pchR, pvdA*	*phzM, phzS*	*aprA, lasA*	*toxA, plcH*	*lasI, rhlI*	*xcpP, xcpQ, xcpR, xcpS, xcpT, xcpU, xcpV, xcpW, xcpX, xcpY, xcpZ*
204	ST 639	*waaA, waaC, waaF, waaG, waaP, chpA, chpB, chpC, chpD, chpE, fimT, fimU, fimV, pilB, pilD, pilE, pilF, pilG, pilH, pilI, pilK, pilM, pilN, pilO, pilP, pilQ, pilR, pilS, pilT, pilU, pilV, pilW, pilX pilY1, pilY2, pilC like, xcpA/pilD*	*alg44, alg8, algA, algB, algC, algD, algE, algF, algG, algI, algJ, algK, algL, algQ, algR, algU, algW, algX, algZ, mucA, mucB, mucC*	*rhlA, rhlB*	*fptA, pchA, pchB, pchC, pchD, pchE, pchF, pchG, pchH, pchI, pchR, pvdA*	*phzM, phzS*	*aprA, lasA*	*toxA, plcH*	*rhlI*	*xcpP, xcpQ, xcpR, xcpS, xcpT, xcpU, xcpV, xcpW, xcpX, xcpY, xcpZ*

Table A9. *Cont.*

Sample ID	ST-Type	Pathogenicity Factor Groups								
		Adherence	Anti-Phagocytosis	Biosurfactant	Iron Uptake	Pigment	Protease	Toxin	Regulation	Secretion System
208	ST 132	*waaA, waaC, waaF, waaG, waaP, chpA, chpB, chpC, chpD, chpE, fimT, fimU, fimV, pilA like, pilB, pilD, pilE, pilF, pilG, pilH, pilI, pilK, pilM, pilN, pilO, pilP, pilQ, pilR, pilS, pilT, pilU, pilV, pilW, pilX pilY1, pilY2, pilC, xcpA/pilD*	*alg44, alg8, algA, algB, algC, algD, algE, algF, algG, algI, algJ, algK, algL, algQ, algR, algU, algW, algX, algZ, mucA, mucB, mucC*	*rhlA, rhlB*	*fptA, pchA, pchB, pchC, pchD, pchE, pchF, pchG, pchH, pchI, pchR, pvdA*	*phzM, phzS*	*aprA, lasA*	*toxA, plcH*	*lasI, rhlI*	*xcpP, xcpQ, xcpR, xcpS, xcpT, xcpU, xcpV, xcpW, xcpX, xcpY, xcpZ*
218	ST 856	*waaA, waaC, waaF, waaG, waaP, chpA, chpB, chpC, chpD, chpE, fimT, fimU, fimV, pilB, pilD, pilE, pilF, pilG, pilH, pilI, pilK, pilM, pilN, pilO, pilP, pilQ, pilR, pilS, pilT, pilU, pilV, pilW, pilX pilY1, pilY2, pilC like, xcpA/pilD*	*alg44, alg8, algA, algB, algC, algD, algE, algF, algG, algI, algJ, algK, algL, algQ, algR, algU, algW, algX, algZ, mucA, mucB, mucC*	*rhlA, rhlB*	*fptA, pchA, pchB, pchC, pchD, pchE, pchF, pchG, pchH, pchI, pchR, pvdA*	*phzM, phzS*	*aprA, lasA*	*toxA, plcH*	*lasI, rhlI*	*xcpP, xcpQ, xcpR, xcpS, xcpT, xcpU, xcpV, xcpW, xcpX, xcpY, xcpZ*
229	ST 270	*waaA, waaC, waaF, waaG, waaP, wzy, wzz, chpA, chpB, chpC, chpD, chpE, fimT, fimU, fimV, pilB, pilD, pilE, pilF, pilG, pilH, pilI, pilK, pilM, pilN, pilO, pilP, pilQ, pilR, pilS, pilT, pilU, pilV, pilW, pilX pilY1, pilY2, pilC like, xcpA/pilD*	*alg44, alg8, algA, algB, algC, algD, algE, algF, algG, algI, algJ, algK, algL, algP/algR3, algQ, algR, algU, algW, algX, algZ, mucA, mucB, mucC*	*rhlA, rhlB*	*fptA, fpvA pchA, pchB, pchC, pchD, pchE, pchF, pchG, pchH, pchI, pchR, pvdA, pvdD, pvdE*	*phzM, phzS*	*aprA, lasA*	*toxA, plcH*	*lasI, rhlI*	*xcpP, xcpQ, xcpR, xcpS, xcpT, xcpU, xcpV, xcpW, xcpX, xcpY, xcpZ*

Table A9. *Cont.*

Sample ID	ST-Type	Pathogenicity Factor Groups									
		Adherence	Anti-Phagocytosis	Biosurfactant	Iron Uptake	Pigment	Protease	Toxin	Regulation	Secretion System
233	ST 3227	*waaA, waaC, waaF, waaG, waaP, chpA, chpB, chpC, chpD, chpE, fimV, pilB, pilD, pilF, pilG, pilH, pilI, pilK, pilM, pilN, pilO, pilP, pilQ, pilR, pilS, pilT, pilU, pilC like, xcpA/pilD*	*alg44, alg8, algA, algB, algC, algD, algE, algF, algG, algI, algJ, algK, algL, algP/algR3, algQ, algR, algU, algW, algX, algZ, mucA, mucB, mucC*	*rhlA, rhlB*	*fptA, pchA, pchB, pchC, pchD, pchE, pchF, pchG, pchH, pchI, pchR, pvdA*	*phzM, phzS*	*aprA, lasA*	*toxA, plcH*	*lasI, rhlI*	*xcpP, xcpQ, xcpR, xcpS, xcpT, xcpU, xcpV, xcpW, xcpX, xcpY, xcpZ*
236	ST 266	*waaA, waaC, waaF, waaG, waaP, chpA, chpB, chpC, chpD, chpE, fimT, fimU, fimV, pilB, pilD, pilE, pilF, pilG, pilH, pilI, pilK, pilM, pilN, pilO, pilP, pilQ, pilR, pilS, pilT, pilU, pilV, pilW, pilX pilY1, pilY2, pilC like, xcpA/pilD*	*alg44, alg8, algA, algB, algC, algD, algE, algF, algG, algI, algJ, algK, algL, algQ, algR, algU, algW, algX, algZ, mucA, mucB, mucC*	*rhlA, rhlB*	*fptA, pchA, pchB, pchC, pchD, pchE, pchF, pchG, pchH, pchI, pchR, pvdA*	*phzM, phzS*	*aprA, lasA*	*toxA, plcH*	*lasI, rhlI*	*xcpP, xcpQ, xcpR, xcpS, xcpT, xcpU, xcpV, xcpW, xcpX, xcpY, xcpZ*
238	ST 3589	*waaA, waaC, waaF, waaG, waaP, chpA, chpB, chpC, chpD, chpE, fimT, fimU, fimV, pilA, pilB, pilD, pilE, pilF, pilG, pilH, pilI, pilK, pilM, pilN, pilO, pilP, pilQ, pilR, pilS, pilT, pilU, pilV, pilW, pilX pilY1, pilY2, pilC like, xcpA/pilD*	*alg44, alg8, algA, algB, algC, algD, algE, algF, algG, algI, algJ, algK, algL, algQ, algR, algU, algW, algX, algZ, mucA, mucB, mucC*	*rhlA, rhlB*	*fptA, pchA, pchB, pchC, pchD, pchE, pchF, pchG, pchH, pchI, pchR, pvdA*	*phzM, phzS*	*aprA, lasA*	*plcH*	*lasI, rhlI*	*xcpP, xcpQ, xcpS, xcpT, xcpU, xcpV, xcpW, xcpX, xcpY, xcpZ*

Table A9. *Cont.*

Sample ID	ST-Type	Pathogenicity Factor Groups								
		Adherence	Anti-Phagocytosis	Biosurfactant	Iron Uptake	Pigment	Protease	Toxin	Regulation	Secretion System
242	ST 3590	*waaA, waaC, waaF, waaG, waaP, chpA, chpB, chpC, chpD, chpE, fimT, fimU, fimV, pilB, pilD, pilE, pilF, pilG, pilH, pilI, pilK, pilM, pilN, pilO, pilP, pilQ, pilR, pilS, pilT, pilU, pilV, pilW, pilX pilY1, pilY2, pilC like, xcpA/pilD*	*alg44, alg8, algA, algB, algC, algD, algE, algF, algG, algI, algJ, algK, algL, algP/algR3 like, algQ, algR, algU, algW, algX, algZ, mucA, mucB, mucC*	*rhlA, rhlB*	*fptA, fpvA pchA, pchB, pchC, pchD, pchE, pchF, pchG, pchH, pchI, pchR, pvdA, pvdE*	*phzM, phzS*	*aprA, lasA*	*toxA, plcH*	*lasI, rhlI*	*xcpP, xcpQ, xcpS, xcpT, xcpU, xcpV, xcpW, xcpX, xcpY, xcpZ*
243	ST 3590	*waaA, waaC, waaF, waaG, waaP, chpA, chpB, chpC, chpD, chpE, fimT, fimU, fimV, pilB, pilD, pilE, pilF, pilG, pilH, pilI, pilK, pilM, pilN, pilO, pilP, pilQ, pilR, pilS, pilT, pilU, pilV, pilW, pilX pilY1, pilY2, pilC like, xcpA/pilD*	*alg44, alg8, algA, algB, algC, algD, algE, algF, algG, algI, algJ, algK, algL, algP/algR3, algQ, algR, algU, algW, algX, algZ, mucA, mucB, mucC*	*rhlA, rhlB*	*fptA, fpvA pchA, pchB, pchC, pchD, pchE, pchF, pchG, pchH, pchI, pchR, pvdA, pvdE*	*phzM, phzS*	*aprA, lasA*	*toxA, plcH*	*lasI, rhlI*	*xcpP, xcpQ, xcpS, xcpT, xcpU, xcpV, xcpW, xcpX, xcpY, xcpZ*
272	ST 2033	*waaA, waaC, waaF, waaG, waaP, chpA, chpB, chpC, chpD, chpE, fimT, fimU, fimV, pilA, pilB, pilD, pilE, pilF, pilG, pilH, pilI, pilK, pilM, pilN, pilO, pilP, pilQ, pilR, pilS, pilT, pilU, pilV, pilW, pilX pilY1, pilY2, pilC, xcpA/pilD*	*alg44, alg8, algA, algB, algC, algD, algE, algF, algG, algI, algJ, algK, algL, algQ, algR, algU, algW, algX, algZ, mucA, mucB, mucC*	*rhlA, rhlB*	*fptA, pchA, pchB, pchC, pchD, pchE, pchF, pchG, pchH, pchI, pchR, pvdA*	*phzM, phzS*	*aprA, lasA*	*toxA, plcH*	*lasI, rhlI*	*xcpP, xcpQ, xcpR, xcpS, xcpT, xcpU, xcpV, xcpW, xcpX, xcpY, xcpZ*

Table A9. *Cont.*

Sample ID	ST-Type	Pathogenicity Factor Groups								
		Adherence	Anti-Phagocytosis	Biosurfactant	Iron Uptake	Pigment	Protease	Toxin	Regulation	Secretion System
274	ST 2033	*waaA, waaC, waaF, waaG, waaP, chpA, chpB, chpC, chpD, chpE, fimT, fimU, fimV, pilA, pilB, pilD, pilE, pilF, pilG, pilH, pilI, pilK, pilM, pilN, pilO, pilP, pilQ, pilR, pilS, pilT, pilU, pilV, pilW, pilX pilY1, pilY2, pilC, xcpA/pilD*	*alg44, alg8, algA, algB, algC, algD, algE, algF, algG, algI, algJ, algK, algL, algQ, algR, algU, algW, algX, algZ, mucA, mucB, mucC*	*rhlA, rhlB*	*fptA, pchA, pchB, pchC, pchD, pchE, pchF, pchG, pchH, pchI, pchR, pvdA*	*phzM, phzS*	*aprA, lasA*	*toxA, plcH*	*lasI, rhlI*	*xcpP, xcpQ, xcpR, xcpS, xcpT, xcpU, xcpV, xcpW, xcpX, xcpY, xcpZ*
278	ST 988	*waaA, waaC, waaF, waaG, waaP, chpA, chpB, chpC, chpD, chpE, fimT, fimU, fimV, pilB, pilD, pilE, pilF, pilG, pilH, pilI, pilK, pilM, pilN, pilO, pilP, pilQ, pilR, pilS, pilT, pilU, pilV, pilW, pilX pilY1, pilY2, pilC like, xcpA/pilD*	*alg44, alg8, algA, algB, algC, algD, algE, algF, algG, algI, algJ, algK, algL, algQ, algR, algU, algW, algX, algZ, mucA, mucB, mucC*	*rhlA, rhlB*	*fptA, pchA, pchB, pchC, pchD, pchE, pchF, pchG, pchH, pchI, pchR, pvdA*	*phzM, phzS*	*aprA, lasA*	*toxA, plcH*	*lasI, rhlI*	*xcpP, xcpQ, xcpR, xcpS, xcpT, xcpU, xcpV, xcpW, xcpX, xcpY, xcpZ*

Table A9. *Cont.*

Sample ID	ST-Type	Pathogenicity Factor Groups								
		Adherence	Anti-Phagocytosis	Biosurfactant	Iron Uptake	Pigment	Protease	Toxin	Regulation	Secretion System
282	ST 554	*waaA, waaC, waaF, waaG, waaP, wzy, wzz, chpA, chpB, chpC, chpD, chpE, fimT, fimU, fimV, pilA like, pilB, pilD, pilE, pilF, pilG, pilH, pilI, pilK, pilM, pilN, pilO, pilP, pilQ, pilR, pilS, pilT, pilU, pilV, pilW, pilX pilY1, pilY2, pilC, xcpA/pilD*	*alg44, alg8, algA, algB, algC, algD, algE, algF, algG, algI, algJ, algK, algL, algQ, algR, algU, algW, algX, algZ, mucA, mucB, mucC*	*rhlA, rhlB*	*fptA, pchA, pchB, pchC, pchD, pchE, pchF, pchG, pchH, pchI, pchR, pvdA*	*phzM, phzS*	*aprA, lasA*	*toxA, plcH*	*lasI, rhlI*	*xcpP, xcpQ, xcpR, xcpS, xcpT, xcpU, xcpV, xcpW, xcpX, xcpY, xcpZ*
285	ST 554	*waaA, waaC, waaF, waaG, waaP, wzy, wzz, chpA, chpB, chpC, chpD, chpE, fimT, fimU, fimV, pilA like, pilB, pilD, pilE, pilF, pilG, pilH, pilI, pilK, pilM, pilN, pilO, pilP, pilQ, pilR, pilS, pilT, pilU, pilV, pilW, pilX pilY1, pilY2, pilC, xcpA/pilD*	*alg44, alg8, algA, algB, algC, algD, algE, algF, algG, algI, algJ, algK, algL, algQ, algR, algU, algW, algX, algZ, mucA, mucB, mucC*	*rhlA, rhlB*	*fptA, pchA, pchB, pchC, pchD, pchE, pchF, pchG, pchH, pchI, pchR, pvdA*	*phzM, phzS*	*aprA, lasA*	*toxA, plcH*	*lasI, rhlI*	*xcpP, xcpQ, xcpR, xcpS, xcpT, xcpU, xcpV, xcpW, xcpX, xcpY, xcpZ*

Antibiotics **2021**, *10*, 339

Table A9. *Cont.*

Sample ID	ST-Type	Pathogenicity Factor Groups									
		Adherence	Anti-Phagocytosis	Biosurfactant	Iron Uptake	Pigment	Protease	Toxin	Regulation	Secretion System
289	ST 1485	*waaA, waaC, waaF, waaG, waaP, chpA, chpB, chpC, chpD, chpE, fimT, fimU, fimV, pilB, pilD, pilE, pilF, pilG, pilH, pilI, pilK, pilM, pilN, pilO, pilP, pilQ, pilR, pilS, pilT, pilU, pilV, pilW, pilX pilY1, pilY2, pilC like, xcpA/pilD*	*alg44, alg8, algA, algB, algC, algD, algE, algF, algG, algI, algJ, algK, algL, algP/algR3, algQ, algR, algU, algW, algX, algZ, mucA, mucB, mucC*	*rhlA, rhlB*	*fptA, fpvA pchA, pchB, pchC, pchD, pchE, pchF, pchG, pchH, pchI, pchR, pvdA, pvdD, pvdE*	*phzM, phzS*	*aprA, lasA*	*toxA, plcH*	*lasI, rhlI*	*xcpP, xcpQ, xcpR, xcpS, xcpT, xcpU, xcpV, xcpW, xcpX, xcpY, xcpZ*
290	ST 1485	*waaA, waaC, waaF, waaG, waaP, chpA, chpB, chpC, chpD, chpE, fimT, fimU, fimV, pilB, pilD, pilE, pilF, pilG, pilH, pilI, pilK, pilM, pilN, pilO, pilP, pilQ, pilR, pilS, pilT, pilU, pilV, pilW, pilX pilY1, pilY2, pilC like, xcpA/pilD*	*alg44, alg8, algA, algB, algC, algD, algE, algF, algG, algI, algJ, algK, algL, algP/algR3, algQ, algR, algU, algW, algX, algZ, mucA, mucB, mucC*	*rhlA, rhlB*	*fptA, fpvA pchA, pchB, pchC, pchD, pchE, pchF, pchG, pchH, pchI, pchR, pvdA, pvdD, pvdE*	*phzM, phzS*	*aprA, lasA*	*toxA, plcH*	*lasI, rhlI*	*xcpP, xcpQ, xcpR, xcpS, xcpT, xcpU, xcpV, xcpW, xcpX, xcpY, xcpZ*
296	ST 235	*waaA, waaC, waaF, waaG, waaP, chpA, chpB, chpC, chpD, chpE, fimT, fimU, fimV, pilB, pilD, pilE, pilF, pilG, pilH, pilI, pilK, pilM, pilN, pilO, pilP, pilQ, pilR, pilS, pilT, pilU, pilW, pilX pilY1, pilY2, pilC, xcpA/pilD*	*alg44, alg8, algA, algB, algC, algD, algE, algF, algG, algI, algJ, algK, algL, algQ, algR, algU, algW, algX, algZ, mucA, mucB, mucC*	*rhlA, rhlB*	*fptA, pchA, pchB, pchC, pchD, pchE, pchF, pchG, pchH, pchI, pchR, pvdA*	*phzM, phzS*	*aprA, lasA*	*toxA, plcH*	*lasI, rhlI*	*xcpP, xcpQ, xcpR, xcpS, xcpT, xcpU, xcpV, xcpW like, xcpX, xcpY, xcpZ*

Table A9. *Cont.*

Sample ID	ST-Type	Pathogenicity Factor Groups								
		Adherence	Anti-Phagocytosis	Biosurfactant	Iron Uptake	Pigment	Protease	Toxin	Regulation	Secretion System
298	ST 3227	*waaA, waaC, waaF, waaG, waaP, chpA, chpB, chpC, chpD, chpE, fimV, pilB, pilD, pilF, pilG, pilH, pilI, pilK, pilM, pilN, pilO, pilP, pilQ, pilR, pilS, pilT, pilU, pilV, pilC, xcpA/pilD*	*alg44, alg8, algA, algB, algC, algD, algE, algF, algG, algI, algJ, algK, algL, algQ, algR, algU, algW, algX, algZ, mucA, mucB, mucC*	*rhlA, rhlB*	*fptA, pchA, pchB, pchC, pchD, pchE, pchF, pchG, pchH, pchI, pchR, pvdA*	*phzM, phzS*	*aprA, lasA*	*toxA, plcH*	*lasI, rhlI*	*xcpP, xcpQ, xcpR, xcpS, xcpT, xcpU, xcpV, xcpW, xcpX, xcpY, xcpZ*
301	ST 3593	*waaA, waaC, waaF, waaG, waaP, wzy, wzz, chpA, chpB, chpC, chpD, chpE, fimT, fimU, fimV, pilB, pilD, pilE, pilF, pilG, pilH, pilI, pilK, pilM, pilN, pilO, pilP, pilQ, pilR, pilS, pilT, pilU, pilV, pilW, pilX pilY1, pilY2, pilC, xcpA/pilD*	*alg44, alg8, algA, algB, algC, algD, algE, algF, algG, algI, algJ, algK, algL, algP/algR3, algQ, algR, algU, algW, algX, algZ, mucA, mucB, mucC*	*rhlA, rhlB*	*fptA, fpvA pchA, pchB, pchC, pchD, pchE, pchF, pchG, pchH, pchI, pchR, pvdA, pvdE*	*phzM, phzS*	*aprA, lasA*	*plcH*	*lasI, rhlI*	*xcpP, xcpQ, xcpR, xcpS, xcpT, xcpU, xcpV, xcpW, xcpX, xcpY, xcpZ*
302	ST 1755	*waaA, waaC, waaF, waaG, waaP, wzy, wzz, chpA, chpB, chpC, chpD, chpE, fimT, fimU, fimV, pilB, pilD, pilE, pilF, pilG, pilH, pilI, pilK, pilM, pilN, pilO, pilP, pilQ, pilR, pilS, pilT, pilU, pilV, pilW, pilX pilY1, pilY2, pilC like, xcpA/pilD*	*alg44, alg8, algA, algB, algC, algD, algE, algF, algG, algI, algJ, algK, algL, algQ, algR, algU, algW, algX, algZ, mucA, mucB, mucC*	*rhlA, rhlB*	*fptA, fpvA pchA, pchB, pchC, pchD, pchE, pchF, pchG, pchH, pchI, pchR, pvdA, pvdE*	*phzM, phzS*	*aprA, lasA*	*toxA, plcH*	*lasI, rhlI*	*xcpP, xcpQ, xcpR, xcpS, xcpT, xcpU, xcpV, xcpW, xcpX, xcpY, xcpZ*

Table A9. *Cont.*

Sample ID	ST-Type	Pathogenicity Factor Groups								
		Adherence	Anti-Phagocytosis	Biosurfactant	Iron Uptake	Pigment	Protease	Toxin	Regulation	Secretion System
309	ST 3592	*waaA, waaC, waaF, waaG, waaP, wzy, wzz, chpA, chpB, chpC, chpD, chpE, fimT, fimU, fimV, pilB, pilD, pilE, pilF, pilG, pilH, pilI, pilK, pilM, pilN, pilO, pilP, pilQ, pilR, pilS, pilT, pilU, pilV, pilW, pilX pilY1, pilY2, pilC, xcpA/pilD*	*alg44, alg8, algA, algB, algC, algD, algE, algF, algG, algI, algJ, algK, algL, algQ, algR, algU, algW, algX, algZ, mucA, mucB, mucC*	*rhlA, rhlB*	*fptA, fpvA pchA, pchB, pchC, pchD, pchE, pchF, pchG, pchH, pchI, pchR, pvdA, pvdE*	*phzM, phzS*	*aprA, lasA*	*plcH*	*lasI, rhlI*	*xcpP, xcpQ, xcpS, xcpT, xcpU, xcpV, xcpW, xcpX, xcpY, xcpZ*
310	ST 532	*waaA, waaC, waaF, waaG, waaP, chpA, chpB, chpC, chpD, chpE, fimV, pilB, pilD, pilF, pilG, pilH, pilI, pilK, pilM, pilN, pilO, pilP, pilQ, pilR, pilS, pilT, pilU, pilC like, xcpA/pilD*	*alg44, alg8, algA, algB, algC, algD, algE, algF, algG, algI, algJ, algK, algL, algQ, algR, algU, algW, algX, algZ, mucA, mucB, mucC*	*rhlA, rhlB*	*fptA, fpvA pchA, pchB, pchC, pchD, pchE, pchF, pchG, pchH, pchI, pchR, pvdA*	*phzM, phzS*	*aprA lasA*	*toxA, plcH*	*lasI, rhlI*	*xcpP, xcpQ, xcpR, xcpS, xcpT, xcpU, xcpV, xcpW, xcpX, xcpY, xcpZ*
312	ST 381	*waaA, waaC, waaF, waaG, waaP, wzy, wzz, chpA, chpB, chpC, chpD, chpE, fimV, pilB, pilD, pilF, pilG, pilH, pilI, pilK, pilM, pilN, pilO, pilP, pilQ, pilR, pilS, pilT, pilU, pilC like, xcpA/pilD*	*alg44, alg8, algA, algB, algC, algD, algE, algF, algG, algI, algJ, algK, algL, algP/algR3, algQ, algR, algU, algW, algX, algZ, mucA, mucB, mucC*	*rhlA, rhlB*	*fptA, pchA, pchB, pchC, pchD, pchE, pchF, pchG, pchH, pchI, pchR, pvdA, pvdD, pvdE*	*phzM, phzS*	*aprA, lasA*	*toxA, plcH*	*lasI, rhlI*	*xcpP, xcpQ, xcpR, xcpS, xcpT, xcpU, xcpV, xcpW, xcpX, xcpY, xcpZ*

Table A10. Analysis of virulence determinants, ordered by strain and MLST type, of the assessed *K. pneumoniae* isolates. ST = Sequence type.

Sample ID	ST-Type	Pathogenicity Factor Groups									
		Adherence	Biofilm Formation	Efflux Pump	Immune Evasion	Iron Uptake	Nutritional Factor	Regulation	Secretion System	Serum Resistance	Toxin
044	ST 327	*fimA, fimB, fimC, fimD, fimE, fimF, fimG, fimH, fimI, fimK*	*mrkA, mrkB, mrkC, mrkD, mrkF, mrkH, mrkI, mrkJ*	*acrA, acrB*	*cpsACP, galF, gnd, ugd, wza like, wzi*	*entA, entB, entC, entD, entE, entF, fepA, fepB, fepC, fepD, fepG, fes, ybdA, iroE like*		*rcsA, rcsB*	*impA/tssA like, sciN/tssJ, tssF, tssG, vasE/tssK, vgrG/tssI, vipA/tssB, vipB/tssC*	*glf, wbbM, wbbN, wbbO, wzm, wzt*	
060	ST 5379	*fimA, fimB, fimC, fimD, fimE, fimF, fimG, fimH, fimI, fimK*	*mrkA, mrkB, mrkC, mrkD, mrkF, mrkH, mrkI, mrkJ*	*acrA, acrB*	*cpsACP, galF, gnd, ugd, wza like, wzi*	*entA, entB, entC, entE, entF, fepA, fepB, fepC, fepD, fepG, fes, ybdA, iroE*		*rcsA, rcsB*	*impA/tssA, sciN/tssJ, tssF, tssG, vasE/tssK, vgrG/tssI, vipA/tssB, vipB/tssC*		
073	ST 39	*fimA, fimB, fimC, fimD, fimE, fimF, fimG, fimH, fimI, fimK*	*mrkA, mrkB, mrkC, mrkD, mrkF, mrkH, mrkI, mrkJ*	*acrA, acrB*	*cpsACP, galF, gnd, ugd, wza like, wzi*	*entA, entB, entC, entE, entF, fepA, fepB, fepC, fepD, fepG, fes, ybdA, iroE, irp1, irp2, ybtA, ybtE, ybtP, ybtQ, ybtS, ybtT, ybtU, ybtX*		*rcsA, rcsB*	*impA/tssA, sciN/tssJ, tle1, tli1, tssF, tssG, vasE/tssK, vgrG/tssI, vipA/tssB, vipB/tssC*	*glf, wbbM, wbbN, wbbO, wzm, wzt*	
100	ST 152	*fimA, fimB, fimC, fimD, fimE, fimF, fimG, fimH, fimI, fimK*	*mrkA, mrkB, mrkC, mrkD, mrkF, mrkH, mrkI, mrkJ*	*acrA, acrB*	*cpsACP, galF, gnd, ugd, wza like, wzi*	*entA, entB, entC, entD, entE, entF, fepA, fepB, fepC, fepD, fepG, fes, ybdA, iroE, irp1, irp2, ybtA, ybtE, ybtP, ybtQ, ybtS, ybtT, ybtU, ybtX*		*rcsA, rcsB*	*impA/tssA like, sciN/tssJ, tssF, tssG*		

Table A10. *Cont.*

Sample ID	ST-Type	Pathogenicity Factor Groups									
		Adherence	Biofilm Formation	Efflux Pump	Immune Evasion	Iron Uptake	Nutritional Factor	Regulation	Secretion System	Serum Resistance	Toxin
102	ST 514	*fimA, fimB, fimC, fimD, fimE, fimF, fimG, fimH, fimI, fimK*	*mrkA, mrkB, mrkC, mrkD, mrkF, mrkI, mrkJ*	*acrA, acrB*	*cpsACP, galF, gnd, manB, manC, ugd, wza, wzi*	*entA, entB, entC, entE, entF, fepA, fepB, fepC, fepD, fepG, fes, ybdA, iroE*		*rcsA, rcsB*	*impA/tssA like, sciN/tssJ, tssF, tssG, vasE/tssK, vgrG/tssI, vipA/tssB, vipB/tssC*	*glf, wbbM, wbbN, wbbO, wzm, wzt*	
124	ST 399	*fimA, fimB, fimC, fimD, fimE, fimF, fimG, fimH, fimI, fimK*	*mrkF, mrkH, mrkJ*	*acrA, acrB*	*cpsACP, galF, ugd, wza like, wzi*	*entA, entB, entC, entE, entF, fepA, fepB, fepC, fepD, fepG, fes, ybdA, iroE*		*rcsA, rcsB*		*glf, wbbM, wbbN, wbbO, wzm, wzt*	
146	ST 4	*fimA, fimC, fimD, fimE, fimF, fimG, fimH, fimI, fimK*	*mrkA, mrkB, mrkC, mrkD, mrkF, mrkH, mrkI, mrkJ*	*acrA, acrB*	*cpsACP, galF, gnd, manB, manC, ugd, wza, wzi*	*entA, entB, entC, entE, entF, fepA, fepB, fepC, fepD, fepG, fes, ybdA, iroE, irp1, irp2, ybtA, ybtE, ybtP, ybtQ, ybtS, ybtT, ybtU, ybtX*		*rcsA, rcsB*	*impA/tssA like, sciN/tssJ, tssF, tssG, vasE/tssK, vgrG/tssI, vipA/tssB, vipB/tssC*	*glf, wbbM, wbbN, wbbO, wzm, wzt*	
177	ST 17	*fimA, fimB, fimC, fimD, fimE, fimF, fimG, fimH, fimI, fimK*	*mrkA, mrkD, mrkF, mrkH, mrkI, mrkJ*	*acrA, acrB*	*cpsACP, galF, gnd, manB, manC, ugd, wza like, wzi*	*entA, entB, entC, entD, entE, entF, fepA, fepB, fepC, fepD, fepG, fes, ybdA, iroE, ybtA, ybtE, ybtP, ybtQ, ybtS, ybtT, ybtU, ybtX*		*rcsA, rcsB*	*impA/tssA, sciN/tssJ, tssF, tssG, vasE/tssK, vgrG/tssI, vipA/tssB, vipB/tssC*	*glf, wbbM, wbbN, wbbO, wzm, wzt*	

Table A10. *Cont.*

Sample ID	ST-Type	Pathogenicity Factor Groups									
		Adherence	Biofilm Formation	Efflux Pump	Immune Evasion	Iron Uptake	Nutritional Factor	Regulation	Secretion System	Serum Resistance	Toxin
181	ST 5380	*fimA, fimB, fimC, fimD, fimE, fimF, fimG, fimH, fimI, fimK*	*mrkA, mrkB, mrkC, mrkD, mrkF, mrkH, mrkI, mrkJ*	*acrA, acrB*	*cpsACP, galF, gmd like, gnd, manB, manC, ugd, wza like, wzi*	*entA, entB, entC, entE, entF, fepA, fepB, fepC, fepD, fepG, fes, ybdA, iroE*		*rcsA, rcsB,*	*impA/tssA, sciN/tssJ, tle1, tli1, tssF, tssG, vasE/tssK, vgrG/tssI, vipA/tssB, vipB/tssC*	*glf, wbbM, wbbN, wbbO, wzm, wzt*	
184	ST 5381	*fimA, fimB, fimC, fimD, fimE, fimF, fimG, fimH, fimI, fimK*	*mrkA, mrkB, mrkC, mrkD, mrkF, mrkH, mrkI, mrkJ*	*acrA, acrB*	*cpsACP, galF, gnd, ugd, wza like, wzi*	*entA, entB, entC, entD like, entE, entF, fepA, fepB, fepC, fepD, fepG, fes, ybdA, iroE like*		*rcsA, rcsB*	*impA/tssA, sciN/tssJ, tssF, tssG, vasE/tssK, vgrG/tssI, vipA/tssB, vipB/tssC*		
199	ST 17	*fimA, fimB, fimC, fimD, fimE, fimF, fimG, fimH, fimI, fimK*	*mrkA, mrkB, mrkC, mrkD, mrkF, mrkH, mrkI, mrkJ*	*acrA, acrB*	*cpsACP, galF, gnd, manB like, manC, ugd, wza, wzi*	*entA, entB, entC, entE, entF, fepA, fepB, fepC, fepD, fepG, fes, ybdA, iroE, irp1, irp2, ybtA, ybtE, ybtP, ybtQ, ybtS, ybtT, ybtU, ybtX*		*rcsA, rcsB*	*impA/tssA, sciN/tssJ, tssF, tssG, vasE/tssK, vgrG/tssI, vipA/tssB, vipB/tssC*		
214	ST 6	*fimA, fimB, fimC, fimD, fimE, fimF, fimG, fimH, fimI, fimK*	*mrkB, mrkC, mrkD, mrkF, mrkH, mrkI,*	*acrA, acrB*	*cpsACP, galF, gnd, manB, manC, ugd, wza like, wzi*	*entA, entB, entC, entE, entF, fepA, fepB, fepC, fepD, fepG, fes, ybdA, iroE, irp1, irp2, ybtA, ybtE, ybtP, ybtQ, ybtS, ybtT, ybtU, ybtX*		*rcsA, rcsB*	*impA/tssA like, sciN/tssJ, tssF, tssG, vasE/tssK, vgrG/tssI, vipA/tssB, vipB/tssC*	*glf, wbbM, wbbN, wbbO, wzm, wzt*	

Table A10. *Cont.*

Sample ID	ST-Type	Pathogenicity Factor Groups									
		Adherence	Biofilm Formation	Efflux Pump	Immune Evasion	Iron Uptake	Nutritional Factor	Regulation	Secretion System	Serum Resistance	Toxin
217	ST 3154	*fimA, fimB, fimC, fimD, fimE, fimF, fimG, fimH, fimI, fimK*	*mrkA, mrkB, mrkC, mrkD, mrkF, mrkH, mrkI, mrkJ*	*acrA, acrB*	*cpsACP, galF, gnd, manB, manC, ugd, wza like, wzi*	*entA, entB, entC, entE, entF, fepA, fepB, fepC, fepD, fepG, fes, ybdA, iroE*		*rcsA, rcsB*	*impA/tssA like, sciN/tssJ, tssF, tssG, vasE/tssK, vgrG/tssI, vipA/tssB, vipB/tssC*	*glf, wbbM, wbbN, wbbO, wzm, wzt*	
220	ST 5382	*fimA, fimB, fimC, fimD, fimE, fimF, fimG, fimH, fimI, fimK*	*mrkA, mrkD, mrkF, mrkH, mrkI, mrkJ*	*acrA, acrB*	*cpsACP, galF, gnd, manB like, manC, ugd, wza like, wzi*	*entA, entB, entC, entE, entF, fepA, fepB, fepC, fepD, fepG, fes, ybdA, iroE,*	*allA, allB, allC, allD, allR, allS*	*rcsA, rcsB*			
234	ST 109	*fimA, fimB, fimC, fimD, fimE, fimF, fimG, fimH, fimI, fimK*	*mrkA, mrkB, mrkC, mrkD, mrkF, mrkH, mrkI, mrkJ*	*acrA, acrB*	*cpsACP, galF, gnd, manB, manC, ugd, wza like, wzi*	*entA, entB, entC, entE, entF, fepA, fepB, fepC, fepD, fepG, fes, ybdA, iroE*		*rcsA, rcsB,*	*impA/tssA, sciN/tssJ, tssF, tssG, vasE/tssK, vgrG/tssI, vipA/tssB, vipB/tssC*	*glf, wbbM, wbbN, wbbO, wzm, wzt*	
240	ST 5383	*fimC, fimD, fimF, fimG, fimH, fimI, fimK*	*mrkF, mrkH, mrkI, mrkJ*	*acrA, acrB*	*cpsACP, galF, gnd, ugd, wza like, wzi*	*entA, entB, entC, entE, entF, fepA, fepB, fepC, fepD, fepG, fes, ybdA, iroE*		*rcsA, rcsB*		*glf, wbbM, wbbN, wbbO, wzm, wzt*	
248	ST 5384		*mrkC, mrkD, mrkF, mrkH, mrkI, mrkJ*	*acrA, acrB*	*cpsACP, galF, gnd, manB, manC, ugd, wza like, wzi*	*entA, entB, entC, entE, entF, fepA, fepB, fepC, fepD, fepG, fes, ybdA, iroE*		*rcsA, rcsB*	*vasE/tssK, vgrG/tssI, vipA/tssB, vipB/tssC*	*glf, wbbM, wbbN, wbbO, wzm, wzt*	
252	ST 607	*fimA, fimB, fimC, fimD, fimE, fimF, fimG, fimH, fimI, fimK*	*mrkA, mrkB, mrkC, mrkD, mrkF, mrkH, mrkI, mrkJ*	*acrA, acrB*	*cpsACP, galF, gnd, wza like, wzi*	*entA, entB, entC, entE, entF, fepA, fepB, fepC, fepD, fepG, fes, ybdA, iroE*		*rcsA, rcsB*	*sciN/tssJ, tssF, tssG, vasE/tssK, vgrG/tssI, vipA/tssB, vipB/tssC*	*glf, wbbM, wbbN, wbbO, wzm, wzt*	

Table A10. *Cont.*

Sample ID	ST-Type	Pathogenicity Factor Groups									
		Adherence	Biofilm Formation	Efflux Pump	Immune Evasion	Iron Uptake	Nutritional Factor	Regulation	Secretion System	Serum Resistance	Toxin
267	ST 36	*fimA, fimB, fimC, fimD, fimE, fimF, fimG, fimH, fimI, fimK*	*mrkA, mrkB, mrkC, mrkD, mrkF, mrkH, mrkI, mrkJ*	*acrA, acrB*	*cpsACP, galF, gnd, manB, manC, ugd, wza, wzi*	*entA, entB, entC, entE, entF, fepA, fepB, fepC, fepD, fepG, fes, ybdA, iroE, irp1, irp2, ybtA, ybtE, ybtP, ybtQ, ybtS, ybtT, ybtU, ybtX*		*rcsA, rcsB*	*impA/tssA, sciN/tssJ, tle1, tli1, tssF, tssG, vasE/tssK, vgrG/tssI, vipA/tssB, vipB/tssC*	*glf, wbbM, wbbN, wbbO, wzm, wzt*	
277	ST 530	*fimA, fimB, fimC, fimD, fimE, fimF, fimG, fimH, fimI, fimK*	*mrkA, mrkB, mrkC, mrkD, mrkF, mrkH, mrkI, mrkJ*	*acrA, acrB*	*cpsACP, galF, gnd, manB, manC, ugd, wza, wzi*	*entA, entB, entC, entE, entF, fepA, fepB, fepC, fepD, fepG, fes, ybdA, iroE*		*rcsA, rcsB*	*impA/tssA like, sciN/tssJ, tssF, tssG, vasE/tssK, vgrG/tssI, vipA/tssB, vipB/tssC*	*glf, wbbM, wbbN, wbbO, wzm, wzt*	
279	ST 5385	*fimA, fimB, fimC, fimD, fimE, fimF, fimG, fimH, fimI, fimK*	*mrkA, mrkB, mrkC, mrkD, mrkF, mrkH, mrkI, mrkJ*	*acrA, acrB*	*cpsACP, galF, gnd, ugd, wza, wzi*	*entA, entB, entC, entE, entF, fepA, fepB, fepC, fepD, fepG, fes, ybdA, iroE*		*rcsA, rcsB,*	*impA/tssA, sciN/tssJ, tle1, tli1, tssF, tssG, vasE/tssK, vgrG/tssI, vipA/tssB, vipB/tssC*	*wzm, wzt*	

Table A11. Details on the strain-specific short-read archive (SRA) accession numbers.

Sample ID	Percentage of Good Targets (SeqSphere+)	Average Coverage (Assembled) (SeqSphere+)	Approximated Genome Size (Megabases) (SeqSphere+)	Species (Kraken2)	Sequence Type	Complex Type (SeqSphere+)	SRA Accession
Iso00017	99.4	105	6.7	*Pseudomonas aeruginosa*	381	1791	SRR13617317
Iso00022	99.4	102	6.9	*Pseudomonas aeruginosa*	2483	1792	SRR13617316
Iso00032	99.2	106	6.6	*Pseudomonas aeruginosa*	3587	1793	SRR13617305
Iso00041	99.4	97	5.0	*Escherichia coli*	2 (Pasteur)	11349	SRR13617294
Iso00044	99.7	116	5.1	*Klebsiella pneumoniae*	327	5462	SRR13617283
Iso00049	98.7	94	5.2	*Escherichia coli*	3 (Pasteur)	11350	SRR13617272
Iso00060	99.6	112	5.3	*Klebsiella pneumoniae*	5379	5463	SRR13617261
Iso00068	99.6	109	4.9	*Escherichia coli*	632 (Pasteur)	11351	SRR13617250
Iso00069	99.6	104	6.8	*Pseudomonas aeruginosa*	360	1794	SRR13617239
Iso00073	99.4	104	5.8	*Klebsiella pneumoniae*	39	5464	SRR13617236
Iso00081	98.7	108	6.6	*Pseudomonas aeruginosa*	244	1795	SRR13617315
Iso00082	99.4	112	6.3	*Pseudomonas aeruginosa*	514	1796	SRR13617314
Iso00088	97.8	105	6.8	*Pseudomonas aeruginosa*	1682	1797	SRR13617313
Iso00099	99.4	106	6.6	*Pseudomonas aeruginosa*	244	1798	SRR13617312
Iso00100	99.2	108	5.5	*Klebsiella pneumoniae*	152	5465	SRR13617311
Iso00102	99.2	111	5.4	*Klebsiella pneumoniae*	514	5466	SRR13617310
Iso00106	99.5	110	6.4	*Pseudomonas aeruginosa*	1521	1799	SRR13617309
Iso00114	99.4	105	6.7	*Pseudomonas aeruginosa*	244	1800	SRR13617308
Iso00117	99.2	95	5.3	*Escherichia coli*	4 (Pasteur)	11352	SRR13617307
Iso00124	99.4	112	5.3	*Klebsiella pneumoniae*	399	5467	SRR13617306
Iso00137	99.4	110	6.4	*Pseudomonas aeruginosa*	3014	1801	SRR13617304
Iso00144	99.6	109	6.5	*Pseudomonas aeruginosa*	245	1802	SRR13617303
Iso00146	99.4	110	5.5	*Klebsiella pneumoniae*	4	5468	SRR13617302
Iso00147	99.5	108	6.6	*Pseudomonas aeruginosa*	245	1802	SRR13617301

Table A11. *Cont.*

Sample ID	Percentage of Good Targets (SeqSphere+)	Average Coverage (Assembled) (SeqSphere+)	Approximated Genome Size (Megabases) (SeqSphere+)	Species (Kraken2)	Sequence Type	Complex Type (SeqSphere+)	SRA Accession
Iso00149	99.6	104	6.9	*Pseudomonas aeruginosa*	381	1803	SRR13617300
Iso00152	99.4	98	5.2	*Escherichia coli*	22 (Pasteur)	11353	SRR13617299
Iso00153	98.5	111	6.4	*Pseudomonas aeruginosa*	704	?	SRR13617298
Iso00154	99.4	102	7.0	*Pseudomonas aeruginosa*	244	1805	SRR13617297
Iso00157	99.6	114	6.3	*Pseudomonas aeruginosa*	2616	1806	SRR13617296
Iso00160	99.2	115	6.2	*Pseudomonas aeruginosa*	170	1807	SRR13617295
Iso00162	99.1	111	6.5	*Pseudomonas aeruginosa*	274	1808	SRR13617293
Iso00176	99.0	98	5.1	*Escherichia coli*	132 (Pasteur)	11354	SRR13617292
Iso00177	99.6	108	5.5	*Klebsiella pneumoniae*	17	5469	SRR13617291
Iso00180	99.8	110	6.5	*Pseudomonas aeruginosa*	856	1809	SRR13617290
Iso00181	99.9	107	5.6	*Klebsiella pneumoniae*	5380	5470	SRR13617289
Iso00183	99.5	107	6.7	*Pseudomonas aeruginosa*	244	1795	SRR13617288
Iso00184	98.3	104	5.6	*Klebsiella variicola* subsp. *variicola*	5381	5471	SRR13617287
Iso00186	98.7	113	6.3	*Pseudomonas aeruginosa*	3588	1810	SRR13617286
Iso00190	99.7	114	6.3	*Pseudomonas aeruginosa*	871	1811	SRR13617285
Iso00195	99.5	111	6.5	*Pseudomonas aeruginosa*	988	1812	SRR13617284
Iso00196	99.5	101	7.1	*Pseudomonas aeruginosa*	2475	1813	SRR13617282
Iso00198	99.6	112	6.4	*Pseudomonas aeruginosa*	2476	1814	SRR13617281
Iso00199	99.4	108	5.6	*Klebsiella pneumoniae*	17	5472	SRR13617280
Iso00204	99.5	104	6.9	*Pseudomonas aeruginosa*	639	1815	SRR13617279
Iso00208	99.7	109	6.5	*Pseudomonas aeruginosa*	132	1816	SRR13617278
Iso00214	99.7	108	5.5	*Klebsiella pneumoniae*	6	5473	SRR13617277
Iso00217	99.8	104	5.7	*Klebsiella pneumoniae*	3154	5474	SRR13617276
Iso00218	99.7	109	6.5	*Pseudomonas aeruginosa*	856	1809	SRR13617275

Table A11. Cont.

Sample ID	Percentage of Good Targets (SeqSphere+)	Average Coverage (Assembled) (SeqSphere+)	Approximated Genome Size (Megabases) (SeqSphere+)	Species (Kraken2)	Sequence Type	Complex Type (SeqSphere+)	SRA Accession
Iso00220	97.8	110	5.4	Klebsiella quasipneumoniae subsp. similipneumoniae	5382	5475	SRR13617274
Iso00221	99.0	94	5.1	Escherichia coli	132 (Pasteur)	11354	SRR13617273
Iso00222	99.0	96	5.1	Escherichia coli	132 (Pasteur)	11354	SRR13617271
Iso00225	99.1	99	5.2	Escherichia coli	506 (Pasteur)	11355	SRR13617270
Iso00229	99.6	109	6.5	Pseudomonas aeruginosa	270	1817	SRR13617269
Iso00233	97.8	114	6.1	Pseudomonas aeruginosa	3227	1818	SRR13617268
Iso00234	99.7	111	5.5	Klebsiella pneumoniae	109	5476	SRR13617267
Iso00236	99.7	112	6.4	Pseudomonas aeruginosa	266	1819	SRR13617266
Iso00238	98.7	108	6.6	Pseudomonas aeruginosa	3589	1820	SRR13617265
Iso00240	98.9	112	5.4	Klebsiella pneumoniae	5383	5477	SRR13617264
Iso00242	98.9	111	6.4	Pseudomonas aeruginosa	3590	1821	SRR13617263
Iso00243	98.9	111	6.4	Pseudomonas aeruginosa	3590	1821	SRR13617262
Iso00245	99.3	107	4.8	Escherichia coli	2 (Pasteur)	11356	SRR13617260
Iso00248	97.2	108	5.5	Klebsiella quasivariicola	5384	5478	SRR13617259
Iso00252	99.6	112	5.3	Klebsiella pneumoniae	607	5479	SRR13617258
Iso00267	99.6	103	5.7	Klebsiella pneumoniae	36	5480	SRR13617257
Iso00270	99.2	100	4.9	Escherichia coli	2 (Pasteur)	11358	SRR13617256
Iso00272	99.5	109	6.5	Pseudomonas aeruginosa	2033	1822	SRR13617255
Iso00274	99.4	109	6.5	Pseudomonas aeruginosa	2033	1822	SRR13617254
Iso00277	99.4	109	5.5	Klebsiella pneumoniae	530	5481	SRR13617253
Iso00278	99.6	110	6.5	Pseudomonas aeruginosa	988	1823	SRR13617252
Iso00279	99.7	111	5.5	Klebsiella pneumoniae	5385	5482	SRR13617251
Iso00282	99.3	108	6.6	Pseudomonas aeruginosa	554	1824	SRR13617249
Iso00285	99.3	109	6.5	Pseudomonas aeruginosa	554	1824	SRR13617248

Table A11. *Cont.*

Sample ID	Percentage of Good Targets (SeqSphere+)	Average Coverage (Assembled) (SeqSphere+)	Approximated Genome Size (Megabases) (SeqSphere+)	Species (Kraken2)	Sequence Type	Complex Type (SeqSphere+)	SRA Accession
Iso00289	99.6	112	6.3	*Pseudomonas aeruginosa*	1485	1825	SRR13617247
Iso00290	99.7	113	6.3	*Pseudomonas aeruginosa*	1485	1825	SRR13617246
Iso00296	99.7	106	6.7	*Pseudomonas aeruginosa*	235	1826	SRR13617245
Iso00298	97.8	116	6.1	*Pseudomonas aeruginosa*	3227	1818	SRR13617244
Iso00299	99.3	108	4.6	*Escherichia coli*	1018 (Pasteur)	11357	SRR13617243
Iso00301	98.6	112	6.3	*Pseudomonas aeruginosa*	3593	1827	SRR13617242
Iso00302	99.6	113	6.3	*Pseudomonas aeruginosa*	1755	1828	SRR13617241
Iso00309	98.6	109	6.5	*Pseudomonas aeruginosa*	3592	1829	SRR13617240
Iso00310	99.3	105	6.8	*Pseudomonas aeruginosa*	532	1830	SRR13617238
Iso00312	99.4	106	6.7	*Pseudomonas aeruginosa*	381	1791	SRR13617237

References

1. Lai, P.S.; Bebell, L.M.; Meney, C.; Valeri, L.; White, M.C. Epidemiology of antibiotic-resistant wound infections from six countries in Africa. *BMJ Glob. Health* **2018**, *2* (Suppl. 4), e000475. [CrossRef]
2. Mama, M.; Abdissa, A.; Sewunet, T. Antimicrobial susceptibility pattern of bacterial isolates from wound infection and their sensitivity to alternative topical agents at Jimma University Specialized Hospital, South-West Ethiopia. *Ann. Clin. Microbiol. Antimicrob.* **2014**, *13*, 14. [CrossRef] [PubMed]
3. Moremi, N.; Mushi, M.F.; Fidelis, M.; Chalya, P.; Mirambo, M.; Mshana, S.E. Predominance of multi-resistant gram-negative bacteria colonizing chronic lower limb ulcers (CLLUs) at Bugando Medical Center. *BMC Res. Notes* **2014**, *7*, 211. [CrossRef]
4. Kassam, N.A.; Damian, D.J.; Kajeguka, D.; Nyombi, B.; Kibiki, G.S. Spectrum and antibiogram of bacteria isolated from patients presenting with infected wounds in a Tertiary Hospital, northern Tanzania. *BMC Res. Notes* **2017**, *10*, 757. [CrossRef]
5. Janssen, H.; Janssen, I.; Cooper, P.; Kainyah, C.; Pellio, T.; Quintel, M.; Monnheimer, M.; Groß, U.; Schulze, M.H. Antimicrobial-Resistant Bacteria in Infected Wounds, Ghana, 2014. *Emerg. Infect. Dis.* **2018**, *24*, 916–919. [CrossRef] [PubMed]
6. Kazimoto, T.; Abdulla, S.; Bategereza, L.; Juma, O.; Mhimbira, F.; Weisser, M.; Utzinger, J.; von Müller, L.; Becker, S.L. Causative agents and antimicrobial resistance patterns of human skin and soft tissue infections in Bagamoyo, Tanzania. *Acta Trop.* **2018**, *186*, 102–106. [CrossRef]
7. Høiby, N.; Ciofu, O.; Johansen, H.K.; Song, Z.J.; Moser, C.; Jensen, P.Ø.; Molin, S.; Givskov, M.; Tolker-Nielsen, T.; Bjarnsholt, T. The clinical impact of bacterial biofilms. *Int. J. Oral. Sci.* **2011**, *3*, 55–65. [CrossRef] [PubMed]
8. Percival, S.L.; Hill, K.E.; Williams, D.W.; Hooper, S.J.; Thomas, D.W.; Costerton, J.W. A review of the scientific evidence for biofilms in wounds. *Wound Repair Regen.* **2012**, *20*, 647–657. [CrossRef]
9. Rahim, K.; Saleha, S.; Zhu, X.; Huo, L.; Basit, A.; Franco, O.L. Bacterial Contributionin Chronicity of Wounds. *Microb. Ecol.* **2017**, *73*, 710–721. [CrossRef]
10. Xu, Z.; Hsia, H.C. The Impact of Microbial Communities on Wound Healing: A Review. *Ann. Plast. Surg.* **2018**, *81*, 113–123. [CrossRef] [PubMed]
11. Mulcahy, L.R.; Isabella, V.M.; Lewis, K. *Pseudomonas aeruginosa* biofilms in disease. *Microb. Ecol.* **2014**, *68*, 1–12. [CrossRef] [PubMed]
12. Serra, R.; Grande, R.; Butrico, L.; Rossi, A.; Settimio, U.F.; Caroleo, B.; Amato, B.; Gallelli, L.; de Franciscis, S. Chronic wound infections: The role of *Pseudomonas aeruginosa* and *Staphylococcus aureus*. *Expert. Rev. Anti. Infect. Ther.* **2015**, *13*, 605–613. [CrossRef] [PubMed]
13. Kirketerp-Møller, K.; Jensen, P.Ø.; Fazli, M.; Madsen, K.G.; Pedersen, J.; Moser, C.; Tolker-Nielsen, T.; Høiby, N.; Givskov, M.; Bjarnsholt, T. Distribution, organization, and ecology of bacteria in chronic wounds. *J. Clin. Microbiol.* **2008**, *46*, 2717–2722. [CrossRef] [PubMed]
14. Kohler, J.E.; Hutchens, M.P.; Sadow, P.M.; Modi, B.P.; Tavakkolizadeh, A.; Gates, J.D. *Klebsiella pneumoniae* necrotizing fasciitis and septic arthritis: An appearance in the Western hemisphere. *Surg. Infect.* **2007**, *8*, 227–232. [CrossRef]
15. Dana, A.N.; Bauman, W.A. Bacteriology of pressure ulcers in individuals with spinal cord injury: What we know and what we should know. *J. Spinal Cord. Med.* **2015**, *38*, 147–160. [CrossRef] [PubMed]
16. Washington, M.A.; Barnhill, J.C.; Duff, M.A.; Griffin, J. Recovery of Bacteria and Fungi from a Leg Wound. *J. Spec. Oper. Med.* **2015**, *15*, 113–116. [PubMed]
17. Vyas, K.S.; Wong, L.K. Detection of Biofilm in Wounds as an Early Indicator for Risk for Tissue Infection and Wound Chronicity. *Ann. Plast. Surg.* **2016**, *76*, 127–131. [CrossRef]
18. Wetterslev, M.; Rose-Larsen, K.; Hansen-Schwartz, J.; Steen-Andersen, J.; Møller, K.; Møller-Sørensen, H. Mechanism of injury and microbiological flora of the geographical location are essential for the prognosis in soldiers with serious warfare injuries. *Dan. Med. J.* **2013**, *60*, A4704. [PubMed]
19. Yun, H.C.; Murray, C.K.; Roop, S.A.; Hospenthal, D.R.; Gourdine, E.; Dooley, D.P. Bacteria recovered from patients admitted to a deployed U.S. military hospital in Baghdad, Iraq. *Mil. Med.* **2006**, *171*, 821–825. [CrossRef] [PubMed]
20. Micheel, V.; Hogan, B.; Rakotoarivelo, R.A.; Rakotozandrindrainy, R.; Razafimanatsoa, F.; Razafindrabe, T.; Rakotondrainiarivelo, J.P.; Crusius, S.; Poppert, S.; Schwarz, N.G.; et al. Identification of nasal colonization with β-lactamase-producing *Enterobacteriaceae* in patients, health care workers and students in Madagascar. *Eur. J. Microbiol. Immunol.* **2015**, *5*, 116–125. [CrossRef]
21. Frickmann, H.; Podbielski, A.; Kreikemeyer, B. Resistant Gram-Negative Bacteria and Diagnostic Point-of-Care Options for the Field Setting during Military Operations. *Biomed. Res. Int.* **2018**, *2018*, 9395420. [CrossRef]
22. McBride, M.E.; Duncan, W.C.; Knox, J.M. Physiological and environmental control of Gram negative bacteria on skin. *Br. J. Dermatol.* **1975**, *93*, 191–199. [CrossRef] [PubMed]
23. Krumkamp, R.; Oppong, K.; Hogan, B.; Strauss, R.; Frickmann, H.; Wiafe-Akenten, C.; Boahen, K.G.; Rickerts, V.; McCormick Smith, I.; Groß, U.; et al. Spectrum of antibiotic resistant bacteria and fungi isolated from chronically infected wounds in a rural district hospital in Ghana. *PLoS ONE* **2020**, *15*, e0237263. [CrossRef]
24. Codjoe, F.S.; Donkor, E.S.; Smith, T.J.; Miller, K. Phenotypic and Genotypic Characterization of Carbapenem-Resistant Gram-Negative Bacilli Pathogens from Hospitals in Ghana. *Microb. Drug. Resist.* **2019**, *25*, 1449–1457. [CrossRef] [PubMed]

25. Del Barrio-Tofiño, E.; López-Causapé, C.; Oliver, A. Pseudomonas aeruginosa epidemic high-risk clones and their association with horizontally-acquired β-lactamases: 2020 update. *Int. J. Antimicrob. Agents* **2020**, *56*, 106196. [CrossRef]

26. Wyres, K.L.; Lam, M.M.C.; Holt, K.E. Population genomics of *Klebsiella pneumoniae*. *Nat. Rev. Microbiol.* **2020**, *18*, 344–359. [CrossRef]

27. Denamur, E.; Clermont, O.; Bonacorsi, S.; Gordon, D. The population genetics of pathogenic *Escherichia coli*. *Nat. Rev. Microbiol.* **2021**, *19*, 37–54. [CrossRef] [PubMed]

28. Eibach, D.; Belmar Campos, C.; Krumkamp, R.; Al-Emran, H.M.; Dekker, D.; Boahen, K.G.; Kreuels, B.; Adu-Sarkodie, Y.; Aepfelbacher, M.; Park, S.E.; et al. Extended spectrum beta-lactamase producing *Enterobacteriaceae* causing bloodstream infections in rural Ghana, 2007–2012. *Int. J. Med. Microbiol.* **2016**, *306*, 249–254. [CrossRef]

29. Eibach, D.; Dekker, D.; Gyau Boahen, K.; Wiafe Akenten, C.; Sarpong, N.; Belmar Campos, C.; Berneking, L.; Aepfelbacher, M.; Krumkamp, R.; Owusu-Dabo, E.; et al. Extended-spectrum beta-lactamase-producing *Escherichia coli* and *Klebsiella pneumoniae* in local and imported poultry meat in Ghana. *Vet. Microbiol.* **2018**, *217*, 7–12. [CrossRef] [PubMed]

30. Labi, A.K.; Bjerrum, S.; Enweronu-Laryea, C.C.; Ayibor, P.K.; Nielsen, K.L.; Marvig, R.L.; Newman, M.J.; Andersen, L.P.; Kurtzhals, J.A.L. High Carriage Rates of Multidrug-Resistant Gram-Negative Bacteria in Neonatal Intensive Care Units from Ghana. *Open Forum Infect. Dis.* **2020**, *7*, ofaa109. [CrossRef]

31. Agyepong, N.; Govinden, U.; Owusu-Ofori, A.; Amoako, D.G.; Allam, M.; Janice, J.; Pedersen, T.; Sundsfjord, A.; Essack, S. Genomic characterization of multidrug-resistant ESBL-producing *Klebsiella pneumoniae* isolated from a Ghanaian teaching hospital. *Int. J. Infect. Dis.* **2019**, *85*, 117–123. [CrossRef] [PubMed]

32. Falgenhauer, L.; Imirzalioglu, C.; Oppong, K.; Akenten, C.W.; Hogan, B.; Krumkamp, R.; Poppert, S.; Levermann, V.; Schwengers, O.; Sarpong, N.; et al. Detection and Characterization of ESBL-Producing *Escherichia coli* from Humans and Poultry in Ghana. *Front. Microbiol.* **2019**, *9*, 3358. [CrossRef] [PubMed]

33. Donkor, E.S.; Horlortu, P.Z.; Dayie, N.T.; Obeng-Nkrumah, N.; Labi, A.K. Community acquired urinary tract infections among adults in Accra, Ghana. *Infect. Drug. Resist.* **2019**, *12*, 2059–2067. [CrossRef] [PubMed]

34. Obeng-Nkrumah, N.; Labi, A.K.; Blankson, H.; Awuah-Mensah, G.; Oduro-Mensah, D.; Anum, J.; Teye, J.; Kwashie, S.D.; Bako, E.; Ayeh-Kumi, P.F.; et al. Household cockroaches carry CTX-M-15-, OXA-48- and NDM-1-producing enterobacteria, and share beta-lactam resistance determinants with humans. *BMC Microbiol.* **2019**, *19*, 272. [CrossRef] [PubMed]

35. Poirel, L.; Corvec, S.; Rapoport, M.; Mugnier, P.; Petroni, A.; Pasteran, F.; Faccone, D.; Galas, M.; Drugeon, H.; Cattoir, V.; et al. Identification of the novel narrow-spectrum beta-lactamase SCO-1 in *Acinetobacter* spp. from Argentina. *Antimicrob. Agents Chemother.* **2007**, *51*, 2179–2184. [CrossRef]

36. Münch, J.; Hagen, R.M.; Müller, M.; Kellert, V.; Wiemer, D.F.; Hinz, R.; Schwarz, N.G.; Frickmann, H. Colonization with Multidrug-Resistant Bacteria—On the Efficiency of Local Decolonization Procedures. *Eur. J. Microbiol. Immunol.* **2017**, *7*, 99–111. [CrossRef]

37. Cornelis, P.; Dingemans, J. *Pseudomonas aeruginosa* adapts its iron uptake strategies in function of the type of infections. *Front. Cell. Infect. Microbiol.* **2013**, *3*, 75. [CrossRef] [PubMed]

38. Hasannejad-Bibalan, M.; Jafari, A.; Sabati, H.; Goswami, R.; Jafaryparvar, Z.; Sedaghat, F.; Sedigh Ebrahim-Saraie, H. Risk of type III secretion systems in burn patients with *Pseudomonas aeruginosa* wound infection: A systematic review and meta-analysis. *Burns* **2020**. [CrossRef]

39. Rumbaugh, K.P.; Griswold, J.A.; Hamood, A.N. Contribution of the regulatory gene *lasR* to the pathogenesis of *Pseudomonas aeruginosa* infection of burned mice. *J. Burn Care Rehabil.* **1999**, *20 Pt 1*, 42–49. [CrossRef]

40. Colmer-Hamood, J.A.; Dzvova, N.; Kruczek, C.; Hamood, A.N. In Vitro Analysis of Pseudomonas aeruginosa Virulence Using Conditions That Mimic the Environment at Specific Infection Sites. *Prog. Mol. Biol. Transl. Sci.* **2016**, *142*, 151–191.

41. Andrews, S. FastQC: A Quality Control Tool for High Throughput Sequence Data. 2010. Available online: http://www.bioinformatics.babraham.ac.uk/projects/fastqc (accessed on 15 December 2020).

42. Wood, D.E.; Lu, J.; Langmead, B. Improved metagenomic analysis with Kraken 2. *Genome Biol.* **2019**, *20*, 257. [CrossRef] [PubMed]

43. Jünemann, S.; Sedlazeck, F.J.; Prior, K.; Albersmeier, A.; John, U.; Kalinowski, J.; Mellmann, A.; Goesmann, A.; von Haeseler, A.; Stoye, J.; et al. Updating benchtop sequencing performance comparison. *Nature Biotechnol.* **2013**, *31*, 294–296. [CrossRef] [PubMed]

44. Prjibelski, A.; Antipov, D.; Meleshko, D.; Lapidus, A.; Korobeynikov, A. Using SPAdes de novo assembler. *Curr. Protoc. Bioinform.* **2020**, *70*, e102. [CrossRef] [PubMed]

45. Seemann, T. Abricate, Github. Available online: https://github.com/tseemann/abricate (accessed on 22 March 2021).

46. Feldgarden, M.; Brover, V.; Haft, D.H.; Prasad, A.B.; Slotta, D.J.; Tolstoy, I.; Tyson, G.H.; Zhao, S.; Hsu, C.H.; McDermott, P.F.; et al. Validating the AMRFinder Tool and Resistance Gene Database by Using Antimicrobial Resistance Genotype-Phenotype Correlations in a Collection of Isolates. *Antimicrob. Agents Chemother.* **2019**, *63*, e00483-19. [CrossRef]

47. Chen, L.; Zheng, D.; Liu, B.; Yang, J.; Jin, Q. VFDB 2016: Hierarchical and refined dataset for big data analysis—10 years on. *Nucleic Acids Res.* **2016**, *44*, D694–D697. [CrossRef]

48. Bortolaia, V.; Kaas, R.S.; Ruppe, E.; Roberts, M.C.; Schwarz, S.; Cattoir, V.; Philippon, A.; Allesoe, R.L.; Rebelo, A.R.; Florensa, A.F.; et al. ResFinder 4.0 for predictions of phenotypes from genotypes. *J. Antimicrob. Chemother.* **2020**, *75*, 3491–3500. [CrossRef]

Article

Clonal Lineages, Antimicrobial Resistance, and PVL Carriage of *Staphylococcus aureus* Associated to Skin and Soft-Tissue Infections from Ambulatory Patients in Portugal

Carolina Ferreira [1] , Sofia Santos Costa [1], Maria Serrano [1], Ketlyn Oliveira [1], Graça Trigueiro [2], Constança Pomba [3,4] and Isabel Couto [1,*]

[1] Global Health and Tropical Medicine (GHTM), Instituto de Higiene e Medicina Tropical (IHMT), Universidade Nova de Lisboa (UNL), Rua da Junqueira 100, 1349-008 Lisboa, Portugal; carolinaf@ihmt.unl.pt (C.F.); scosta@ihmt.unl.pt (S.S.C.); a21000870@ihmt.unl.pt (M.S.); mmm0012@ihmt.unl.pt (K.O.)

[2] Laboratório de Análises Clínicas Dr. Joaquim Chaves, Av. General Norton de Matos, 71 R/C, 1495-148 Algés, Portugal; graca.trigueiro@jcs.pt

[3] CIISA, Centre of Interdisciplinary Research in Animal Health, Faculty of Veterinary Medicine, University of Lisbon, Avenida da Universidade Técnica, 1300-477 Lisboa, Portugal; cpomba@fmv.ulisboa.pt

[4] GeneVet, Laboratório de Diagnóstico Molecular Veterinário, Rua Quinta da Nora Loja 3B, 2790-140 Carnaxide, Portugal

* Correspondence: icouto@ihmt.unl.pt; Tel.: +351-21-3652652; Fax: +351-21-3632105

Abstract: *Staphylococcus aureus* (*S. aureus*) is a leading cause of skin and soft-tissue infections (SSTIs) in the community. In this study, we characterized a collection of 34 *S. aureus* from SSTIs in ambulatory patients in Portugal and analyzed the presence of Panton–Valentine leucocidin (PVL)-encoding genes and antibiotic-resistance profile, which was correlated with genetic determinants, plasmid carriage, and clonal lineage. Nearly half of the isolates (15, 44.1%) were methicillin-resistant *Staphylococcus aureus* (MRSA) and/or multidrug resistant (MDR). We also detected resistance to penicillin (33/34, 97.1%), fluoroquinolones (17/34, 50.0%), macrolides and lincosamides (15/34, 44.1%), aminoglycosides (6/34, 17.6%), and fusidic acid (2/34, 5.9%), associated with several combinations of resistance determinants (*blaZ*, *erm*(A), *erm*(C), *msr*(A), *mph*(C), *aacA-aphD*, *aadD*, *aph*(3′)-*IIIa*, *fusC*), or mutations in target genes (*fusA*, *grlA/gyrA*). The collection presented a high genetic diversity (Simpson's index of 0.92) with prevalence of clonal lineages CC5, CC22, and CC8, which included the MRSA and also most MDR isolates (CC5 and CC22). PVL-encoding genes were found in seven isolates (20.6%), three methicillin-susceptible *Staphylococcus aureus* (MSSA) (ST152-*agr*I and ST30-*agr*III), and four MRSA (ST8-*agr*I). Plasmid profiling revealed seventeen distinct plasmid profiles. This work highlights the high frequency of antimicrobial resistance and PVL carriage in SSTIs-related *S. aureus* outside of the hospital environment.

Keywords: *Staphylococcus aureus*; skin and soft-tissue infections; antibiotic resistance; clonal lineages; plasmids; Panton–Valentine leucocidin

Citation: Ferreira, C.; Costa, S.S.; Serrano, M.; Oliveira, K.; Trigueiro, G.; Pomba, C.; Couto, I. Clonal Lineages, Antimicrobial Resistance, and PVL Carriage of *Staphylococcus aureus* Associated to Skin and Soft-Tissue Infections from Ambulatory Patients in Portugal. *Antibiotics* **2021**, *10*, 345. https://doi.org/10.3390/antibiotics10040345

Academic Editor: Giovanna Batoni

Received: 19 February 2021
Accepted: 19 March 2021
Published: 24 March 2021

Publisher's Note: MDPI stays neutral with regard to jurisdictional claims in published maps and institutional affiliations.

1. Introduction

Staphylococcus aureus (*S. aureus*) is a major human pathogen responsible for a wide range of infections both in hospitals and in the community. It is one of the main causes of severe nosocomial infections such as bacteremia and infective endocarditis and in the community is a frequent cause of skin and soft-tissue infections (SSTIs) [1]. Besides their potential severity, infections caused by *S. aureus* are usually difficult to treat due to the frequent acquisition of antimicrobial resistance determinants. In the last decades, there has been an emergence and dissemination of methicillin-resistant *S. aureus* (MRSA) as well as of multidrug-resistant (MDR) strains [2,3]. Consequently, MRSA are now included in the

World Health Organization (WHO) list of high-priority bacteria for development of new drugs [4].

S. aureus is the most frequent pathogen associated with SSTIs, which can range from minor or superficial infections such as impetigo to life-threating infections such as necrotizing fasciitis [5]. Topical antibiotics that are often used for the prevention or treatment of milder infections include mupirocin, fusidic acid, neomycin, and bacitracin [6,7]. The use of some of these topical antibiotics is particularly relevant in the community/ambulatory settings, where they may not require medical prescription. Other antibiotics for systemic use, such as clindamycin, trimethoprim-sulfamethoxazole, tetracyclines, and linezolid, are also indicated for treatment of severe forms of SSTIs caused by *S. aureus* [6,8]. The frequency of antibiotic-resistant *S. aureus* isolates associated with SSTIs is rising worldwide [9–12], particularly to fusidic acid and mupirocin, which is probably linked with the widespread use of these antibiotics [3].

Resistance to antibiotics in *S. aureus* can be mediated by several mechanisms, such as antibiotic modification or degradation, target mutation, or antibiotic efflux. Resistance to penicillins can occur by inactivation of the antibiotic molecule through the action of the β-lactamase BlaZ. The *blaZ* gene occurs frequently in *S. aureus* clinical isolates. Resistance to penicillins and other β-lactams, with the exception of fifth-generation cephalosporins, is mediated by the acquisition of the *mecA* gene, which is part of the mobile genetic element SCC*mec* (staphylococcal cassette chromosome *mec*) and encodes for an additional penicillin-binding protein, PBP2a, with low affinity for the β-lactam antibiotics [3]. Resistance to macrolides and lincosamides can occur through several mechanisms, including the acquisition of rRNA methylases-encoding *erm* genes that methylate the binding site of the antibiotics [3]. Resistance to aminoglycosides is associated with acquisition of several genes, like *aacA-aphD* or *aadD* that encode enzymes that modify the antibiotic molecule rendering it inactive [3]. Resistance to fluoroquinolones is usually linked to the occurrence of mutations in the quinolone-resistant determining region (QRDR) of the *grlA/B* and *gyrA/B* genes that encode the DNA topoisomerase IV and DNA gyrase, respectively. Fluoroquinolone resistance can also be conveyed by overexpression of chromosomally-encoded efflux pump genes such as *norA/B/C* and *mepA* [13]. Resistance to fusidic acid can be achieved by the acquisition of the *fusB/C* genes that encode ribosomal protection proteins or by mutations in the *fusA* gene [3].

S. aureus produces several virulence factors, including toxins, proteins associated with immune evasion, and tissue-degrading enzymes [1]. The cytotoxin Panton–Valentine leucocidin (PVL), encoded by the genes *lukF-PV* and *lukS-PV* carried on bacteriophage φSa2, is a two-component pore-forming protein that has been strongly associated with *S. aureus* isolates causing skin infections in the community and with necrotizing pneumonia [1]. Nevertheless, the role of PVL in *S. aureus* infection pathogenesis is still not fully elucidated [14]. The *S. aureus* accessory gene regulator (*agr*) locus regulates the expression of several virulence factors like cell-wall-associated and extracellular proteins, contributing to infection severity and persistence. The polymorphism of the *agr* locus allows the classification of *S. aureus* in four predominant *agr* types (I to IV), that may differ in terms of infection type, carriage of virulence factors, and temporal patterns of autoinduction [15].

Most antimicrobial resistance and virulence genes of *S. aureus* are located on mobile genetic elements (MGEs) such as plasmids, bacteriophages, pathogenicity islands, transposons, integrative conjugative elements (ICEs), integrons, and staphylococcal chromosome cassettes (SCCs), which make up to 15–20% of its genome [16]. The acquisition of antimicrobial resistance by *S. aureus* is mostly due to horizontal gene transfer (HGT), and plasmids have been identified as one of the main responsible for the dissemination of resistance genes [17].

Several studies have evaluated the main clones of MRSA circulating both in hospitals and in the community in Portugal, a country with a high prevalence of MRSA [18–22]. However, there have been fewer studies focusing on *S. aureus*, both methicillin-susceptible *Staphylococcus aureus* (MSSA) and MRSA causing SSTIs. The aim of this work was to perform a phenotypic and genotypic characterization of a collection of *S. aureus* isolated from SSTIs in ambulatory patients and to assess their virulence determinants and susceptibility to the main antibiotics used in SSTI therapeutics, correlating their resistance profile to genetic determinants and identifying their main mechanisms of dissemination among *S. aureus* strains.

2. Results

2.1. Antimicrobial Susceptibility Profile and Correlation with Resistance Determinants

The antimicrobial susceptibility profile of the 34 isolates is described in Table 1. Resistance to penicillin was detected in 97.1% (33/34) of the isolates, and 44.1% (15/34) were MRSA (*mecA*$^+$ and cefoxitin resistant). We have also observed resistance to the fluoroquinolones ciprofloxacin and moxifloxacin (50.0%, 17/34), erythromycin (44.1%, 15/34), clindamycin (35.3%, 12/34) either constitutive (2.9%, 1/34) or inducible (32.4%, 11/34), kanamycin (17.6%, 6/34), tobramycin (14.7%, 5/34), amikacin (8.8%, 3/34), gentamycin (2.9%, 1/34), and fusidic acid (5.9%, 2/34). Fifteen isolates (44.1%) were MDR, mainly resistant to β-lactams, fluoroquinolones, macrolides, and lincosamides. All isolates were susceptible to tetracyclines, tigecycline, rifampicin, trimethoprim-sulfamethoxazole, linezolid, chloramphenicol, retapamulin, and quinupristin-dalfopristin. All MRSA isolates were susceptible to ceftaroline. Only one isolate was susceptible to all antibiotics tested. Although no breakpoints or epidemiological cut-off values are established by the European Committee on Antimicrobial Susceptibility Testing (EUCAST) for bacitracin or neomycin (for 30 µg discs), one isolate showed no inhibition zone toward each of these topical antibiotics. The presence of antibiotic-resistance determinants was confirmed for all isolates presenting phenotypic resistance (Table 1). The *blaZ* gene was detected in all isolates resistant to penicillin. All isolates showing resistance to cefoxitin harbored the *mecA* gene. Mutations in QRDR regions of *grlA* and *gyrA* genes were found in different combinations in all the representative fluoroquinolone resistant isolates screened. Resistance to macrolides and lincosamides was associated with *erm*(A), *erm*(C), *msr*(A), and/or *mph*(C). Resistance to aminoglycosides was mainly linked to the *aadD* gene. For the two isolates resistant to fusidic acid, one harbored the *fusC* gene, whereas the other carried three mutations in the *fusA* gene.

Table 1. Antimicrobial resistance phenotypes of the 34 *S. aureus* included in this study and correlation with resistance determinants.

Class	Antibiotic	Resistant Isolates (%)	Resistance Determinants (No. Isolates); [Mutations]
β-lactams	PEN	33 (97.1%)	*blaZ* (33)
	CXI	15 (44.1%)	*mecA* (15)
Fluoroquinolones	CIP	17 (50%)	Mutations in GrlA [S80Y, E84G, S80F] and GyrA [S84L, E88K]
	MOX	17 (50%)	
Macrolides/Lincosamides	ERY	15 (44.1%)	*erm*(A) (9), *erm*(C) (7) *msr*(A) (4), *mph*(C) (3)
	CLI	12 (35.3%)	
Aminoglycosides	KAN	6 (17.6%)	*aadD* (4), *aacA-aphD* (1) *aph(3′)-IIIa* (1)
	TOB	5 (14.7%)	
	AMI	3 (8.8%)	
	GEN	1 (2.9%)	
Fusidanes	FUS	2 (5.9%)	*fusC* (1) FusA mutations [A71V, H457Q, G476C]

PEN: penicillin; CXI: cefoxitin; ERY: erythromycin; CLI: clindamycin; CIP: ciprofloxacin; MOX: moxifloxacin; KAN: kanamycin; GEN: gentamycin; TOB: tobramycin; AMI: amikacin; and FUS: fusidic acid.

2.2. Efflux Activity

The presence of increased efflux activity in the 34 isolates was assessed by different approaches. The minimum inhibitory concentrations (MICs) of ethidium bromide (EtBr) for the entire collection ranged from 2 to 16 µg/mL with a unimodal distribution (data not shown). Eleven isolates presented an EtBr MIC of 16 µg/mL, suggesting increased efflux activity in those isolates. In addition, these 11 isolates were also resistant to fluoroquinolones, a class of antibiotics that is substrate of the main efflux pumps in *S. aureus* [13]. To verify the presence of an efflux-mediated resistance in these 11 isolates, the EtBr and ciprofloxacin (CIP) MICs were determined in the presence of the known efflux inhibitors (EIs) thioridazine (TZ) and verapamil (VER) and compared to their original values (Table 2). A significant decrease (four- to eight-fold) in EtBr MICs was observed for all isolates but one, confirming the presence of increased efflux activity in these isolates. However, none of the isolates carried the plasmid-encoded *qacA/B* or *smr* genes, which code for the efflux pumps QacA/B and Smr, respectively, responsible for the extrusion of EtBr and several biocides. These results indicate that the increased efflux activity present in these isolates may be driven by chromosomally-encoded efflux pumps, like NorA, which extrudes EtBr and biocides but also several fluoroquinolones like ciprofloxacin and norfloxacin [13]. The effect of EIs on CIP MICs was less significant, with MIC reductions of two-fold for the majority of the isolates. This result does not exclude the presence of increased efflux activity associated with fluoroquinolone resistance, since these isolates harbor mutations in the QRDR of *grlA* and *gyrA* genes, which are responsible for conferring high-level fluoroquinolone resistance and thus may be hindering the screening of efflux activity associated with resistance to these antibiotics [23].

Table 2. The effect of the efflux inhibitors thioridazine and verapamil on ethidium bromide and ciprofloxacin MICs for selected *S. aureus* isolates.

Isolate	MIC (µg/mL)					
	EtBr	**EtBr + TZ**	**EtBr + VER**	**CIP**	**CIP + TZ**	**CIP + VER**
BIOS-H4	16	4	2	16	16	16
BIOS-H7	16	8	8	512	256	256
BIOS-H8	16	8	4	512	256	256
BIOS-H10	16	8	4	512	256	256
BIOS-H11	16	4	2	256	128	128
BIOS-H14	16	4	2	128	64	64
BIOS-H19	16	4	2	512	256	256
BIOS-H23	16	4	4	256	128	256
BIOS-H24	16	2	2	32	16	16
BIOS-H31	16	4	2	512	**128**	**128**
BIOS-H33	16	4	2	512	256	256

MIC: minimum inhibitory concentration; EtBr: ethidium bromide; CIP: ciprofloxacin; TZ: thioridazine; VER: verapamil. Bold-type numbers indicate MIC reductions $\geq$ four-fold in the presence of EIs when compared to the original MIC values.

2.3. Main Clonal Lineages and Genetic Diversity of the S. aureus Isolates

Analysis of *Sma*I-macrorestriction profiles revealed the presence of 15 pulsed-field gel electrophoresis (PFGE) types (A to O) and 18 subtypes (Figure 2) among the collection studied. The three most common profiles, PFGE types G, N, and E, are represented by seven, five, and four isolates, respectively. An isolate representative of each PFGE type was selected for typing by multilocus sequence typing (MLST). Fourteen sequence types (STs) were identified belonging to 10 clonal complexes. The clonal complexes identified were CC5 (ST5, ST105, and the newly identified ST6531, which is a single-locus variant (SLV) of ST5), CC8 (ST8, ST72), CC152 (ST152), CC30 (ST30), CC7 (ST7), CC97 (ST97), CC15 (ST15), CC25 (ST25), CC22 (ST22), and CC45 (ST278). We also detected a newly identified singleton, ST6564. In general, each ST identified was associated with a single PFGE type, except for ST5 (CC5), associated with PFGE types F and D and ST8 (CC8) associated with PFGE types C and E. The most common PFGE types were linked to ST105 (CC5), ST22 (CC22), and ST8 (CC8). The Simpson's index of diversity (SID), calculated based upon the PFGE *Sma*I-macrorestriction profiles, revealed a highly diverse *S. aureus* population (SID = 0.92, CI: 0.87–0.98).

2.4. Correlation of Strain Lineage with agr Typing and PVL Carriage

The *agr* typing of the *S. aureus* isolates identified *agr*I as the predominant type, which was detected in 20 out of the 34 isolates (58.8%), followed by *agr* type II, identified in 13/34 (38.2%) isolates, and *agr* type III, observed in a single isolate (1/34, 2.9%). No isolate of *agr* type IV was identified. The PVL-encoding genes *lukS-lukF* were detected in seven isolates (20.6%), corresponding to three MSSA (3/19, 15.8%) and four MRSA (4/15, 26.7%), all classified as *agr* type I or III.

As shown in Figure 2, an association was observed between *S. aureus* clonal lineages ST8 (CC8), ST25 (CC25), ST22 (CC22), ST7 (CC7), ST278 (CC45), ST97 (CC97) and ST152 (CC152) and *agr* type I, whereas clonal complexes CC5 and CC15 were linked to *agr* type II, and the singe isolate harboring *agr* type III belonged to ST30 (CC30). The newly identified singleton ST6564 belongs to *agr* type I.

Carriage of PVL was associated with MRSA belonging to ST8 (CC8) and MSSA assigned to ST152 (CC152) or ST30 (CC30).

2.5. Correlation of Strain Lineage with Antimicrobial Resistance and Plasmid Profiles

Analysis of the methicillin resistance status and clonal lineage showed that the MRSA isolates identified in the collection were restricted to the clonal complexes CC22, CC8 (ST8), and CC5 (ST5 and ST105) (Figure 1). Most isolates from ST22 and clonal lineages of the CC5 presented MDR phenotypes (Figure 1).

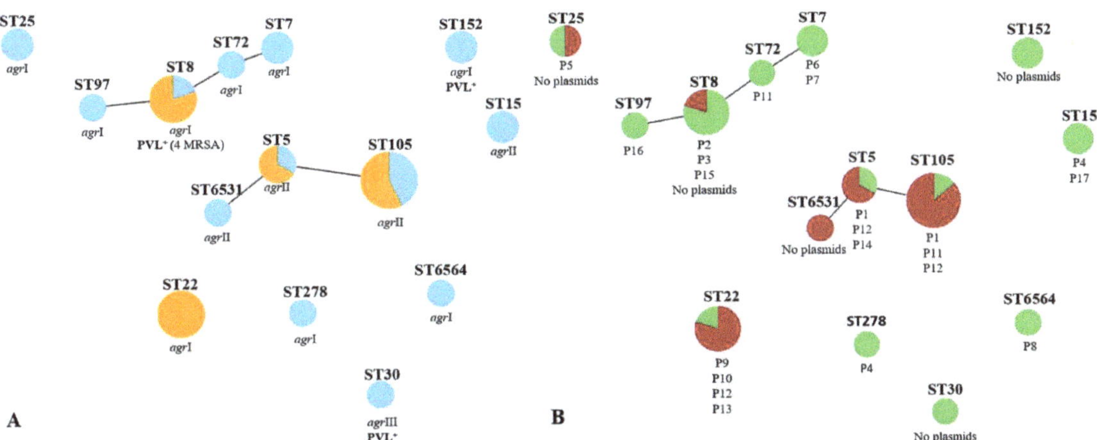

Figure 1. Relation of clonal lineages identified amongst the *S. aureus* associated with SSTIs in ambulatory patients determined using PHYLOViZ software and correlation with (**A**) methicillin resistance status, PVL carriage, and *agr* type; and (**B**) MDR phenotypes and plasmid profile. In panel (**A**), MRSA isolates are displayed in orange whereas MSSA isolates are shown in light blue. In panel (**B**), MDR isolates are presented in red, while non-MDR isolates are shown in green.

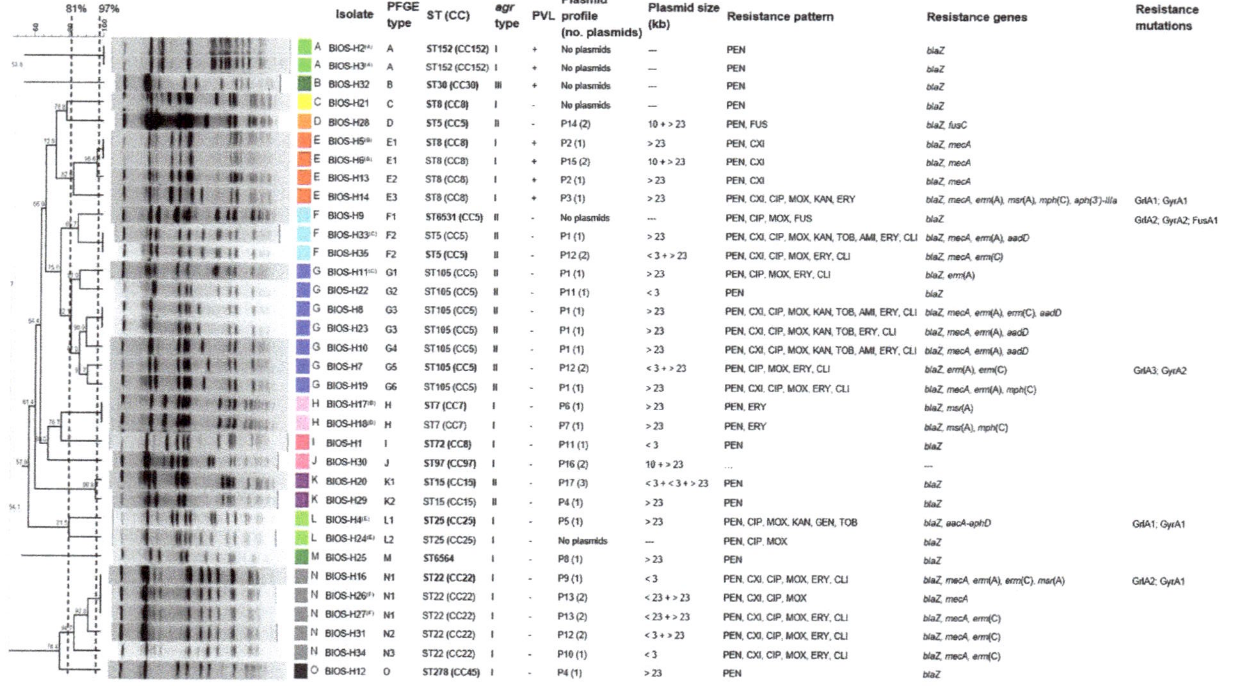

Isolate	PFGE type	ST (CC)	*agr* type	PVL	Plasmid profile (no. plasmids)	Plasmid size (kb)	Resistance pattern	Resistance genes	Resistance mutations
A BIOS-H2[A]	A	ST152 (CC152)	I	+	No plasmids	—	PEN	*blaZ*	
A BIOS-H3[A]	A	ST152 (CC152)	I	+	No plasmids	—	PEN	*blaZ*	
B BIOS-H32	B	ST30 (CC30)	III	+	No plasmids	—	PEN	*blaZ*	
C BIOS-H21	C	ST8 (CC8)	I	-	No plasmids	—	PEN	*blaZ*	
D BIOS-H28	D	ST5 (CC5)	II	-	P14 (2)	10 + > 23	PEN, FUS	*blaZ, fusC*	
E BIOS-H5[B]	E1	ST8 (CC8)	I	+	P2 (1)	> 23	PEN, CXI	*blaZ, mecA*	
E BIOS-H6[B]	E1	ST8 (CC8)	I	+	P15 (2)	10 + > 23	PEN, CXI	*blaZ, mecA*	
E BIOS-H13	E2	ST8 (CC8)	I	+	P2 (1)	> 23	PEN, CXI	*blaZ, mecA*	
E BIOS-H14	E3	ST8 (CC8)	I	+	P3 (1)	> 23	PEN, CXI, CIP, MOX, KAN, ERY	*blaZ, mecA, erm(A), msr(A), mph(C), aph(3')-IIIa*	GrlA1; GyrA1
F BIOS-H9	F1	ST6531 (CC5)	II	-	No plasmids	—	PEN, CIP, MOX, FUS	*blaZ*	GrlA2; GyrA2; FusA1
F BIOS-H33[C]	F2	ST5 (CC5)	II	-	P1 (1)	> 23	PEN, CXI, CIP, MOX, KAN, TOB, AMI, ERY, CLI	*blaZ, mecA, erm(A), aadD*	
F BIOS-H35	F2	ST5 (CC5)	II	-	P12 (2)	< 3 + > 23	PEN, CXI, CIP, MOX, ERY, CLI	*blaZ, mecA, erm(C)*	
G BIOS-H11[C]	G1	ST105 (CC5)	II	-	P1 (1)	> 23	PEN, CIP, MOX, ERY, CLI	*blaZ, erm(A)*	
G BIOS-H22	G2	ST105 (CC5)	II	-	P11 (1)	< 3	PEN	*blaZ*	
G BIOS-H8	G3	ST105 (CC5)	II	-	P1 (1)	> 23	PEN, CXI, CIP, MOX, KAN, TOB, AMI, ERY, CLI	*blaZ, mecA, erm(A), erm(C), aadD*	
G BIOS-H23	G3	ST105 (CC5)	II	-	P1 (1)	> 23	PEN, CXI, CIP, MOX, KAN, TOB, ERY, CLI	*blaZ, mecA, erm(A), aadD*	
G BIOS-H10	G4	ST105 (CC5)	II	-	P1 (1)	> 23	PEN, CXI, CIP, MOX, KAN, TOB, AMI, ERY, CLI	*blaZ, mecA, erm(A), aadD*	
G BIOS-H7	G5	ST105 (CC5)	II	-	P12 (2)	< 3 + > 23	PEN, CIP, MOX, ERY, CLI	*blaZ, erm(A), erm(C)*	GrlA3; GyrA2
G BIOS-H19	G6	ST105 (CC5)	II	-	P1 (1)	> 23	PEN, CXI, CIP, MOX, ERY, CLI	*blaZ, mecA, erm(A), mph(C)*	
H BIOS-H17[D]	H	ST7 (CC7)	I	-	P6 (1)	> 23	PEN, ERY	*blaZ, msr(A)*	
H BIOS-H18[D]	H	ST7 (CC7)	I	-	P7 (1)	> 23	PEN, ERY	*blaZ, msr(A), mph(C)*	
I BIOS-H1	I	ST72 (CC8)	I	-	P11 (1)	< 3	PEN	*blaZ*	
J BIOS-H30	J	ST97 (CC97)	I	-	P16 (2)	10 + > 23	…	---	
K BIOS-H20	K1	ST15 (CC15)	II	-	P17 (3)	< 3 + < 3 + > 23	PEN	*blaZ*	
K BIOS-H29	K2	ST15 (CC15)	II	-	P4 (1)	> 23	PEN	*blaZ*	
L BIOS-H4[E]	L1	ST25 (CC25)	I	-	P5 (1)	> 23	PEN, CIP, MOX, KAN, GEN, TOB	*blaZ, aacA-aphD*	GrlA1; GyrA1
L BIOS-H24[E]	L2	ST25 (CC25)	I	-	No plasmids	—	PEN, CIP, MOX	*blaZ*	
M BIOS-H25	M	ST6564	I	-	P8 (1)	> 23	PEN	*blaZ*	
N BIOS-H16	N1	ST22 (CC22)	I	-	P9 (1)	< 3	PEN, CXI, CIP, MOX, ERY, CLI	*blaZ, mecA, erm(A), erm(C), msr(A)*	GrlA2; GyrA1
N BIOS-H26[F]	N1	ST22 (CC22)	I	-	P13 (2)	< 23 + > 23	PEN, CXI, CIP, MOX	*blaZ, mecA*	
N BIOS-H27[F]	N1	ST22 (CC22)	I	-	P13 (2)	< 23 + > 23	PEN, CXI, CIP, MOX, ERY, CLI	*blaZ, mecA, erm(C)*	
N BIOS-H31	N2	ST22 (CC22)	I	-	P12 (2)	< 3 + > 23	PEN, CXI, CIP, MOX, ERY, CLI	*blaZ, mecA, erm(C)*	
N BIOS-H34	N3	ST22 (CC22)	I	-	P10 (1)	< 3	PEN, CXI, CIP, MOX, ERY, CLI	*blaZ, mecA, erm(C)*	
O BIOS-H12	O	ST278 (CC45)	I	-	P4 (1)	> 23	PEN	*blaZ*	

Figure 2. *Sma*I-PFGE macrorestriction profile analysis of the *S. aureus* isolates associated with SSTIs in ambulatory patients and corresponding clonal lineages as determined by MLST and their correlation with PVL carriage and *agr* types, plasmid profiles, and phenotypic and genotypic resistance traits. The pairs of isolates recovered from different anatomical sites of the same patient are marked by (A) to (F), where each letter corresponds to a different patient. The dendrogram was built using Bionumerics and the UPGMA algorithm, using Dice coefficient, and an optimization of 0.5% and tolerance of band of 1%. The dashed lines correspond to the similarity criteria for considering isolates belonging to the same PFGE type ($\geq$81%) or subtype ($\geq$97%). Isolates sharing the same PFGE type or subtype were considered as belonging to the same sequence type (ST). The isolates subjected to MLST are indicated in bold-type. Each plasmid profile corresponds to a unique pattern of undigested and/or *Eco*RI-digested plasmids. CC: clonal complex; ST: sequence type; PFGE: pulsed-field gel electrophoresis; PVL: Panton–Valentine leucocidin; PEN: penicillin; CXI: cefoxitin; ERY: erythromycin; CLI: clindamycin; CIP: ciprofloxacin; MOX: moxifloxacin; KAN: kanamycin; GEN: gentamycin; TOB: tobramycin; AMI: amikacin; and FUS: fusidic acid. Resistance mutations: GrlA1: S80Y; GrlA2: S80F; GrlA3: S80Y, E84G; GyrA1: S84L; GyrA2: E88K; FusA1: A71V, H547Q, G476C.

The majority of the *S. aureus* isolates studied carried plasmids (28/34, 82.4%), with 19 isolates (55.9%) carrying one plasmid, eight isolates (23.5%) carrying two plasmids, and only one isolate (2.9%) carrying three plasmids. Large plasmids ($\geq$23 kb) were present in most isolates (24/34, 70.6%), alone or in combination with medium or smaller plasmids (10 kb or $\leq$3 kb). Isolates with large plasmids harbor, in general, a higher number of resistance determinants than those carrying small or no plasmids (Figure 2).

Seventeen plasmid profiles were identified, designated P1 to P17 (Figure 2). For strains carrying a single plasmid, these profiles were defined after restriction with *Eco*RI–profiles P1 to P10. The most frequent profile, P1, is represented by a single large plasmid (>23 kb), identified in six isolates, five of which belonging to ST105 (CC5). All isolates with this plasmid profile are MDR and carry several resistance genes. The second most frequent profile, P12, is shared by three isolates belonging to CC5 and CC22. Isolates of the same clonal complex show a high variety of plasmid profiles. For example, isolates of CC5, CC8, and CC22 have four different plasmids profiles each (Figure 2).

Of the six pairs of isolates recovered from two anatomical sites of the same patient, only one pair was assigned to two distinct PFGE types. Four pairs of isolates were indistinguishable by PFGE, while the remaining pair included subtypes of the same PFGE type. However, different phenotypical or genotypical trait(s) were observed within each pair except one (Figure 2). Isolates of three pairs differed in plasmid content, while two pairs of isolates differed in terms of resistance profile and/or resistance determinants. Another pair of isolates displayed different resistance profile and determinants although sharing the same plasmid profile.

3. Discussion

S. aureus is a leading cause of bacterial infections not only in healthcare settings but also in the community, many of which are caused by MRSA and MDR strains [2]. According to the most recent data of EARS-Net (European Antimicrobial Resistance Surveillance Network), in 2019, the prevalence of MRSA in bloodstream infections in Portugal was 34.8%. Even though this value has been decreasing over the last decade, it was still the fifth highest registered in Europe [24].

A high frequency of antibiotic resistance was observed in this collection. All isolates except one were resistant to at least one class of antibiotics, mainly β-lactams, and nearly half (44.1%) were MDR, which was unexpected considering they were not from hospitalized patients. However, these isolates were collected from ambulatory patients who could have been under antibiotic therapy or could have had recent contact with hospitals and that might explain the high rates of resistance observed. The 44.1% rate of MRSA identified is higher than the MRSA rates reported in the community (21.6%) in Portugal [19] and in children affected by SSTIs attending a pediatric emergency in Lisbon area (8.6%) [25] in years close to the year of collection of these isolates and is closer to the values observed in hospitals (47.4%) for 2014 [26]. On the other hand, the fact that these patients used laboratory services suggests that these may reflect more complex infections, which may explain the high frequencies of resistance observed [27]. Of the 15 MRSA, 11 (73.3%) were also MDR and the most common pattern was resistance to β-lactams, fluoroquinolones, macrolides, and lincosamides, which is a profile frequently observed in hospital-acquired MRSA (HA-MRSA) [19,20]. Previous studies have shown that there is a high prevalence of HA-MRSA strains in the community in our country, due to dissemination of these strains from the hospital [19–21]. The molecular analysis revealed that most of the isolates studied (16/34, 47.1%) presented genetic backgrounds related to hospital-associated lineages, such as CC5 and CC22, which were the predominant HA-MRSA lineages in Portugal during this period, identified in nosocomial or community isolates [18,19]. This finding, together with the use of community laboratory services to treat possible resilient and complex infections, may explain the high frequency of resistance observed in this collection. Regarding MSSA strains, only four strains showed an MDR profile (4/19, 21%). This observed rate of MDR strains is higher than previously reported for other MSSA collected from the community in Portugal [25,28].

With the exception of ST278 and the two new STs, all the other strain lineages identified in this work have been found in other studies in Portugal, with ST8 being the most frequent CA-MRSA clone, while ST30 and ST72 were the most prevalent MSSA clones [18,19,29]. ST278 belongs to CC45 and has been reported in the USA as a MSSA clone [30,31]. Strains of CC45 are prevalent in Portugal [18,19,29], but as far as we know, ST278 has not been yet reported in our country. We have also identified two new STs, ST6531, a SLV of ST5, and the singleton ST6564.

Although the *S. aureus* studied were isolated from patients with SSTIs, a low frequency of resistance was observed toward topical antibiotics, particularly to neomycin and fusidic acid, which are some of the most commonly used for the treatment of SSTIs in the community [7]. Only two isolates (5.9%) were resistant to fusidic acid, and only one isolate (2.9%) did not show inhibition zone to neomycin or bacitracin. The current rates of resistance to fusidic acid reported in the literature for SSTIs-associated *S. aureus* vary geographically, ranging from over 30% in Africa [32] to much lower rates, 2 to 6% in Asia or South America [33,34]. These low levels of resistance to fusidic acid are similar to the ones detected in other contemporary studies in Portugal [19]. The rates of resistance to neomycin and bacitracin in our collection are lower than the ones reported for other CA-MRSA from SSTIs [34]. However, a higher frequency of resistance was detected toward clindamycin (35.3%), an antibiotic also recommended for topical treatment of these infections, in comparison with other CA-MRSA from SSTIs [35,36].

Antibiotic-resistance determinants were identified in all isolates presenting phenotypic resistance (Table 1). The distribution of the fusidic-acid-resistance determinants in *S. aureus* reported in the literature is variable. While some studies report that *fusB* and *fusC* are the most prevalent genes [37], others report *fusA* mutations as the most common mechanism of fusidic-acid resistance [38]. In this study, only two isolates were resistant to fusidic acid. One of these had three mutations in the *fusA* gene, two of which (A71V and H457Q) already associated with resistance to this antibiotic [39–41], while the third mutation found, G476C, was described for the first time in this work and could also be contributing to fusidic-acid resistance. The other isolate resistant to fusidic acid carried the *fusC* gene. This is an MSSA that belongs to ST5. Several studies have shown that *fusC* gene can be located in SSC*mec* cassettes, with or without *mecA* gene [42,43].

Screening for mutations in fluoroquinolone-resistant representative isolates identified several patterns of mutations in the QRDR regions of GrlA and GyrA (namely, GrlA S80Y, GyrA S84L; GrlA S80F, GyrA E88K; GrlA S80Y E84G, GyrA E88K; and GrlA S80F GyrA S84L) already associated with high level resistance to these antibiotics [44,45]. These patterns of QRDR mutations were also detected in an earlier study from *S. aureus* clinical isolates in Lisbon [46]. The GrlA S80F and GyrA S84L mutations are the most commonly described in the literature [47–51] and are characteristic of ST22 and some ST8 lineages [51]. In our study, only one isolate carried both mutations and belonged to ST22. The GyrA S84L mutation was also found in one isolate of ST8. The GrlA E84G and S80Y and GyrA E88K mutations are also described in some studies [50,51] but appear to be less frequent. Besides these mutations, the activity of chromosomally-encoded MDR efflux pumps might also be contributing to fluoroquinolone resistance. A subset of fluoroquinolone-resistant isolates presented increased efflux activity of EtBr, a common substrate of MDR efflux pumps like NorA/B/C and MepA, which also extrude fluoroquinolones. No significant reduction in CIP MICs was observed in the presence of EIs, yet the effect of these compounds may be potentially hindered by the presence of QRDR mutations. The absence in this collection of the plasmid-encoded efflux pump genes *qacA/B* or *smr* indicates that the higher EtBr efflux activity detected is probably due to the overexpression of chromosomal efflux pump genes such as *norA/B/C* or *mepA* [13,46]. In the future this, hypothesis can be confirmed be quantifying the expression levels of these genes by RT-qPCR.

Plasmid profiling revealed a high proportion of plasmid-bearing isolates (82.4%) and a high diversity of plasmids, with 17 different profiles identified distributed amongst 14 clonal lineages. Most isolates carried a large plasmid, potentially associated with determinants for resistance to β-lactams, macrolides, lincosamides, and aminoglycosides (Figure 2). These results are similar to the ones found in a previous study that analyzed the plasmid content of a collection of 53 *S. aureus* isolated from a hospital in Lisbon between 2006 and 2007 [52]. The proportion of plasmid-bearing isolates in that study was 83%, and most isolates carried a large plasmid that was frequently associated with resistance to β-lactams, macrolides, and lincosamides. Other studies have also demonstrated that large plasmids are quite common in *S. aureus* and that they can carry several resistance determinants associated with resistance to the classes of antibiotics mentioned above [16].

The occurrence of PVL is linked to the bacteriophage φSa2 and generally associated with community-acquired MRSA (CA-MRSA), being traditionally considered a marker for the identification of CA-MRSA isolates [53], although some CA-MRSA strains do not produce this toxin. Its prevalence in HA-MRSA isolates, albeit lower, has been documented in several countries [54]. PVL is also strongly linked with *S. aureus* isolates collected from skin infections [53,54]. The overall rate of 20.6% of PVL-positive isolates in our set of *S. aureus* associated with SSTIs is lower than the ones reported from children with SSTI attending a pediatric emergency (37%) [25] but higher than the ones reported for other MSSA, CA-MRSA, or HA-MRSA collections in Portugal [19,22,55,56], albeit most of these other collections are not exclusively associated with skin infections. PVL carriage in our set of MRSA isolates appears restricted to the ST8-*agr*I clonal lineage, as found in a previous study by Tavares and colleagues [19]. Interestingly, the single MSSA ST8 isolate of our collection did not harbor PVL. The PVL-positive MSSA detected in our collection belong to the genetic backgrounds ST30-*agr*III and ST152-*agr*I, different from the ones reported in that earlier study [19]. However, PVL-positive ST30 isolates were also detected in children with SSTIs attending a pediatric emergency [25]. The ST8 and ST30 clonal lineages were frequently encountered in isolates from the community and less frequently associated with nosocomial isolates [19,20].

In this study, we performed a phenotypic and genotypic characterization of a collection of *S. aureus* isolated from SSTIs in ambulatory patients. Although this can be considered a relatively small sample, this is a convenience collection that represents the diversity of the population affected by SSTIs in an ambulatory setting over a five-months period where the only condition criteria for inclusion of the *S. aureus* isolates was to be SSTI-related. The genetic diversity of this collection was demonstrated by the high value of the Simpson's index (SID of 0.92).

4. Materials and Methods

4.1. Bacterial Isolates

The study comprised a collection of 34 *S. aureus* isolates associated with SSTIs of 28 ambulatory patients. Of the 34 isolates, 31 were collected from wounds (legs, *n* = 17; foot, *n* = 5; armpit, *n* = 3; ear, *n* = 1; and from unidentified sites, *n* = 4), and three were collected from ulcers. Six pairs of isolates (*n* = 12) were collected from different anatomical sites (right/left leg, *n* = 6; right/left armpit, *n* = 2; ear/leg, *n* = 2; and unidentified sites, *n* = 2) of six patients. The isolates were collected between February and June of 2014 at a community clinical diagnostic laboratory in Lisbon, Portugal. All isolates were grown in tryptic soy broth (TSB) (Oxoid™, Hampshire, UK), with shaking or tryptic soy agar (TSA) (Oxoid™) at 37 °C. Species identification was confirmed by amplification of the *nuc* gene following the protocol described by Poulsen and colleagues [57], using the primers described in Table S1 of Supplementary Data.

4.2. Antimicrobial Susceptibility Testing

Antimicrobial susceptibility was determined for a panel of 24 antibiotics by disk diffusion in Mueller-Hinton agar (MHA, Oxoid™), according to the EUCAST guidelines [58].

Antibiotics discs were obtained from Oxoid™. The following antibiotic discs (antibiotic content per disc) were used: penicillin (PEN, 1 U), oxacillin (OXA, 1 µg), cefoxitin (CXI, 30 µg), ceftaroline (CPT, 5 µg), ciprofloxacin (CIP, 5 µg), moxifloxacin (MOX, 5 µg), gentamicin (GEN, 10 µg), kanamycin (KAN, 30 µg), tobramycin (TOB, 10 µg), neomycin (NEO, 30 µg), amikacin (AMI, 30 µg), tetracycline (TET, 30 µg), minocycline (MIN, 30 µg), tigecycline (TIG, 15 µg), chloramphenicol (CHL, 30 µg), erythromycin (ERY, 15 µg), clindamycin (CLI, 2 µg), quinupristin/dalfopristin (QD, 15 µg), linezolid (LIN, 10 µg), trimethoprim-sulfamethoxazole (TRS, 25 µg), rifampicin (RIF, 5 µg), bacitracin (BAC, 10 U), fusidic acid (FUS, 10 µg), and mupirocin (MUP, 200 µg). The D-zone test was performed for detection of inducible clindamycin resistance, and the penicillin inhibition zone was examined to detect production of β-lactamases. Susceptibility testing to ceftaroline was performed for MRSA isolates only. Susceptibility to retapamulin (RET) was evaluated by determination of MICs by the two-fold microdilution method with cation-adjusted Mueller-Hinton broth (CAMHB, Oxoid™), according to the Clinical and Laboratory Standards Institute (CLSI) guidelines [59]. Retapamulin was acquired in powder form from Sigma-Aldrich (St. Louis, MO, USA), dissolved in dimethyl sulfoxide, and diluted in water with 10% β-cyclodextrin [60]. The reference strain *S. aureus* ATCC®29213™ was used as quality control. Isolates resistant to one antibiotic of at least three classes of antibiotics were considered multidrug resistant [61].

4.3. Detection of Resistance Genes by PCR

Total DNA was extracted from each isolate by the boiling method as described by Alexopoulou and colleagues [62]. All isolates were screened by PCR for the presence of the resistance genes *mecA* and *blaZ* and plasmid-encoded efflux pump genes *qacA/B* and *smr* (reduced susceptibility to biocides and EtBr). Isolates presenting phenotypic resistance to antibiotics were also screened for the presence of the genes *erm*(A), *erm*(B), *erm*(C), *msr*(A), *mph*(C), *vga*(A), *vga*(C) (resistance to macrolides, lincosamides, and streptogramins), *aadD*, *aph*(3′)-*IIIa*, *aacA-aphD* (resistance to aminoglycosides), *fusB*, and *fusC* (resistance to fusidic acid) using the primers described in Table S1 of Supplementary Data.

4.4. Screening of Mutations in grlA, gyrA, and fusA Genes

Mutations in the QRDRs of *grlA* and *gyrA* genes associated with fluoroquinolone resistance were screened for representative isolates, chosen according to their PFGE types. Mutations in the *fusA* gene were screened for isolates presenting resistance to fusidic acid. The primers used for amplification and sequencing of *grlA*, *gyrA*, and *fusA* genes are described in Table S1. Amplification products were purified using the kit NZYGelpure (NZYTech, Lisboa, Portugal) and sequenced. Sequences were analyzed using the programs SnapGene Viewer (GSL Biotech; available at snapgene.com) and blastx (NCBI, Bethesda, MD, USA).

4.5. Evaluation of Efflux Activity

The presence of increased efflux activity was evaluated by (i) determining the EtBr MIC [63] and (ii) determination of EtBr and CIP MICs in the presence of the EIs TZ and VER [23,46]. MICs of EtBr, CIP, TZ, and VER (Sigma-Aldrich) were determined by the two-fold broth microdilution method. Briefly, from overnight cultures, a cellular suspension equivalent to McFarland 0.5 was prepared in CAMHB and aliquoted in 96-well plates containing two-fold dilutions of the compound to be tested. Plates were incubated at 37 °C for 18 h, and the MIC registered as the lowest concentration of compound that inhibited visible growth. EtBr and CIP MICs were then redetermined in the presence of TZ and VER at 12.5 µg/mL and 400 µg/mL, respectively, corresponding to a subinhibitory concentration (1/2 MIC) [23]. The 96-well plates were prepared as described previously, except for the addition of a 0.01 mL aliquot of TZ or VER to each well prior to inoculation of the plate. Each assay was performed in duplicate. A four-fold, or higher, decrease in MICs values in the presence of EIs is indicative of inhibition of efflux activity [23].

4.6. Plasmid DNA Extraction and Profiling

Plasmid DNA of each isolate was extracted with the kit NZYMiniprep (NZYTech), adding 35 μg/mL of lysostaphin (Sigma-Aldrich) in the cell lysis step with buffer A1, followed by an incubation at 37 °C for 90 min. For isolates carrying a single plasmid, plasmid DNA was digested with 10 U of the enzyme *Eco*RI (NZYTech). The reaction mixture was incubated at 37 °C for 90 min and inactivated at 65 °C for 20 min. Restriction profiles were analyzed by 1% (*w/v*) agarose gel electrophoresis for 90 min.

4.7. Detection of lukSF Genes

The presence of the determinants *lukF-PV* and *lukS-PV* encoding PVL was screened by PCR, using the primers described in Table S1 of the Supplementary Data.

4.8. Molecular Typing

All isolates were characterized by PFGE. *Sma*I-PFGE was performed as previously described [64], and macrorestriction profiles were analyzed with the Bionumerics software v 7.6 using the Dice coefficient and dendrograms built based on the UPGMA algorithm, considering a band tolerance of 1% and an optimization of 0.5%. Isolates presenting macrorestriction profiles with a similarity ≥81% or ≥97% were considered as belonging to the same PFGE type or subtype, respectively [65]. The genetic diversity of the collection was calculated, based on PFGE types, by Simpson's index of diversity with a confidence interval of 95% [66].

A subset of isolates representative of each PFGE type was further analyzed by MLST. Isolates sharing the same PFGE type or subtype were considered as belonging to the same ST. Internal fragments of the seven housekeeping genes *arcC, aroE, glpF, gmk, pta, tpi*, and *yqiL* were amplified by PCR and sequenced using the primers and conditions previously described [67,68]. Allelic profiles and STs were obtained from MLST database (PubMLST.org (accessed on 28 December 2020). New alleles and ST profiles were submitted to PubMLST for validation and allele/ST assignment. The relationship between clonal lineages were inferred with the PHYLOViZ freeware using the goeBurst algorithm [69].

agr typing of all isolates was performed according to the protocol described by Lina and colleagues [70]. The set or primers used for *agr* typing is described in Table S1.

5. Conclusions

This work demonstrates a high prevalence of antibiotic resistance in *S. aureus* of SSTIs from outside the hospital environment, correlating it with the presence of several antibiotic-resistance determinants and a high prevalence of PVL-positive isolates, assigned to three MSSA (ST152-*agr*I and ST30-*agr*III) and four MRSA (ST8-*agr*I) isolates. This study also highlights the phenotypic and genotypic variability that may be present in *S. aureus* isolates causing infection in distinct anatomical sites of the same patient. The high diversity of plasmids identified in this collection demonstrates the important role these MGEs have in the transmission of antimicrobial resistance in *S. aureus* and the relevance of studying these elements to further prevent the dissemination of MDR strains.

Supplementary Materials: The following is available online at https://www.mdpi.com/2079-6382/10/4/345/s1, Table S1: Primers used in this study.

Author Contributions: Conceptualization, S.S.C. and I.C.; Funding acquisition, S.S.C. and I.C.; Investigation, C.F., S.S.C., M.S. and K.O.; Methodology, S.S.C. and I.C.; Project administration, S.S.C. and I.C.; Resources, G.T., C.P. and I.C.; Supervision, I.C.; Validation, S.S.C. and I.C.; Visualization, C.F. and S.S.C.; Writing—original draft, C.F. and S.S.C.; Writing—review and editing, C.F., S.S.C., M.S., K.O., G.T., C.P. and I.C. All authors have read and agreed to the published version of the manuscript.

Funding: This work was supported by Project BIOSAFE funded by FEDER through the Programa Operacional Factores de Competitividade—COMPETE and by Fundação para a Ciência e a Tecnologia (FCT, Portugal)—Grant LISBOA-01-0145-FEDER-030713, PTDC/CAL-EST/30713/2017 and by FCT through funds to GHTM (UID/04413/2020) and CIISA Project (UID/CVT/00276/2020).

Institutional Review Board Statement: Ethics approval was not required for this study.

Informed Consent Statement: Not applicable.

Data Availability Statement: All relevant data have been provided in the paper. Raw data can be provided by the authors upon reasonable request.

Conflicts of Interest: The authors declare no conflict of interest.

References

1. Turner, N.A.; Sharma-Kuinkel, B.K.; Maskarinec, S.A.; Eichenberger, E.M.; Shah, P.P.; Carugati, M.; Holland, T.L.; Fowler, V.G., Jr. Methicillin-resistant *Staphylococcus aureus*: An overview of basic and clinical research. *Nat. Rev. Microbiol.* **2019**, *17*, 203–218. [CrossRef]
2. Lee, A.S.; de Lencastre, H.; Garau, J.; Kluytmans, J.; Malhotra-Kumar, S.; Peschel, A.; Harbarth, S. Methicillin-resistant *Staphylococcus aureus*. *Nat. Rev. Dis. Primers* **2018**, *4*, 18033. [CrossRef]
3. Foster, T. Antibiotic resistance in *Staphylococcus aureus*. Current status and future prospects. *FEMS Microbiol. Rev.* **2017**, *41*, 430–449. [CrossRef]
4. Tacconelli, E.; Carrara, E.; Savoldi, A.; Harbarth, S.; Mendelson, M.; Monnet, D.L.; Pulcini, C.; Kahlmeter, G.; Kluytmans, J.; Carmeli, Y.; et al. Discovery, research, and development of new antibiotics: The WHO priority list of antibiotic-resistant bacteria and tuberculosis. *Lancet Infect. Dis.* **2018**, *18*, 318–327. [CrossRef]
5. Olaniyi, R.; Pozzi, C.; Grimaldi, L.; Bagnoli, F. *Staphylococcus aureus*-associated skin and soft tissue infections: Anatomical localization, epidemiology, therapy and potential prophylaxis. *Curr. Top. Microbiol. Immunol.* **2016**, *409*, 199–227. [CrossRef]
6. Liu, C.; Bayer, A.; Cosgrove, S.E.; Daum, R.S.; Fridkin, S.K.; Gorwitz, R.J.; Kaplan, S.L.; Karchmer, A.W.; Levine, D.P.; Murray, B.E.; et al. Clinical practice guidelines by the Infectious Diseases Society of America for the treatment of methicillin-resistant *Staphylococcus aureus* infections in adults and children. *Clin. Infect. Dis.* **2011**, *52*, e18–e55. [CrossRef] [PubMed]
7. Esposito, S.; Bassetti, M.; Bonnet, E.; Bouza, E.; Chan, M.; De Simone, G.; Dryden, M.; Gould, I.; Lye, D.C.; Saeed, K.; et al. Hot topics in the diagnosis and management of skin and soft-tissue infections. *Int. J. Antimicrob. Agents* **2016**, *48*, 19–26. [CrossRef] [PubMed]
8. Montravers, P.; Snauwaert, A.; Welsch, C. Current guidelines and recommendations for the management of skin and soft tissue infections. *Curr. Opin. Infect. Dis.* **2016**, *29*, 131–138. [CrossRef]
9. Lee, G.C.; Dallas, S.D.; Wang, Y.; Olsen, R.J.; Lawson, K.A.; Wilson, J.; Frei, C.R. Emerging multidrug resistance in community-associated *Staphylococcus aureus* involved in skin and soft tissue infections and nasal colonization. *J. Antimicrob. Chemother.* **2017**, *72*, 2461–2468. [CrossRef]
10. Davey, R.X.; Tong, S.Y.C. The epidemiology of *Staphylococcus aureus* skin and soft tissue infection in the southern Barkly region of Australia's Northern Territory in 2017. *Pathology* **2019**, *51*, 308–312. [CrossRef]
11. Stefanaki, C.; Ieronymaki, A.; Matoula, T.; Caroni, C.; Polythodoraki, E.; Chryssou, S.E.; Kontochristopoulos, G.; Antoniou, C. Six-year retrospective review of hospital data on antimicrobial resistance profile of *Staphylococcus aureus* isolated from skin infections from a single institution in Greece. *Antibiotics* **2017**, *6*, 39. [CrossRef]
12. Watanabe, S.; Ohnishi, T.; Yuasa, A.; Kiyota, H.; Iwata, S.; Kaku, M.; Watanabe, A.; Sato, J.; Hanaki, H.; Manabe, M.; et al. The first nationwide surveillance of antibacterial susceptibility patterns of pathogens isolated from skin and soft-tissue infections in dermatology departments in Japan. *J. Infect. Chemother.* **2017**, *23*, 503–511. [CrossRef] [PubMed]
13. Costa, S.S.; Viveiros, M.; Amaral, L.; Couto, I. Multidrug efflux pumps in *Staphylococcus aureus*: An update. *Open Microbiol. J.* **2013**, *7*, 59–71. [CrossRef] [PubMed]
14. Lakhundi, S.; Zhang, K. Methicillin-Resistant *Staphylococcus aureus*: Molecular characterization, evolution, and epidemiology. *Clin. Microbiol. Rev.* **2018**, *31*. [CrossRef] [PubMed]
15. Wang, B.; Muir, T.W. Regulation of virulence in *Staphylococcus aureus*: Molecular mechanisms and remaining puzzles. *Cell Chem. Biol.* **2016**, *23*, 214–224. [CrossRef]
16. Haaber, J.; Penadés, J.R.; Ingmer, H. Transfer of antibiotic resistance in *Staphylococcus aureus*. *Trends Microbiol.* **2017**, *25*, 893–905. [CrossRef]
17. Jensen, S.O.; Lyon, B.R. Genetics of antimicrobial resistance in *Staphylococcus aureus*. *Future Microbiol.* **2009**, *4*, 565–582. [CrossRef]
18. Faria, N.; Miragaia, M.; de Lencastre, H.; The Multi Laboratory Project Collaborators. Massive dissemination of methicillin resistant *Staphylococcus aureus* in bloodstream infections in a high MRSA prevalence country: Establishment and diversification of EMRSA-15. *Microbial. Drug Resist.* **2013**, *19*, 483–490. [CrossRef]
19. Tavares, A.; Miragaia, M.; Rolo, J.; Coelho, C.; de Lencastre, H. High prevalence of hospital-associated methicillin-resistant *Staphylococcus aureus* in the community in Portugal: Evidence for the blurring of community–hospital boundaries. *Eur. J. Clin. Microbiol. Infect. Dis.* **2013**, *32*, 1269–1283. [CrossRef]
20. Espadinha, D.; Faria, N.A.; Miragaia, M.; Lito, L.M.; Melo-Cristino, J.; de Lencastre, H.; Médicos Sentinela Network. Extensive dissemination of methicillin-resistant *Staphylococcus aureus* (MRSA) between the hospital and the community in a country with a high prevalence of nosocomial MRSA. *PLoS ONE* **2013**, *8*, e59960. [CrossRef]

21. Conceição, T.; Diamantino, F.; Coelho, C.; de Lencastre, H.; Aires-de-Sousa, M. Contamination of public buses with MRSA in Lisbon, Portugal: A possible transmission route of major MRSA clones within the community. *PLoS ONE* **2013**, *8*, e77812. [CrossRef] [PubMed]

22. Conceição, T.; Martins, H.; Rodrigues, S.; de Lencastre, H.; Aires-de-Sousa, M. *Staphylococcus aureus* nasal carriage among homeless population in Lisbon, Portugal. *Eur. J. Clin. Microbiol. Infect. Dis.* **2019**, *38*, 2037–2044. [CrossRef]

23. Costa, S.S.; Junqueira, E.; Palma, C.; Viveiros, M.; Melo-Cristino, J.; Amaral, L.; Couto, I. Resistance to antimicrobials mediated by efflux pumps in *Staphylococcus aureus*. *Antibiotics* **2013**, *2*, 83–99. [CrossRef] [PubMed]

24. European Centre for Disease Prevention and Control. *Surveillance of Antimicrobial Resistance in Europe 2019*; ECDC: Stockholm, Sweden, 2020.

25. Conceição, T.; Aires-de-Sousa, M.; Pona, N.; Brito, M.J.; Barradas, C.; Coelho, R.; Sardinha, T.; Sancho, L.; de Sousa, G.; Machado, M.C.; et al. High prevalence of ST121 in community-associated methicillin-susceptible *Staphylococcus aureus* lineages responsible for skin and soft tissue infections in Portuguese children. *Eur. J. Clin. Microbiol. Infect. Dis.* **2011**, *30*, 293–297. [CrossRef]

26. European Centre for Disease Prevention and Control. *Antimicrobial Resistance Surveillance in Europe 2014. Annual Report of the European Antimicrobial Resistance Surveillance Network (EARS-NET)*; ECDC: Stockholm, Sweden, 2015.

27. Livermore, D.M.; Wain, J. Revolutionising bacteriology to improve treatment outcomes and antibiotic stewardship. *Infect. Chemother.* **2013**, *45*, 1–10. [CrossRef] [PubMed]

28. Rolo, J.; Miragaia, M.; Turlej-Rogacka, A.; Empel, J.; Bouchami, O.; Faria, N.A.; Tavares, A.; Hryniewicz, W.; Fluit, A.C.; de Lencastre, H.; et al. High genetic diversity among community-associated *Staphylococcus aureus* in Europe: Results from a multicenter study. *PLoS ONE* **2012**, *7*, e34768. [CrossRef] [PubMed]

29. Tavares, A.; Faria, N.A.; de Lencastre, H.; Miragaia, M. Population structure of methicillin-susceptible *Staphylococcus aureus* (MSSA) in Portugal over a 19-year period (1192-2011). *Eur. J. Clin. Microbiol. Infect. Dis.* **2014**, *33*, 423–432. [CrossRef]

30. Sun, J.; Yang, M.; Sreevatsan, S.; Bender, J.B.; Singer, R.S.; Knutson, T.P.; Marthaler, D.G.; Davies, P.R. Longitudinal study of *Staphylococcus aureus* colonization and infection in a cohort of swine veterinarians in the United States. *BMC Infect. Dis.* **2017**, *17*, 690. [CrossRef]

31. David, M.Z.; Siegel, J.D.; Henderson, J.; Leos, G.; Lo, K.; Iwuora, J.; Taylor, A.R.; Zychowski, D.L.; Porsa, E.; Boyle-Vavra, S.; et al. Hand and nasal carriage of discordant *Staphylococcus aureus* isolates among urban jail detainees. *J. Clin. Microbiol.* **2014**, *52*, 3422–3425. [CrossRef]

32. Rasslan, O.S.E.; Khater, W.S.; Elnour, S.S.A.; Asaad, M.K. Fusidic acid resistance among *Staphylococcus aureus* causing community acquired skin and soft tissue infections. *Egypt J. Med. Lab. Sci.* **2016**, *25*, 21–31.

33. Liu, Y.; Geng, W.; Yang, Y.; Wang, C.; Zheng, Y.; Shang, Y.; Wu, D.; Li, X.; Wang, L.; Yu, S.; et al. Susceptibility to and resistance determinants of fusidic acid in *Staphylococcus aureus* isolated from Chinese children with skin and soft tissue infections. *FEMS Immunol. Med. Microbiol.* **2012**, *64*, 212–218. [CrossRef]

34. Bessa, G.R.; Quinto, V.P.; Machado, D.C.; Lipnharski, C.; Weber, M.B.; Bonamigo, R.R.; D'Azevedo, P.A. *Staphylococcus aureus* resistance to topical antimicrobials in atopic dermatitis. *Anais Bras. Dermatol.* **2016**, *91*, 604–610. [CrossRef]

35. Stein, M.; Komerska, J.; Prizade, M.; Sheinberg, B.; Tasher, D.; Somekh, E. Clindamycin resistance among *Staphylococcus aureus* strains in Israel: Implications for empirical treatment of skin and soft tissue infections. *Int. J. Infect. Dis.* **2016**, *46*, 18–21. [CrossRef]

36. Vicetti, M.C.P.; Mejias, A.; Leber, A.; Sanchez, P.J. A decade of antimicrobial resistance in *Staphylococcus aureus*: A single center experience. *PLoS ONE* **2019**, *14*, e0212029. [CrossRef]

37. Mclaws, F.B.; Larsen, A.R.; Skov, R.L.; Chopra, I.; O'Neill, A.J. Distribution of fusidic acid resistance determinants in methicillin-resistant *Staphylococcus aureus*. *Antimicrob. Agents Chemother.* **2011**, *55*, 1173–1176. [CrossRef]

38. Chen, H.J.; Hung, W.; Tseng, S.; Tsai, J.; Hsueh, P.; Teng, L. Fusidic acid resistance determinants in *Staphylococcus aureus* clinical isolates. *Antimicrob. Agents Chemother.* **2010**, *54*, 4985–4991. [CrossRef]

39. Besier, S.; Ludwig, A.; Brade, V.; Wichelhaus, T.A. Molecular analysis of fusidic acid resistance in *Staphylococcus aureus*. *Mol. Microbiol.* **2003**, *47*, 463–469. [CrossRef]

40. Castanheira, M.; Watters, A.A.; Bell, J.M.; Turnidge, J.; Jones, R.N. Fusidic acid resistance rates and prevalence of resistance mechanisms among *Staphylococcus* spp. isolated in North America and Australia, 2007–2008. *Antimicrob. Agents Chemother.* **2010**, *54*, 3614–3617. [CrossRef] [PubMed]

41. Lannergård, T.; Norstrom, T.; Hughes, D. Genetic determinants of resistance to fusidic acid among clinical bacteremia isolates of *Staphylococcus aureus*. *Antimicrob. Agents Chemother.* **2009**, *53*, 2059–2065. [CrossRef] [PubMed]

42. Ellington, M.J.; Reuterb, S.; Harrisb, S.R.; Holden, M.T.G.; Cartwright, E.J.; Greaves, D.; Gerver, S.M.; Hope, R.; Brown, N.M.; Török, M.E.; et al. Emergent and evolving antimicrobial resistance cassettes in community-associated fusidic acid and meticillin-resistant *Staphylococcus aureus*. *Int. J. Antimicrob. Agents* **2015**, *45*, 477–484. [CrossRef] [PubMed]

43. Baines, S.L.; Howden, B.P.; Heffernan, H.; Stinear, T.P.; Carter, G.P.; Seeman, T.; Kwong, J.C.; Ritchie, S.R.; Williamson, D.A. Rapid emergence and evolution of *Staphylococcus aureus* clones harboring *fusC*-containing staphylococcal cassette chromosome elements. *Antimicrob. Agents Chemother.* **2016**, *60*, 2359–2365. [CrossRef]

44. Schmitz, F.J.; Jones, M.E.; Hofmann, B.; Hansen, B.; Scheuring, S.; Luckefahr, M.; Fluit, A.; Verhoef, J.; Hadding, U.; Heinz, H.P.; et al. Characterization of *grlA*, *grlB*, *gyrA*, and *gyrB* Mutations in 116 unrelated isolates of *Staphylococcus aureus* and effects of mutations on ciprofloxacin MIC. *Antimicrob. Agents Chemother.* **1998**, *42*, 1249–1252. [CrossRef]

45. Fitzgibbon, J.E.; John, J.F.; Delucia, J.L.; Dubin, D.T. Topoisomerase mutations in trovafloxacin-resistant *Staphylococcus aureus*. *Antimicrob. Agents Chemother.* **1998**, *42*, 2122–2124. [CrossRef]

46. Costa, S.S.; Falcão, C.; Viveiros, M.; Machado, D.; Martins, M.; Melo-Cristino, J.; Amaral, L.; Couto, I. Exploring the contribution of efflux on the resistance to fluoroquinolones in clinical isolates of *Staphylococcus aureus*. *BMC Microbiol.* **2011**, *11*, 241. [CrossRef]

47. Sierra, J.M.; Marco, F.; Ruiz, J.; de Anta, M.T.J.; Vila, J. Correlation between the activity of different fluoroquinolones and the presence of mechanisms of quinolone resistance in epidemiologically related and unrelated strains of methicillin-susceptible and -resistant *Staphylococcus aureus*. *Clin. Microbiol. Infect.* **2002**, *8*, 781–790. [CrossRef]

48. Trong, H.N.; Prunier, A.L.; Leclercq, R. Hypermutable and fluoroquinolone-resistant clinical isolates of *Staphylococcus aureus*. *Antimicrob. Agents Chemother.* **2005**, *49*, 2098–2101. [CrossRef] [PubMed]

49. Sanfilippo, C.M.; Hesje, C.K.; Haas, W.; Morris, T.W. Topoisomerase mutations that are associated with high-level resistance to earlier fluoroquinolones in *Staphylococcus aureus* have less effect on the antibacterial activity of besifloxacin. *Chemotherapy* **2011**, *57*, 363–371. [CrossRef]

50. Marasa, B.S.; Iram, S.; Sung, k.; Kweon, O.; Cerniglia, C.E.; Khan, S. Molecular characterization of fluoroquinolone resistance of methicillin–resistant clinical *Staphylococcus aureus* isolates from Rawalpindi, Pakistan. *Med. Res. Arch.* **2015**, *2*. [CrossRef]

51. Fuzi, M.; Szabo, D.; Csercsik, R. Double-serine fluoroquinolone resistance mutations advance major international clones and lineages of various multi-drug resistant bacteria. *Front. Microbiol.* **2017**, *8*, 2261. [CrossRef]

52. Costa, S.S.; Palma, C.; Kladec, K.; Fessler, A.T.; Viveiros, M.; Melo-Cristino, J.; Schwarz, S.; Couto, I. Plasmid-borne antimicrobial resistance of *Staphylococcus aureus* isolated in a hospital in Lisbon, Portugal. *Microb. Drug Resist.* **2016**, *22*, 617–626. [CrossRef]

53. David, M.Z.; Daum, R.S. Community-associated methicillin-resistant *Staphylococcus aureus*: Epidemiology and clinical consequences of an emerging epidemic. *Clin. Microbiol. Rev.* **2010**, *23*, 616–687. [CrossRef]

54. Shallcross, L.J.; Fragaszy, E.; Johnson, A.M.; Hayward, A.C. The role of the Panton-Valentine leucocidin toxin in staphylococcal disease: A systematic review and meta-analysis. *Lancet Infect. Dis.* **2013**, *13*, 43–54. [CrossRef]

55. Castro, A.; Komora, N.; Ferreira, V.; Lira, A.; Mota, M.; Silva, J.; Teixeira, P. Prevalence of *Staphylococcus aureus* from nares and hands on health care professionals in a Portuguese Hospital. *J. Appl. Microbiol.* **2016**, *121*, 831–839. [CrossRef] [PubMed]

56. Mottola, C.; Semedo-Lemsaddek, T.; Mendes, J.J.; Melo-Cristino, J.; Tavares, L.; Cavaco-Silva, P.; Oliveira, M. Molecular typing, virulence traits and antimicrobial resistance of diabetic foot staphylococci. *J. Biomed. Sci.* **2016**, *23*, 33. [CrossRef]

57. Poulsen, A.B.; Skov, R.; Pallesen, L.V. Detection of methicillin resistance in coagulase-negative staphylococci and in staphylococci directly from simulated blood cultures using the EVIGENE MRSA Detection Kit. *J. Antimicrob. Chemother.* **2003**, *51*, 419–421. [CrossRef]

58. European Committee on Antimicrobial Susceptibility Testing. *Breakpoint Tables for Interpretation of MICs and Zone Diameters*; Version 10.0; 2020; Available online: http://www.eucast.org (accessed on 28 December 2020).

59. Clinical and Laboratory Standards Institute. *M100 Performance Standards for Antimicrobial Suscetibility Testing*, 30th ed.; Clinical and Laboratory Standards Institute: Annapolis, MD, USA, 2020.

60. Traczewski, M.M.; Brown, S.D. Proposed MIC and disk diffusion microbiological cutoffs and spectrum of activity of retapamulin, a novel topical antibiotic. *Antimicrob. Agents Chemother.* **2008**, *52*, 3863–3867. [CrossRef] [PubMed]

61. Tenover, F.C. Mechanisms of antimicrobial resistance in bacteria. *Am. J. Infect. Control* **2006**, *34*, S3–S10. [CrossRef]

62. Alexopoulou, K.; Foka, A.; Petinaki, E.; Jelastopulu, E.; Dimitracopoulos, G.; Spiliopoulou, I. Comparison of two commercial methods with PCR restriction fragment length polymorphism of the *tuf* gene in the identification of coagulase-negative staphylococci. *Lett. Appl. Microbiol.* **2006**, *43*, 450–454. [CrossRef]

63. Patel, D.; Kosmidis, C.; Seo, S.M.; Kaatz, G.W. Ethidium bromide MIC screening for enhanced efflux pump gene expression or efflux activity in *Staphylococcus aureus*. *Antimicrob. Agents Chemother.* **2010**, *54*, 5070–5073. [CrossRef]

64. Chung, M.; de Lencastre, H.; Matthews, P.; Tomasz, A.; Adamsson, I.; Aires-de-Sousa, M.; Camou, T.; Cocuzza, T.; Corso, A.; Couto, I.; et al. Molecular typing of methicillin-resistant *Staphylococcus aureus* by pulsed-field gel electrophoresis: Comparison of results obtained in a multilaboratory effort using identical protocols and MRSA strains. *Microb. Drug Resist.* **2000**, *6*, 189–198. [CrossRef]

65. Carriço, J.A.; Pinto, F.R.; Simas, C.; Nunes, S.; Sousa, N.G.; Frazão, N.; de Lencastre, H.; Almeida, J.S. Assessment of band-based similarity coefficients for automatic type and subtype classification of microbial isolates analyzed by pulsed-field gel electrophoresis. *J. Clin. Microbiol.* **2005**, *43*, 5483–5490. [CrossRef]

66. Carriço, J.A.; Silva-Costa, C.; Melo-Cristino, J.; Pinto, F.R.; de Lencastre, H.; Almeida, J.S.; Ramirez, M. Illustration of a common framework for relating multiple typing methods by application to macrolide-resistant *Streptococcus pyogenes*. *J. Clin. Microbiol.* **2006**, *44*, 2524–2532. [CrossRef]

67. Enright, M.C.; Day, N.P.; Davies, C.E.; Peacock, S.J.; Spratt, B.G. Multilocus sequence typing for characterization of methicillin-resistant and methicillin-susceptible clones of *Staphylococcus aureus*. *J. Clin. Microbiol.* **2000**, *38*, 1008–1015. [CrossRef]

68. Crisóstomo, M.I.; Westh, H.; Tomasz, A.; Chung, M.; Oliveira, D.C.; de Lencastre, H. The evolution of methicillin resistance in *Staphylococcus aureus*: Similarity of genetic backgrounds in historically early methicillin susceptible and resistant isolates and contemporary epidemic clones. *Proc. Natl. Acad. Sci. USA* **2001**, *98*, 9865–9870. [CrossRef]

69. Francisco, A.P.; Vaz, C.; Monteiro, P.T.; Melo-Cristino, J.; Ramirez, M.; Carriço, J.A. PHYLOViZ: Phylogenetic inference and data visualization for sequence based typing methods. *BMC Bioinform.* **2012**, *13*, 87. [CrossRef]
70. Lina, G.; Boutite, F.; Tristan, A.; Bes, M.; Etienne, J.; Vandenesch, F. Bacterial competition for human nasal cavity colonization: Role of Staphylococcal *agr* alleles. *Appl. Environ. Microbiol.* **2003**, *69*, 18–23. [CrossRef]

Article

Nisin Influence on the Antimicrobial Resistance Ability of Canine Oral Enterococci

Eva Cunha [1] , Rita Janela [1], Margarida Costa [1], Luís Tavares [1], Ana Salomé Veiga [2] and Manuela Oliveira [1,*]

[1] CIISA—Centro de Investigação Interdisciplinar em Sanidade Animal, Faculdade de Medicina Veterinária, Universidade de Lisboa, Av. da Universidade Técnica, 1300-477 Lisboa, Portugal; evacunha@fmv.ulisboa.pt (E.C.); ritajanela@msn.com (R.J.); margaridaxavicosta@hotmail.com (M.C.); ltavares@fmv.ulisboa.pt (L.T.)

[2] Instituto de Medicina Molecular, Faculdade de Medicina, Universidade de Lisboa, Avenida Professor Egas Moniz, 1649-028 Lisboa, Portugal; aveiga@medicina.ulisboa.pt

* Correspondence: moliveira@fmv.ulisboa.pt; Tel.: +351-213652800

Received: 20 November 2020; Accepted: 8 December 2020; Published: 10 December 2020

Abstract: Periodontal disease (PD) is one of the most common diseases in dogs. Although previous studies have shown the potential of the antimicrobial peptide nisin for PD control, there is no information regarding its influence in the development of antimicrobial resistance or horizontal gene transfer (HGT). Nisin's mutant prevention concentration (MPC) and selection window (MSW) were determined for a collection of canine oral enterococci. Isolates recovered after the determination of the MPC values were characterized for their antimicrobial profile and its nisin minimum inhibitory and bactericidal concentrations. The potential of *vanA* HGT between *Enterococcus faecium* CCGU36804 and nine clinical canine staphylococci and enterococci was evaluated. Nisin MPC values ranged from 400 to more than 600 μg/mL. In comparison with the original enterococci collection, the isolates recovered after the determination of the nisin MPC showed increased resistance towards amoxicillin/clavulanate (5%), vancomycin (5%), enrofloxacin (10%), gentamicin (10%) and imipenem (15%). The HGT of *vanA* gene was not observed. This work showed that nisin selective pressure may induce changes in the bacteria's antimicrobial resistance profile but does not influence horizontal transfer of *vanA* gene. To our knowledge, this is the first report of nisin's MPC and MSW determination regarding canine enterococci.

Keywords: nisin; mutant prevention concentration; mutant selection window; antimicrobial susceptibility testing; horizontal gene transfer

1. Introduction

Periodontal disease is one of the most widespread inflammatory diseases in dogs [1,2], that results from the establishment of a polymicrobial biofilm (dental plaque) on the teeth surface and a subsequent local inflammatory response [3]. Recently, we proposed the use of a new nisin biogel as a promising strategy to control this disease [4]. Nisin is an antimicrobial peptide, active mainly against Gram-positive bacteria, including multi-drug-resistant bacteria [4–7], with demonstrated potential for medical application [8]. However, considering that antimicrobial resistance is a major public health problem, any antimicrobial compound under investigation for clinical purposes should be characterized for the mechanisms responsible for resistance development and its environmental persistence [9,10]. Bacteria can become resistant to antimicrobial compounds by acquisition of resistance genes through horizontal gene transfer (HGT), but resistance can also result from the accumulation of mutations that decrease susceptibility [11,12]. Therapeutic protocols that favor mutant subpopulations may facilitate resistance development when compared with regimens that suppress mutant formation [11]. Thus, it is important to optimize the antimicrobial concentrations needed to prevent the selection and amplification of resistant mutants [13]. In this context, the mutant selection window (MSW) hypothesis, described by Zhao and Drlica, postulates that single-step resistant mutant subpopulations, although naturally present, are selectively enriched and amplified when drug concentrations fall within a specific range [14–16]. The MSW comprises a range of concentrations between the minimal inhibitory concentration (MIC) and the mutant prevention concentration (MPC) [11]. The MIC is the lowest drug concentration that inhibits the multiplication of the majority of susceptible cells, while MPC is the drug concentration that blocks the growth of the least susceptible, first step mutant, when a high inoculum is applied [11,14–19].

As described, resistance dissemination can also occur by HGT, which plays an important role in the emergence of new pathogens [12,20]. The dental plaque biofilm is a perfect environment for the transfer of resistance and virulence genes between bacteria [21]. Present in the oral cavity of dogs with PD, commensal enterococci have a high genome plasticity, being capable of acquiring, conserving, and disseminating genetic determinants, such as resistance genes, easily becoming opportunistic pathogens and being associated with PD-systemic consequences [22,23]. In fact, vancomycin resistance associated with the *vanA* gene is one of the most important antimicrobial resistance determinants associated with vancomycin-resistant *Enterococcus faecium*, considered by WHO as a high priority pathogen [23,24]. This gene is usually present in the Tn1546 transposon, harbored in a plasmid, being transferred by HGT [23,25,26]. Several studies demonstrated that vancomycin resistant enterococci (VRE) can transfer *vanA* to other bacteria, such as staphylococci, which are commensals of the skin and mucosa of animals and humans [23,26]. This transfer ability was associated by some authors to the presence of a pSK41-like plasmids in the recipients [27,28]. Furthermore, a continuous antimicrobial pressure due to the presence of sub inhibitory concentrations of antibiotics in the environment may contribute to the mobilization of acquired resistance genes, being essential to understand how the application of new antimicrobial compounds, such as antimicrobial peptides, can interfere with this phenomenon [26,29,30].

In this work, we determined the MSW of nisin from a previously characterized collection of enterococci obtained from the oral cavity of dogs with PD [31]. The isolates recovered after the determination of the MPC were collected and used to re-evaluate nisin's inhibitory (MIC) and bactericidal (MBC) concentrations, as well as their antimicrobial susceptibility profile against 11 antimicrobials relevant for veterinary medicine and public health. The influence of subinhibitory concentrations of nisin in the horizontal transfer of *vanA* gene from *Enterococcus faecium* to canine staphylococci and enterococci was also evaluated.

2. Results

2.1. Determination of the Nisin Mutant Prevention Concentration (MPC)

Nisin MPC values were determined regarding a previously characterized canine enterococci collection, obtained from the oral cavity of dogs with PD [31]. These values, in combination with the previously determined MIC values [1], allowed us to define the MSW of nisin towards the isolates under study. In addition, the relationship between these values, expressed as the ratio MPC/MIC, was determined, which can be used to compare antimicrobial agents for their ability to select resistant mutants [14].

It was possible to determine the MPC values for 85% ($n = 17$) of the strains used, with the exception of strains B28d, B29c and B32a, which presented an MPC higher than 600 µg/mL. The nisin MPC values for the 17 strains ranged from 400 to 600 µg/mL, with an average MPC of 447.06 ± 84.84 µg/mL (Table 1). Considering the nisin MPC/MIC, the resulting MPC values were 15 to 39 times higher than the previously determined MIC values [4].

Table 1. Minimum inhibitory concentration (MIC), mutant prevention concentration (MPC), and MIC/MPC ratio of nisin against the enterococci collection obtained from the oral cavity of dogs with periodontal disease (PD).

Isolates ID	MIC (μg/mL) [1]	MPC (μg/mL)	MPC/MIC Ratio
M2b	12.75	400	31
M2c	15.75	400	25
M3b	14.75	400	27
M3d	15.75	400	25
M4a	21.50	600	28
M4c	26.75	400	15
M15b	19.25	600	31
M15d	15.25	600	39
M21a	12.50	400	32
M21c	16.00	400	25
M23a	12.50	400	32
M23c	12.50	400	32
M25a	12.50	400	32
M25c	12.50	400	32
M28a	10.50	400	38
M28d	8.50	>600	-
M29b	12.50	400	32
M29c	12.50	>600	-
M32a	17.50	>600	-
M32b	16.25	600	37
Average	14.90	447.06	32
SD	4.10	84.84	-

ID—identification, MIC—minimum inhibitory concentration, MPC—mutant prevention concentration, SD—standard deviation.

2.2. Antimicrobial Susceptibility Testing

The antimicrobial susceptibility profiles of the original enterococci collection were compared with the ones obtained with the bacterial isolates recovered from the MPC protocol plates with the highest nisin concentrations. It was possible to observe that none of the isolates were susceptible to all the antibiotics tested, being in fact resistant to more than one compound. In the original enterococci collection, resistance levels ranged from 0% (imipenem and amoxicillin/clavulanate) to 100% (cefotaxime and gentamicin-10 μg), while for the isolates recovered from the MPC protocol, the resistance levels varied between 5% (amoxicillin/clavulanate) and 100% (cefotaxime and gentamicin−10 μg). When compared with the original isolates, the MPC recovered isolates presented an increased resistance towards amoxicillin/clavulanate (5%), vancomycin (5%), imipenem (15%), enrofloxacin (10%) and gentamicin-120 μg (10%) (Table 2).

According to the definitions proposed by Magiorakos and collaborators (2012) [32], which indicate that a multidrug-resistant *Enterococcus* spp. is non-susceptible to at least one agent in three or more antimicrobial categories, in our study, 15 isolates (75%) in the original collection and 18 isolates (90%) in the MPC recovered collection exhibited a multidrug-resistance profile.

Table 2. Representation of the resistance levels of the enterococci from the original collection, obtained from the oral cavity of dogs with PD, and the isolates recovered in the MPC protocol, determined by disc diffusion method according to Clinical and Laboratory Standards Institute (CLSI) guidelines.

Antibiotic	Resistant		Intermediate		Susceptible	
	Number of Original Isolates	Number of MPC Recovered Isolates	Number of Original Isolates	Number of MPC Recovered Isolates	Number of Original Isolates	Number of MPC Recovered Isolates
Ampicillin	3	3	0	0	17	17
Amoxicillin/clavulanate	0	1	3	2	17	17
Vancomycin	2	3	9	8	9	9
Imipenem	0	3	6	5	14	12
Cefotaxime	20	20	0	0	0	0
Enrofloxacin	16	18	4	2	0	0
Ciprofloxacin	11	11	9	9	0	0
Tetracycline	19	19	0	1	1	0
Doxycycline	17	17	2	2	1	1
Gentamicin 10 µg	20	20	0	0	0	0
Gentamicin 120 µg	4	6	0	0	16	14
Streptomycin	15	14	0	0	5	6

2.3. Determination of Nisin's Minimum Inhibitory Concentration (MIC) and Minimum Bactericidal Concentration (MBC)

The isolates recovered after determination of the nisin MPC values were used to evaluate the effect of nisin selective pressure on the nisin 's MIC and MBC values, by comparison with the values from the original collection [1].

MIC values of nisin regarding the isolates of the original collection and the isolates recovered in the MPC protocol are presented in Table 3.

Concerning the isolates recovered after the determination of the MPC, nisin MIC values ranged from 18.75 to 81.25 µg/mL, with an average value of 48.41 ± 21.62 µg/mL. MIC values were higher than 100 µg/mL for three isolates. MBC values ranged from 37.50 to 92.19 µg/mL, with an average value of 60.46 ± 19.40 µg/mL. An MBC value higher than 100 µg/mL was observed for eight isolates.

Nisin MIC results obtained against the MPC recovered isolates were higher and statistically different (p-value < 0.05) when compared with the nisin MIC values of the original collection. Concerning the MBC values, no statistical difference was observed between the results regarding the isolates from the two collections; however, most isolates recovered in the MPC protocol showed higher MBC values when compared with the original ones (65%, n = 13/20).

Table 3. MIC, MBC, and MBC/MIC ratio of nisin regarding the enterococci collection recovered after MPC protocol and the original enterococci collection obtained from the oral cavity of dogs with PD.

Isolates ID	MIC (µg/mL)		MBC (µg/mL)		MBC/MIC Ratio	
	MPC Recovered Isolates	Original Isolates [1]	MPC Recovered Isolates	Original Isolates [1]	MPC Recovered Isolates	Original Isolates
M2b	29.17	12.75	45.83	73.00	1.57	5.73
M2c	29.17	15.75	41.67	85.50	1.43	5.43
M3b	39.58	14.75	43.75	60.25	1.11	4.08
M3d	60.42	15.75	>100	82.25	-	5.22

Table 3. *Cont.*

Isolates ID	MIC (µg/mL)		MBC (µg/mL)		MBC/MIC Ratio	
	MPC Recovered Isolates	Original Isolates [1]	MPC Recovered Isolates	Original Isolates [1]	MPC Recovered Isolates	Original Isolates
M4a	>100	21.50	>100	98.50	-	4.58
M4c	>100	26.75	>100	>100	-	-
M15b	79.69	19.25	>100	77.00	-	4.00
M15d	>100	15.25	>100	86.50	-	5.67
M21a	76.56	12.50	>100	59.75	-	4.48
M21c	64.06	16.00	90.63	46.25	1.41	2.89
M23a	56.25	12.50	92.19	64.50	1.64	5.16
M23c	70.31	12.50	76.56	54.25	1.09	4.34
M25a	43.75	12.50	75.00	91.25	1.71	7.30
M25c	62.50	12.50	>100	72.25	-	5.78
M28a	81.25	10.50	>100	48.50	-	4.62
M28d	34.38	8.50	51.56	37.50	1.50	4.41
M29b	27.08	12.50	43.75	41.00	1.62	3.28
M29c	18.75	12.50	64.58	39.25	3.44	3.14
M32a	20.83	17.50	37.50	79.25	1.80	4.53
M32b	29.17	16.25	62.50	69.25	2.14	4.26
Average	48.41	14.90	60.46	66.63	1.71	4.70
SD	21.62	4.10	19.40	18.57	0.62	1.05

ID—identification, MIC—minimum inhibitory concentration, MBC—minimum bactericidal concentration, MPC—mutant prevention concentration, SD—standard deviation.

2.4. Nisin's Influence on vanA Horizontal Gene Transfer (HGT)

Bacterial acquired resistance by HGT is an important form of resistance dissemination [12]. The *vanA* gene is responsible for vancomycin resistance and its transfer between enterococci and staphylococci is well documented [25]. To evaluate nisin's influence in *vanA* transfer between enterococci and staphylococci, first we assessed the presence or absence of *vanA* in the isolates under study. In the case of staphylococci, we also evaluated the presence of *mecA* gene, associated with methicillin resistance, and of the pSK41-like plasmid, that may prompt *vanA* transfer [23,28].

In the initial PCR screening, none of the isolates from the oral enterococci collection (*n* = 20) presented the *vanA* gene. In addition, none of the six staphylococci obtained from canine skin lesions presented the *vanA* and *mecA* genes or the pSK41-like plasmid. Then, three isolates from the enterococci collection, M3b, M23a and M29b, were selected to participate in the HGT protocol, based on their strong capacity of biofilm production [31].

Afterwards, two mating rounds aiming to promote HGT of the *vanA* gene from *E. faecium* CCUG 36804 to nine clinical enterococci and staphylococci were performed. One round was performed in the absence of antimicrobial environmental pressure, and another in the presence of subinhibitory concentrations of nisin. None of the two mating experiments allowed the development of transconjugants in the MSA plates supplemented with rifampicin and vancomycin. All isolates recovered from the SBA and the MSA plates supplemented with rifampicin were submitted to PCR analysis, the results of which confirmed the absence of the *vanA* gene.

3. Discussion

Antimicrobial resistance is considered one of the major health treats of our time [10,24]. Misuse, overuse, and improper antimicrobial dosage promote a selective environmental pressure to bacteria, favoring resistance development [9,10]. Nisin, commonly used as a preservative in the food industry, is showing relevance in the biomedical field, being a promising antimicrobial agent to be used for the control of canine PD [4,33]. Despite the low resistance rate associated with this antimicrobial peptide, a few cases of nisin resistance have been reported, reinforcing the need to unveil related mechanisms, to evaluate its influence on dental plaque bacterial interaction, and to adopt correct doses to prevent the emergence and amplification of nisin resistant strains [8,34,35].

The MSW hypothesis allows us to define a range of concentrations that can promote mutant's development, which is useful to evaluate dose regimens [14]. In the present work, the MPC values of nisin against a collection of enterococci from the oral cavity of dogs with PD were determined. The values obtained were up to 39 times higher than the previously determined MIC values for the same isolates [4]. Similar results were also obtained for daptomycin against *Enterococcus faecalis*, with MPC values being 2 to 32 times the MIC values [19]. In fact, several authors have used this methodology to evaluate multiple antimicrobials and bacteria [36–43]. All these studies revealed high MPC values in comparison with the MIC values for the same microorganism; however, a high variation was observed between bacteria and drugs. For example, a MPC/MIC ratio of 48 to 72 was obtained for fosfomycin against *Escherichia coli*, while for *Pseudomonas aeruginosa* the ratio was 28 to 57; likewise, orbifloxacin presented a MPC/MIC ratio against *E. coli* of 4 to 32, while for *P. aeruginosa* the ratio was 16 to 64 [41,43]. In fact, Gianvecchio and collaborators (2019) suggested that MPC values present high variability for a given bacterial strain–antimicrobial combination, and should be understood as a range with confidence intervals, contrasting with MIC values [43,44].

To better understand the effect of nisin selective pressure over 72 h, as promoted in the determination of the MPC, antimicrobial susceptibility profiling along with nisin's MIC and MBC determination were performed on the isolates obtained in the MPC protocol and results were compared with the original collection.

Considering isolates' antimicrobial susceptibility profile, differences in the resistance profile of the isolates recovered after MPC protocol were observed when compared with those of the original isolates, specifically concerning amoxicillin/clavulanate, vancomycin, imipenem, enrofloxacin and gentamycin (120 μg). These results suggest a possible influence of nisin in increasing antimicrobial resistance. Cross resistance between nisin and antimicrobials is rare; however, there are some reports describing its occurrence [34,35,45]. Cross resistance may occur regarding antimicrobials that present a similar mode of action, or when the resistance mechanisms are related [35]. Nisin acts by binding to the lipid II, present in the bacterial membrane, which leads to pore formation and inhibition of peptidoglycan synthesis [8]. Considering that, parallel mechanisms may be observed in resistance to vancomycin, an antimicrobial that also acts on lipid II but in a different location, or in resistance to antimicrobials that act on the bacterial wall, such carbapenems or aminopenicillins [5,46]. Resistance to nisin is usually related to proteolytic degradation (by nisinase and nisin resistant protein); however, there are descriptions suggesting that resistance can also arise from mutations that induce changes in the membrane and cell wall composition, such as cell wall thickening, increased positive charges, the presence of penicilin binding proteins and modifications of membrane phospholipid and fatty acid composition [8,34,35,47]. Other nisin resistance mechanisms described so far are related to ABC transporters and multiple regulatory networks [34].

In addition, Drlica (2003) showed that the mutants derived from the MPC protocol are expected to develop mechanisms that inactivate the antimicrobial agent, including efflux or degradation systems [11]. These mechanisms may explain the increased resistance towards enrofloxacin and gentamicin, that act by inhibition of nucleic acid and protein synthesis, respectively [12].

Considering the MIC and MBC determinations, MIC values were higher and statistically different (p-value $\leq$ 0.05) towards all the recovered isolates in comparison with those of the original collection, while MBC values were higher regarding 65% of the recovered isolates in comparison with the originals. These results suggest that incubation in the presence of nisin leads to a reduction in the inhibitory activity, in spite of its bactericidal activity being maintained towards most isolates (60%, Table 3). According to Levinson and collaborators (2009), an MBC/MIC ratio lower than four indicates that the antimicrobial agent is bactericidal [48]. As such, nisin presented a bactericidal activity towards 42% of the isolates from the original collection [4]. On the other hand, nisin presented a bactericidal activity against all isolates recovered from the MPC protocol, except for isolates with nisin MBC values higher than 100 µg/mL.

Animals' oral cavities present a high bacterial concentration and diversity [21]. Located at the teeth surface, dental plaque is a highly complex polymicrobial biofilm where bacteria easily interact and act as reservoirs of transferable resistance genes [21]. *Enterococcus* spp. are known to be a central hub for resistance gene acquisition, conservation, and dissemination [23]. Classified by the WHO as a high priority pathogen, vancomycin-resistant *Enterococcus faecium*, along with other enterococcal species, are opportunistic pathogens frequently associated with nosocomial infections, and capable of transferring relevant genes to other bacterial species such as *Staphylococcus aureus*, *E. coli* and *Listeria* spp. [23,24,26]. In this work, a protocol aiming to promote the horizontal transfer of *vanA* from *E. faecium* to *Staphylococcus* spp. and *Enterococcus faecalis* clinical isolates was established. This gene is linked to vancomycin and teicoplanin resistance in enterococci, being harbored in a mobile genetic element, allowing its transfer to other bacteria [26]. Mating experiments performed in the absence of nisin selective pressure did not allow the transfer of *vanA* gene. Several studies demonstrated that *vanA* transfer may be facilitated by some molecules, such as pheromone-inducible surface proteins, or be related to the presence of specific plasmids, such as *S. aureus* pSK41 [28,49–51]. Although two of our enterococci recipients (M3b and M23a) were able to express an aggregation substance—more specifically, a pheromone-inducible surface protein that facilitates conjugative exchange [4,31,49,52]—no transfer occurred. None of the staphylococcal recipients presented the pSK41 plasmid, which may have influenced the results. Nevertheless, *vanA* gene horizontal transfer is a complex process which is not yet fully understood.

It is known that the use of antimicrobials can enhance gene transfer between bacteria [26]. In order to evaluate the influence of nisin in HGT, mating experiments were performed in the presence of this antimicrobial peptide at subinhibitory concentration [4]. None of the recipients presented the *vanA* gene after the mating experiments, reinforcing the potential of nisin to be used in the clinical setting, more precisely in veterinary medicine for canine PD control.

To our knowledge, this is the first report of nisin's MPC and MSW determination regarding canine enterococci and of its influence on gene transfer between enterococci and staphylococci. This approach is an important step in the development of new antimicrobial compounds, allowing to understand their potential influence in resistance evolution.

4. Materials and Methods

4.1. Bacterial Collection

A collection of 20 oral enterococci obtained from the oral cavity of dogs diagnosed with PD, previously characterized regarding clonality, antimicrobial resistance and virulence profiles, were used as bacterial models [31]. From these 20 isolates, 17 correspond to strains belonging to the species *Enterococcus faecalis*, and the remaining 3 to *Enterococcus faecium* [4].

For the HGT protocol, one *Staphylococcus aureus* and five *Staphylococcus pseudintermedius* obtained from canine skin lesions and an *Enterococcus faecium* reference strain (CCUG 36804, *vanA* positive) were used.

Enterococcus faecalis ATCC® 29212, *Staphylococcus aureus* ATCC® 25293, a *Staphylococcus aureus mecA* positive strain kindly provided by Dr. Birgit Strommenger, Robert Koch Institute, Germany, and a *Staphylococcus aureus* RN4220 pGO1 positive strain kindly provided by Dr. Alex O'Neill, University of Leeds, were included as controls.

4.2. Nisin Preparation

A nisin stock solution (1000 µg/mL) was obtained by dissolving 1 g of nisin powder (2.5% purity, 1000 IU/mg, Sigma-Aldrich, St. Louis, MO, USA) in 25 mL of HCl (0.02 M) (Merck, Darmstadt, Germany) [4]. Then, the stock solution was filtered using a 0.22 µm Millipore filter, and serial dilutions were prepared in distilled sterile water. Solutions were kept at 4 °C during the study.

4.3. Determination of the Nisin Mutant Prevention Concentration (MPC)

To determine the MPC of nisin against the canine oral enterococci collection [31], a modified version of the protocol described by Sinel and collaborators (2016) was performed [19]. Briefly, each isolate was spread onto three brain heart infusion (BHI) agar plates (VWR, Leuven, Belgium) and incubated for 24 h at 37 °C. Afterwards, all the bacterial lawn developed in the three BHI plates was resuspended in 450 µL of BHI broth and further incubated at 37 °C for 20 min, to achieve a bacterial suspension of 10^{10} CFU/mL, which was confirmed by viable cell count. Then, an aliquot of 50 µL of this bacterial suspension was inoculated onto Mueller–Hinton (MH) agar plates (Oxoid, Hampshire, UK), supplemented with two-fold concentration increments of nisin ranging from 6.25 to 40× the MIC value of 14.9 µg/mL [4]. Thus, the MH agar plates series contained 6.25, 15, 25, 50, 100, 200, 400, 600 µg/mL of nisin, previously determined. Plates were incubated at 37 °C for 72 h and observed daily for detection of colony growth.

MPC was defined as the lowest concentration of nisin that prevented the growth of any resistant mutant subpopulations after a 72 h incubation period [19]. It was also possible to establish the mutant selection window (MSW) of nisin for the collection of oral enterococci isolates from dogs, a value defined as the antimicrobial concentration ranging between the MIC and MPC values [11,15]. In addition, colonies grown in MH plates with the higher nisin concentration were isolated and kept at −80 °C in a solution of buffered peptone water with 20% glycerol. These isolates were classified as MPC recovered isolates and further used for antimicrobial profiling and for the determination of nisin's MIC and MBC.

4.4. Antimicrobial Susceptibility Testing

Antimicrobial susceptibility profiling was performed regarding the original clinical isolates and those recovered after the MPC protocol to determine if incubation in the presence of nisin interferes with the susceptibility profiles. Using the disk diffusion method, the susceptibility profile regarding a total of 11 different antibiotics, (MASTDISCS® AST, Mast Group, Liverpool, UK), presented in Table 4, was determined in accordance with the Clinical and Laboratory Standards Institute (CLSI) guidelines [53,54]. For that, a 0.5 MacFarland bacterial suspension was prepared for each isolate. Afterwards, the inoculum was evenly spread over the surface of a MH agar plate and the disks impregnated with the antimicrobial agents were placed over the surface of the agar plate. Plates were then incubated at 37 °C under aerobic conditions for 18 h or, in the case of vancomycin, 24 h. After incubation, the inhibition zone diameters were measured and compared with the CLSI standard breakpoints established in VETS01-S2 and M100S, allowing to define the antimicrobial profile (resistant, intermediate or susceptible) of each isolate regarding the antimicrobial agents tested [53,54]. Quality control was performed using the reference strain *Staphylococcus aureus* ATCC® 25293.

Antibiotics were selected based on their relevance to veterinary medicine, as well as to public health. Specifically, CN-120 µg and S-300 µg were included to detect high-level aminoglycoside resistance in *Enterococcus* spp., whereas IMI and VA were chosen due to their importance to public health [53–55].

Table 4. Antimicrobial agents used in the antimicrobial susceptibility test, grouped by mechanism of action, class and concentration [46].

Mechanism of Action	Antimicrobial Class	Antimicrobial Drug	Concentration (µg Per Disk)
Inhibition of cell-wall synthesis	Aminopenicillins	Ampicillin (AMP)	10
		Amoxicillin/Clavulanate * (AMX)	30
	Glycopeptides	Vancomycin (VA)	30
	Carbapenems	Imipenem (IMI)	10
	Cephalosporins	Cefotaxime (CTX)	30
Inhibition of nucleic acid synthesis	Fluoroquinolones	Enrofloxacin (ENR)	5
		Ciprofloxacin (CIP)	5
Inhibition of protein synthesis	Tetracyclines	Tetracycline (T)	30
		Doxycycline (DTX)	30
	Aminoglycosides	Gentamicin (CN)	10/120
		Streptomycin (S)	300

* Beta lactamase inhibitor.

4.5. Determination of Nisin's Minimum Inhibitory Concentration (MIC) and Minimum Bactericidal Concentration (MBC)

Nisin MIC determination was performed regarding the isolates recovered after the MPC protocol using the broth microdilution method, to assess their current susceptibility to nisin, as previously described by Cunha and collaborators (2018) [4]. Briefly, a 96-well microplate was filled with 20 µL of nisin solution at different concentrations (final nisin concentrations ranged from 1.25 to 100 µg/mL) and 180 µL of 10^6 CFU/mL bacterial suspensions of each isolate. A negative control with only tryptic soy broth (TSB) medium (VWR, Leuven, Belgium) and positive controls with bacterial suspensions were also included.

The 96-well microplates were incubated at 37 °C for 24 h, after which bacterial growth was visually assessed in order to determine MIC value. This parameter is defined as the lowest nisin concentration capable of preventing bacterial multiplication in vitro, with no visible growth on the well [4].

Subsequently, after MIC assessment, the MBC was determined. Five microliters of the bacterial suspension from each well with no visible growth were plated onto tryptic soy agar (TSA) plates (VWR, Leuven, Belgium), followed by incubation at 37 °C for 24 h. MBC was defined as the lowest antimicrobial concentration that inhibits bacterial growth after sub-culture of the suspensions on solid unselective media without any antimicrobial agent [6].

These assays were performed in triplicate, on independent days, and 10% of replicates were tested to assure results representability.

4.6. Nisin's Influence on vanA Horizontal Gene Transfer (HGT)

4.6.1. DNA Extraction and Isolates PCR Screening

DNA extraction was performed based on the protocol described by Semedo-Lemsaddek et al. (2016) and Mottola et al. (2016) [22,56]. Then, all canine staphylococci and enterococci were evaluated by multiplex PCR, in order to identify the presence of the gene *vanA* and *mecA* [57]. Two pairs of primers synthesized by STABVIDA® (Lisbon, Portugal), targeting *vanA* (5′ GGGAAAACGACAATTGC 3′ and 3′ GTACAATGCGGCCGTTA 5′) and *mecA* (5′ TCCAGATTACAACTTCACCAGG 3′ and 3′CCACTTCATATCTTGTAACG 5′) were used [56,57]. The PCR mixture had a final volume of 28.5 µL, with 10 µL of Supreme NZYTaq 2× Green Master Mix (NZYtech®, Lisbon, Portugal), 0.5 uM of the *vanA* primer, 0.4 uM of the *mecA* primer, 16.88 µL of PCR-grade water and 5 µL of DNA template.

PCR amplification was completed using the conditions: initial denaturation at 94 °C for 4 min; 10 cycles involving denaturation at 94 °C for 30 s, annealing at 64 °C for 30 s and elongation at 72 °C for 45 s; 25 cycles involving denaturation at 94 °C for 30 s, annealing at 50 °C for 45 s and elongation at 72 °C for 2 min, and a final extension step at 72 °C for 10 min. Electrophoresis (90 V for 45 min) was performed to evaluate the amplified products, using a 1.5% agarose gel (NZYtech®, Lisbon, Portugal) stained with GreenSafe (NZYtech®, Lisbon, Portugal). A molecular weight marker, NZYDNA ladder VI (NZYtech®, Lisbon, Portugal) was also included. Results were visualized by transillumination.

Two positive control strains, *Staphylococcus aureus* 01-00694 (*mecA* positive) and *Enterococcus faecium* CCUG 36,804 (*vanA* positive), were included [56].

In addition, the presence of the pSK41-like plasmid in the 6 canine staphylococci under study was evaluated by PCR, using a primer targeting the *traE* gene (5′ ACAAATGCGTA CTACAGACCCTAAACGA 3′ and 3′GCCCTGCTGTTGCTGTATCCATATT 5′), synthesized by STABVIDA® [28,58].

A PCR mixture composed by 10 µL of Supreme NZYTaq 2× Green Master Mix (NZYtech®, Lisbon, Portugal), 0.4 uM of *traE* primer; 39.2 µL of PCR-grade water and 5 µL of DNA template was used. The PCR amplification was completed using the following conditions: initial denaturation at 94 °C for 2 min; 30 cycles involving denaturation at 95 °C for 15 s, annealing at 55 °C for 90 s and elongation at 72 °C for 90 s, and a final extension step at 72 °C for 7 min [28].

An electrophoresis (90 V for 45 min) was performed to evaluate the amplified products, using a 1.5% agarose gel (NZYtech®, Lisbon, Portugal) stained with GreenSafe (NZYtech®, Lisbon, Portugal). A molecular weight marker, NZYDNA ladder VII (NZYtech®, Lisbon, Portugal), was also included. Results were visualized by transillumination.

A positive control strain, *Staphylococcus aureus* RN4220 (pGO1 positive), was included in the PCR amplification protocol [59].

4.6.2. HGT Protocol

To test if selective pressure due to the presence of subinhibitory concentrations of nisin induces HGT, a protocol adapted from Niederhäusern and collaborators (2011) was developed [25]. Mating experiments were performed in two rounds, using the VRE rifampicin susceptible (Van^r Rif^s) *Enterococcus faecium* CCUG 36804 strain, as donor of the *vanA* gene, and as recipients the 6 canine staphylococci from canine skin lesions, and 3 canine *Enterococcus faecalis* isolates from our collection of enterococci from the oral cavity of dogs with PD, selected according to their strong biofilm forming ability. All recipients were susceptible to vancomycin, and rifampicin resistance was induced (Van^s Rif^r) [25], as it is associated with a point mutation rather than to an acquired gene [60]. After performing a 0.5 MacFarland suspension in 0.9% NaCl for each isolate, 500 µL of the donor suspension and 500 µL of the suspension of one of the recipients were inoculated into 5 mL of TSB and incubated at 37 °C for 18 h. After incubation, 1 mL of the dual bacterial suspension was added to 5 mL of TSB and further incubated for 6 h at 37 °C. Afterwards, 2 mL of each dual suspension were inoculated in TSA plates and incubated for 5 h at 37 °C in a slight movement on a shaker, to promote mating. Then, plates were incubated at 37 °C for 24 h, after which the bacterial suspension that remained at the surface of the agar plates was removed and inoculated in 5 mL of TSB. After an incubation period of 12 h at 37 °C, 100 µL of the suspension was inoculated in Mannitol Salt agar (MSA, PanReac AppliChem, Barcelona, Spain) supplemented with rifampicin (64 µg/mL, PanReac AppliChem, Barcelona, Spain) and vancomycin (8 µg/mL, PanReac AppliChem, Barcelona, Spain), to select for transconjugants. If mating occurred, recombinant isolates developed should be resistant to rifampicin and vancomycin. In addition, the suspension was also inoculated in Manitol Salt Agar (MSA, PanReac AppliChem, Barcelona, Spain) and Slanetz and Bartley agar (SBA, PanReac AppliChem, Barcelona, Spain) supplemented only with rifampicin (64 µg/mL).

The second mating round was performed in the presence of nisin, with all the media used being supplemented with nisin at sub-MIC concentration, 7.45 µg/mL for enterococci [4] and 5.63 µg/mL for staphylococci [6]. All recovered isolates and transconjugants recovered from the supplemented media were submitted to a PCR analysis to detect the presence of the *vanA* gene.

4.7. Statistical Analysis

Data statistical analysis was carried out using Microsoft Excel 2016®. All quantitative data are expressed as means ± standard deviation. Student's *t*-test was used for statistical analysis of the nisin MIC and MBC values regarding the original collection and the collection recovered after MPC protocol. A confidence interval of 95% was considered, with a *p*-value ≤ 0.05 indicating statistical significance.

5. Conclusions

Periodontal disease is a highly prevalent inflammatory disease in dogs, and nisin might be a promising molecule for its control. The study of nisin influence on mutant's development, antimicrobial signatures and transfer of resistance determinants revealed that this compound can influence isolates antimicrobial profiles. MPC and MSW determinations can be an interesting measure to establish more accurate treatment protocols based on appropriate antimicrobial doses. However, the utility of the MSW in the definition of dose regimens must be demonstrated not only in vitro but also in vivo. In addition, this study showed that nisin did not promote horizontal transfer of the *vanA* gene between the isolates tested, which emphasizes its potential to be used in PD control. To our knowledge, this is the first report of nisin's MPC and MSW determination regarding canine enterococci, being a relevant step towards its application in both human and veterinary medicine.

Author Contributions: Conceptualization, M.O., A.S.V. and E.C.; methodology, E.C., M.O., R.J. and M.C.; software, E.C. and R.J.; validation, M.O., L.T. and A.S.V.; formal analysis, E.C.; investigation, E.C. and M.C.; data curation, M.O.; writing—original draft preparation, E.C.; writing—review and editing, M.O., L.T. and A.S.V.; visualization, E.C.; supervision, M.O. and A.S.V.; project administration, M.O.; funding acquisition, M.O. and E.C. All authors have read and agreed to the published version of the manuscript.

Funding: Authors would like to acknowledge the Foundation for Science and Technology (Eva Cunha PhD fellowship SFRH/BD/131384/2017) and to CIISA–Centro de Investigação Interdisciplinar em Sanidade Animal, Faculdade de Medicina Veterinária, Universidade de Lisboa, Project UIDB/00276/2020 (Funded by FCT), for financial support of this work.

Conflicts of Interest: The authors declare no conflict of interest.

References

1. Riggio, M.P.; Lennon, A.; Taylor, D.J.; Bennett, D. Molecular identification of bacteria associated with canine periodontal disease. *Vet. Microbiol* **2011**, *150*, 394–400. [CrossRef] [PubMed]
2. Marshall, M.; Wallis, C.; Milella, L.; Colyer, A.; Tweedie, A.; Harris, S. A longitudinal assessment of periodontal disease in 52 miniature schnauzers. *BMC Vet. Res.* **2014**, *10*, 1–13. [CrossRef] [PubMed]
3. Niemiec, B.A. Periodontal disease. *Top. Companion Anim. M* **2008**, *23*, 72–80. [CrossRef] [PubMed]
4. Cunha, E.; Trovão, T.; Pinheiro, A.; Nunes, T.; Santos, R.; Moreira da Silva, J.; São Braz, B.; Tavares, L.; Veiga, A.S.; Oliveira, M. Potential of two delivery systems for nisin topical application to dental plaque biofilms in dogs. *BMC Vet. Res.* **2018**, *14*, 375. [CrossRef] [PubMed]
5. Field, D.; Cotter, P.D.; Ross, R.P.; Hill, C. Bioengineering of the model lantibiotic nisin. *Bioengineered* **2015**, *6*, 187–192. [CrossRef] [PubMed]
6. Santos, R.; Gomes, D.; Macedo, H.; Barros, D.; Tibério, C.; Veiga, A.S.; Tavares, L.; Castanho, M.; Oliveira, M. Guar gum as a new antimicrobial peptide delivery system against diabetic foot ulcers *Staphylococcus aureus* isolates. *J. Med. Microbiol.* **2016**, *65*, 1092–1099. [CrossRef]
7. Ahmad, V.; Khan, M.S.; Jamal, Q.M.S.; Alzohairy, M.A.; Al Karaawi, M.A.; Siddiqui, U.M. Antimicrobial potential of bacteriocins: In therapy, agriculture and food preservation. *Int. J. Antimicrob. Agents* **2017**, *49*, 1–11. [CrossRef]
8. Shin, J.M.; Gwak, J.W.; Kamarajan, P.; Fenno, J.C.; Rickard, A.H.; Kapila, Y.L. Biomedical applications of nisin. *J. Appl. Microbiol.* **2015**, *120*, 1449–1465. [CrossRef]

9. Köck, R.; Kreienbrock, L.; van Duijkeren, E.; Schwarz, S. Antimicrobial resistance at the interface of human and veterinary medicine. *Vet. Microbiol.* **2017**, *200*, 1–5. [CrossRef]

10. Dadgostar, P. Antimicrobial Resistance: Implications and Costs. *Infect. Drug Resist.* **2019**, *12*, 3903–3910. [CrossRef]

11. Drlica, K. The mutant selection window and antimicrobial resistance. *J. Antimicrob. Chemother.* **2003**, *52*, 11–17. [CrossRef] [PubMed]

12. Munita, J.M.; Arias, C.A. Mechanisms of Antibiotic Resistance. *Microbiol. Spectr.* **2016**, *4*. [CrossRef] [PubMed]

13. Pasquali, F.; Manfreda, G. Mutant prevention concentration of ciprofloxacin and enrofloxacin against *Escherichia coli*, *Salmonella typhimurium* and *Pseudomonas aeruginosa*. *Vet. Microbiol.* **2007**, *119*, 304–310. [CrossRef] [PubMed]

14. Zhao, X.; Drlica, K. Restricting the selection of antibiotic-resistant mutants: A general strategy derived from fluoroquinolone studies. *Clin. Infect. Dis* **2001**, *15*, 147–156. [CrossRef]

15. Zhao, X.; Drlica, K. Restricting the selection of antibiotic-resistant mutant bacteria: Measurement and potential use of the Mutant Selection Window. *J. Infect. Dis.* **2002**, *185*, 561–565. [CrossRef]

16. Drlica, K.; Zhao, X. Mutant selection window hypothesis updated. *Clin. Infect. Dis.* **2007**, *44*, 681–688. [CrossRef]

17. Dong, Y.; Zhao, X.; Domagala, J.; Drlica, K. Effect of fluoroquinolone concentration on selection of resistant mutants of *Mycobacterium bovis* BCG and *Staphylococcus aureus*. *Antimicrob. Agents Chemother.* **1999**, *43*, 1756–1758. [CrossRef]

18. Firsov, A.A.; Vostrov, S.N.; Lubenko, I.Y.; Drlica, K.; Portnoy, Y.A.; Zinner, S.H. In vitro pharmacodynamic evaluation of the mutant selection window hypothesis using four fluoroquinolones against *Staphylococcus aureus*. *Antimicrob. Agents Chemother.* **2003**, *47*, 1604–1613. [CrossRef]

19. Sinel, C.; Jaussaud, C.; Auzou, M.; Giard, J.; Cattoir, V. Mutant prevention concentrations of daptomycin for *Enterococcus faecium* clinical isolates. *Int. J. Antimicrob. Agents* **2016**, *48*, 449–452. [CrossRef]

20. Schiwon, K.; Arends, K.; Rogowski, K.M.; Fürch, S.; Prescha, K.; Sakinc, T.; Van Houdt, R.; Werner, G.; Grohmann, E. Comparison of antibiotic resistance, biofilm formation and conjugative transfer of *Staphylococcus* and *Enterococcus* isolates from international space station and antarctic research station Concordia. *Microb. Ecol.* **2013**, *65*, 638–651. [CrossRef]

21. Roberts, A.; Mullany, P. Oral biofilms: A reservoir of transferable, bacterial, antimicrobial resistance. *Expert Rev. Anti. Infect.* **2010**, *8*, 1441–1450. [CrossRef] [PubMed]

22. Semedo-Lemsaddek, T.; Tavares, M.; Braz, B.S.; Tavares, L.; Oliveira, M. Enterococcal infective endocarditis following periodontal disease in dogs. *PLoS ONE* **2016**, *11*, 1–6. [CrossRef] [PubMed]

23. Werner, G.; Coque, T.M.; Franz, C.; Grohmann, E.; Hegstad, K.; Jensen, L.; van Schaik, W.; Weaver, K. Antibiotic resistant enterococci—Tales of a drug resistance gene trafficker. *Int. J. Med. Microbiol.* **2013**, *303*, 360–379. [CrossRef] [PubMed]

24. World Health Organization (WHO). Global Priority List of Antibiotic-Resistant Bacteria to Guide Research, Discovery, and Development of New Antibiotics. 2017. Available online: https://www.who.int/medicines/publications/WHO-PPL-Short_Summary_25Feb-ET_NM_WHO.pdf (accessed on 10 May 2019).

25. Niederhäusern, S.; Bondi, M.; Messi, P.; Iseppi, R.; Sabia, C.; Manicardi, G.; Anacarso, I. Vancomycin-resistance transferability from VanA enterococci to *Staphylococcus aureus*. *Curr. Microbiol.* **2011**, *62*, 1363–1367. [CrossRef]

26. Sparo, M.; Delpech, G.; Allende, N.G. Impact on public health of the spread of high-level resistance to gentamicin and vancomycin in enterococci. *Front. Microbiol.* **2018**, *9*. [CrossRef]

27. Palmer, K.L.; Kos, V.N.; Gilmore, M.S. Horizontal gene transfer and the genomics of enterococcal antibiotic resistance. *Curr. Opin. Microbiol.* **2010**, *13*, 632–639. [CrossRef]

28. Zhu, W.; Clark, N.; Patel, J.B. pSK41-like plasmid is necessary for Inc18-like *vanA* plasmid transfer from *Enterococcus faecalis* to *Staphylococcus aureus* in vitro. *Antimicrob. Agents Chemother.* **2013**, *57*, 212–219. [CrossRef]

29. Kataoka, Y.; Ito, C.; Kawashima, A.; Ishii, M.; Yamashiro, S.; Harada, K.; Ochi, H.; Sawada, T. Identification and antimicrobial susceptibility of Enterococci isolated from dogs and cats subjected to differing antibiotic pressures. *J. Vet. Med. Sci.* **2013**, *75*, 749–753. [CrossRef]

30. Argudín, M.A.; Deplano, A.; Meghraoui, A.; Dodémont, M.; Heinrichs, A.; Denis, O.; Nonhoff, C.; Roisin, S. Bacteria from animals as a pool of antimicrobial resistance genes. *Antibiotics* **2017**, *6*, 12. [CrossRef]

31. Oliveira, M.; Tavares, M.; Gomes, D.; Touret, T.; São Braz, B.; Tavares, L.; Semedo-Lemsaddek, T. Virulence traits and antibiotic resistance among enterococci isolated from dogs with periodontal disease. *Comp. Immunol. Microbiol. Infect. Dis.* **2016**, *46*, 27–31. [CrossRef]

32. Magiorakos, A.; Srinivasan, A.; Carey, R.; Carmeli, Y.; Falagas, M.; Giske, C.; Harbarth, S.; Hindler, J.F.; Kahlmeter, G.; Olsson-Liljequist, B.; et al. Multidrug-resistant, extensively drug-resistant and pandrug-resistant bacteria: An international expert proposal for interim standard definitions for acquired resistance. *Clin. Microbiol. Infect.* **2012**, *18*, 268–281. [CrossRef] [PubMed]

33. Howell, T.H.; Fiorellini, J.P.; Blackburn, P.; Projan, S.J.; de la Harpe, J.; Williams, R.C. The effect of a mouthrinse based on nisin, a bacteriocin, on developing plaque and gingivitis in beagle dogs. *J. Clin. Periodontol.* **1993**, *20*, 335–339. [CrossRef] [PubMed]

34. Zhou, H.; Fang, J.; Tian, Y.; Yang Lu, X. Mechanisms of nisin resistance in Gram-positive bacteria. *Ann. Microbiol.* **2014**, *64*, 413–420. [CrossRef]

35. Draper, L.; Cotter, P.; Hill, C.; Ross, R. Lantibiotic resistance. *Microbiol Mol. Biol. R* **2015**, *79*, 171–191. [CrossRef] [PubMed]

36. Hansen, G.T.; Zhaob, X.; Drlica, K.; Blondeau, J.M. Mutant prevention concentration for ciprofloxacin and levofloxacin with *Pseudomonas aeruginosa*. *Int. J. Antimicrob. Agents* **2006**, *27*, 120–124. [CrossRef]

37. Cui, J.; Liu, Y.; Chen, L. Mutant prevention concentration of tigecycline for carbapenem-susceptible and -resistant *Acinetobacter baumannii*. *J. Antibiot.* **2010**, *63*, 29–31. [CrossRef]

38. Gebru, E.; Choi, M.J.; Lee, S.J.; Damte, D.; Park, S.C. Mutant-prevention concentration and mechanism of resistance in clinical isolates and enrofloxacin/marbofloxacin-selected mutants of *Escherichia coli* of canine origin. *J. Med. Microbiol.* **2011**, *60*, 1512–1522. [CrossRef]

39. Balaje, R.; Sidhu, P.; Kaur, G.; Rampal, S. Mutant prevention concentration and PK–PD relationships of enrofloxacin for *Pasteurella multocida* in buffalo calves. *Res. Vet. Sci.* **2013**, *95*, 1114–1124. [CrossRef]

40. Berghaus, L.J.; Giguère, S.; Guldbech, K. Mutant prevention concentration and mutant selection window for 10 antimicrobial agents against *Rhodococcus equi*. *Vet. Microbiol.* **2013**, *166*, 670–675. [CrossRef]

41. Shimizu, T.; Harada, K.; Kataoka, Y. Mutant prevention concentration of orbifloxacin: Comparison between *Escherichia coli*, *Pseudomonas aeruginosa*, and *Staphylococcus pseudintermedius* of canine origin. *Acta Vet. Scand.* **2013**, *55*, 37. [CrossRef]

42. Hesje, C.; Drlica, K.; Blondeau, J. Mutant prevention concentration of tigecycline for clinical isolates of *Streptococcus pneumoniae* and *Staphylococcus aureus*. *J. Antimicrob. Chemother.* **2015**, *70*, 494–497. [CrossRef] [PubMed]

43. Pan, A.; Mei, Q.; Ye, Y.; Li, H.; Liu, B.; Li, J. Validation of the mutant selection window hypothesis with fosfomycin against *Escherichia coli* and *Pseudomonas aeruginosa*: An in vitro and in vivo comparative study. *J. Antibiot.* **2017**, *70*, 166–173. [CrossRef] [PubMed]

44. Gianvecchio, C.; Lozano, N.A.; Henderson, C.; Kalhori, P.; Bullivant, A.; Valencia, A.; Su, L.; Bello, G.; Wong, M.; Cook, E.; et al. Variation in mutant prevention concentrations. *Front. Microbiol.* **2019**, *10*. [CrossRef] [PubMed]

45. Martínez, B.; Rodríguez, A. Antimicrobial susceptibility of nisin resistant *Listeria monocytogenes* of dairy origin. *FEMS Microbiol. Lett.* **2005**, *252*, 67–72. [CrossRef]

46. Plumb, D.C. *Plumbs Veterinary Drug Handbook*, 7th ed.; Blackwell Publishing: Ames, IA, USA, 2011.

47. Field, D.; Blake, T.; Mathur, H.; O' Connor, P.M.; Cotter, P.D.; Ross, R.P.; Hill, C. Bioengineering nisin to overcome the nisin resistance protein. *Mol. Microbiol.* **2019**, *111*, 717–731. [CrossRef] [PubMed]

48. Levison, M.E.; Levison, J.H. Pharmacokinetics and Pharmacodynamics of antibacterial agents. *Infect. Dis. Clin. North Am.* **2009**, *23*, 791–819. [CrossRef]

49. Paoletti, C.; Foglia, G.; Princivalli, M.S.; Magi, G.; Guaglianone, E.; Donelli, G.; Pruzzo, C.; Biavasco, F.; Facinelli, B. Co-transfer of *vanA* and aggregation substance genes from *Enterococcus faecalis* isolates in intra- and interspecies matings. *J. Antimicrob. Chemother.* **2007**, *59*, 1005–1009. [CrossRef]

50. Clewell, D.B.; Weaver, K.E.; Dunny, G.M.; Coque, T.M.; Francia, M.V.; Hayes, F. Extrachromosomal and mobile elements in enterococci: Transmission, maintenance, and epidemiology. In *Enterococci from Commensals to Leading Causes of Drug Resistant Infection*; Gilmore, M.S., Clewell, D.B., Ike, Y., Shankar, N., Eds.; Eye and Ear Infirmary: Boston, MA, USA, 2014.

51. Terra, M.R.; Tosoni, N.F.; Furlaneto, M.C.; Furlaneto-Maia, L. Assessment of vancomycin resistance transfer among enterococci of clinical importance in milk matrix. *J. Environ. Sci. Health Part B* **2019**. [CrossRef]

52. Eaton, J.T.; Gasson, M.J. Molecular screening of *Enterococcus* virulence determinants and potential for genetic exchange between food and medical isolates. *Appl. Environ. Microbiol.* **2001**, *67*, 1628–1635. [CrossRef]

53. Clinical and Laboratory Standards Institute (CLSI). *Performance Standards for Antimicrobial Disk and Dilution Susceptibility Tests for Bacteria Isolated from Animals: Second Informational Supplement VET01-S2*; Clinical and Laboratory Standards Institute (CLSI): Wayne, PA, USA, 2013.

54. Clinical and Laboratory Standards Institute (CLSI). *Performance Standards for Antimicrobial Susceptibility Tests: M100S*, 26th ed.; Clinical and Laboratory Standards Institute (CLSI): Wayne, PA, USA, 2016.

55. Davido, B.; Moussiegt, A.; Dinh, A.; Bouchand, F.; Matt, M.; Senard, O.; Deconinck, L.; Espinasse, F.; Lawrence, C.; Fortineau, N.; et al. Germs of thrones—Spontaneous decolonization of Carbapenem-Resistant *Enterobacteriaceae* (CRE) and Vancomycin-Resistant Enterococci (VRE) in Western Europe: Is this myth or reality? *Antimicrob. Resist. Infect. Control* **2018**, *7*. [CrossRef]

56. Mottola, C.; Matias, C.S.; Mendes, J.J.; Melo-Cristino, J.; Tavares, L.; Cavaco-Silva, P.; Oliveira, M. Susceptibility patterns of *Staphylococcus aureus* biofilms in diabetic foot infections. *BMC Microbiol.* **2016**, *16*, 119. [CrossRef] [PubMed]

57. Ramos-Trujillo, E.; Pérez-Roth, E.; Méndez-Alvarez, S.; Claverie-Martín, F. Multiplex PCR for simultaneous detection of enterococcal genes *vanA* and *vanB* and staphylococcal genes *mecA*, *ileS-2* and *femB*. *Int. Microbiol.* **2003**, *6*, 113–115. [CrossRef] [PubMed]

58. Albrecht, V.S.; Zervos, M.J.; Kaye, K.S.; Tosh, P.K.; Arshad, S.; Hayakawa, K.; Kallen, A.J.; McDougal, L.K.; Limbago, B.M.; Guh, A.Y. Prevalence of and risk factors for vancomycin-resistant *Staphylococcus aureus* precursor organisms in southeastern Michigan. *Infect. Control. Hosp. Epidemiol.* **2014**, *35*, 1531–1534. [CrossRef] [PubMed]

59. Caryl, J.A.; O'Neill, A.J. Complete nucleotide sequence of pGO1, the prototype conjugative plasmid from the staphylococci. *Plasmid* **2009**, *62*, 35–38. [CrossRef]

60. Conwell, M.; Daniels, V.; Naughton, P.J.; Dooley, J.S.G. Interspecies transfer of vancomycin, erythromycin and tetracycline resistance among *Enterococcus* species recovered from agrarian sources. *BMC Microbiol.* **2017**, *17*, 19. [CrossRef]

Publisher's Note: MDPI stays neutral with regard to jurisdictional claims in published maps and institutional affiliations.

Article

Salmonella spp. in Pet Reptiles in Portugal: Prevalence and Chlorhexidine Gluconate Antimicrobial Efficacy

João B. Cota *, Ana C. Carvalho, Inês Dias, Ana Reisinho, Fernando Bernardo and Manuela Oliveira

CIISA–Centro de Investigação Interdisciplinar em Sanidade Animal, Faculdade de Medicina Veterinária, Universidade de Lisboa, Av. da Universidade Técnica, 1300-477 Lisboa, Portugal; accarvalho@fmv.ulisboa.pt (A.C.C.); ines.adpdias@gmail.com (I.D.); anareisinho@fmv.ulisboa.pt (A.R.); fbernardo@fmv.ulisboa.pt (F.B.); moliveira@fmv.ulisboa.pt (M.O.)
* Correspondence: joaobcota@fmv.ulisboa.pt; Tel.: +351-213-652-800

Abstract: A fraction of human *Salmonella* infections is associated with direct contact with reptiles, yet the number of reptile-associated Salmonellosis cases are believed to be underestimated. Existing data on *Salmonella* spp. transmission by reptiles in Portugal is extremely scarce. The aim of the present work was to evaluate the prevalence of *Salmonella* spp. in pet reptiles (snakes, turtles, and lizards), as well as evaluate the isolates' antimicrobial resistance and virulence profiles, including their ability to form biofilm in the air-liquid interface. Additionally, the antimicrobial effect of chlorhexidine gluconate on the isolates was tested. *Salmonella* was isolated in 41% of the animals sampled and isolates revealed low levels of antimicrobial resistance. Hemolytic and lypolytic phenotypes were detected in all isolates. The majority (90.63%) of the *Salmonella* isolates were positive for the formation of pellicle in the air-liquid interface. Results indicate chlorhexidine gluconate is an effective antimicrobial agent, against the isolates in both their planktonic and biofilm forms, demonstrating a bactericidal effect in 84.37% of the *Salmonella* isolates. This study highlights the possible role of pet reptiles in the transmission of non-typhoidal *Salmonella* to humans, a serious and increasingly relevant route of exposure in the *Salmonella* public health framework.

Keywords: *Salmonella*; reptiles; isolation; antimicrobial resistance; biofilms; chlorhexidine gluconate; public health

Citation: Cota, J.B.; Carvalho, A.C.; Dias, I.; Reisinho, A.; Bernardo, F.; Oliveira, M. *Salmonella* spp. in Pet Reptiles in Portugal: Prevalence and Chlorhexidine Gluconate Antimicrobial Efficacy. *Antibiotics* **2021**, *10*, 324. https://doi.org/10.3390/antibiotics10030324

Academic Editor: Jonathan Frye

Received: 27 February 2021
Accepted: 17 March 2021
Published: 19 March 2021

Publisher's Note: MDPI stays neutral with regard to jurisdictional claims in published maps and institutional affiliations.

1. Introduction

Salmonella is a well-known food-borne illness etiological agent, reported as the second most common zoonotic agent, causing 91,857 confirmed cases of disease in the European Union during 2018 [1] and an estimated number of 93.8 million cases worldwide annually [2]. The clinical manifestations of human salmonellosis are frequently those associated with a self-limited gastroenteritis, namely nausea, vomiting and diarrhea, but can also include severe complications, including bacteremia and extra-intestinal infections [3]. Though most commonly associated with contaminated food, human salmonellosis can also occur through the contact with infected animals, such as farm animals and pets, including reptiles [4].

In the course of the past years, reptiles have been increasingly regarded as household pets, with their estimated numbers ascending up to 8 million only in the Europe Union in 2019 [5]. *Salmonella* not only can be found in the gastrointestinal tract of healthy reptiles, but also in the environments where those animals are kept [6,7]. *Salmonella enterica* subspecies *enterica* is commonly found in warm-blooded animals, while the remainder subspecies, *salamae, arizonae, diarizonae, houtenae,* and *indica,* along with *Salmonella bongori* are frequently isolated either from reptiles or from the environment [8]. Furthermore, among more than 2500 known *Salmonella* serotypes, over 40% are associated with reptiles and are rarely isolated from other animals, including humans [9]. Although infrequent when compared with food-borne cases, accounting for 6% of all human salmonellosis cases both in the

USA and in Europe [10], reptile-associated salmonellosis (RAS) seems to be more related with more severe clinical scenarios, such as systemic and severe disease development, especially in children, elderly people, and pregnant women [6]. In fact, RAS is a growing public health concern worldwide, with different reports pointing out for its role in disease outbreaks [11,12]. Despite the several RAS cases that have been reported in different European countries [13], there seems to be no available data regarding Portugal.

As observed for non-typhoidal salmonellae of other sources, there has been an increasing focus on antimicrobial resistance in reptile-associated *Salmonella* [14–16] since this feature can impair the success of treatments of both human and veterinary *Salmonella* infections [17]. Antimicrobial resistance can either arise from mutations in chromosomal genes (intrinsic resistance), which are caused by selective pressure, or through the acquisition of antimicrobial resistance determinants encoded in plasmids (extrinsic resistance), by horizontal transfer [18]. The role of reptiles as disseminators of antimicrobial resistant (AMR) *Salmonella* has been suggested [19,20]. Furthermore, *Salmonella* is known to have the ability of producing biofilms in different biotic and abiotic surfaces [21]. Not only are bacterial cells in biofilms more tolerant to antimicrobials when compared with the corresponding planktonic cells [22] but also more resistant to several chemical disinfectants [21].

Chlorhexidine is a biocide widely included in antiseptic products, especially in handwashing and oral products, due to its broad-spectrum efficacy and low irritability [23]. For surgical skin preparations and hand scrub, chlorhexidine is available in 4% solutions, while for wound cleaning is used as a 0.5% concentrated solution [24]. In veterinary care, chlorhexidine gluconate is a common disinfectant. In reptile treatment, chlorhexidine solutions are frequently used for topical application and preoperative scrubs, in concentrations below 2% [25], but there is a lack of clear guidelines regarding the most appropriate concentration to use.

The aim of the present study was to assess the presence of *Salmonella* spp. among the intestinal microbiota of pet reptiles in the Metropolitan area of Lisbon, Portugal, and to characterize those isolates, regarding antimicrobial susceptibility and virulence traits, bringing more information on the role of reptile-associated *Salmonella* on the public health scenario. Additionally, the antimicrobial efficacy of chlorhexidine gluconate against both planktonic cells and biofilms was also evaluated.

2. Results

2.1. Salmonella spp. Isolates

Of the 78 reptiles sampled 32 were identified as *Salmonella* positive (41%), specifically four Ophidians (50%), 14 Saurians (51.9%), and nine Chelonians (20.9%), belonging to 12 different owners (Table 1). Overall, the *Salmonella* recovery rates where higher both in Ophidians and Saurians when comparing with the one recorded in Chelonians ($p = 0.016$). After assessing the biochemical profile using API20E strip tests of the presumptive *Salmonella* isolates, 13 were identified as *Salmonella enterica* subspecies *arizonae* and 19 as *Salmonella* spp. (Table 2).

More than half of all *Salmonella* positive animals (62.5%) were detained by only three owners (E, F, and J). Moreover, owner J alone kept 12 *Salmonella* positive reptiles, more specifically Saurians. *Salmonella* isolates from co-habiting animals belonged to similar species with the exception for the isolates recovered from the animals of owner J, where the majority was identified as *Salmonella enterica* subspecies *arizonae* (10/12) and the remaining as *Salmonella* spp. (2/12) (Table 2). Notably, whenever an owner possessed multiple *Salmonella* positive animals, those animals belonged to the same reptile group.

Table 1. *Salmonella* positive animals, divided by category and species.

Category	Species	Number of Positive Animals
Ophidians	*Pantherophis guttatus guttatus*	2
	Python regius	2
Chelonians	*Centrochelys sulcata*	1
	Chelonoidis carbonaria	1
	Geochelone sulcata	1
	Pseudemys spp.	2
	Sternotherus odoratus	1
	Testudo horsfield	1
	Traquemys scripta elegans	2
Saurians	*Chlamydosaurus kingii*	2
	Ctenosaura quinquecarinata	1
	Gerrhosaurus major	1
	Hydrosaurus amboinensis	1
	Iguana iguana	1
	Physignatus cocincinus	3
	Physignatus lesueurii lesueurii	1
	Pogona vitticeps	8
	Tupinambis rufrescens	1

Table 2. Detailed information regarding the *Salmonella* isolates under study.

Isolate Number	Group	Species	Owner	API20E Result
4	Ophidian	*Python regius*	A	*Salmonella enterica* subsp. *arizonae*
12	Chelonian	*Pseudemys* spp.	B	*Salmonella* spp.
21	Ophidian	*Pantherophis guttatus guttatus*	C	*Salmonella* spp.
26	Chelonian	*Geochelone sulcata*	D	*Salmonella enterica* subsp. *arizonae*
27	Chelonian	*Chelonoidis carbonaria*	D	*Salmonella enterica* subsp. *arizonae*
30	Saurian	*Pogona vitticeps*	E	*Salmonella* spp.
31	Saurian	*Pogona vitticeps*	E	*Salmonella* spp.
32	Saurian	*Pogona vitticeps*	E	*Salmonella* spp.
33	Saurian	*Physignatus cocincinus*	E	*Salmonella* spp.
34	Saurian	*Pogona vitticeps*	E	*Salmonella* spp.
35	Chelonian	*Centrochelys sulcata*	F	*Salmonella* spp.
36	Chelonian	*Testudo horsfield*	F	*Salmonella* spp.
41	Chelonian	*Sternotherus odoratus*	F	*Salmonella* spp.
44	Chelonian	*Pseudemys* spp.	G	*Salmonella* spp.
46	Saurian	*Pogona vitticeps*	H	*Salmonella* spp.
47	Chelonian	*Traquemys scripta elegans*	I	*Salmonella enterica* subsp. *arizonae*
48	Chelonian	*Traquemys scripta elegans*	I	*Salmonella* spp.
50	Saurian	*Ctenosaura quinquecarinata*	J	*Salmonella enterica* subsp. *arizonae*
52	Saurian	*Physignatus cocincinus*	J	*Salmonella enterica* subsp. *arizonae*
53	Saurian	*Physignatus cocincinus*	J	*Salmonella einterica* subsp. *arizonae*
54	Saurian	*Tupinambis rufrescens*	J	*Salmonella enterica* subsp. *arizonae*
55	Saurian	*Pogona vitticeps*	J	*Salmonella* spp.
56	Saurian	*Pogona vitticeps*	J	*Salmonella enterica* subsp. *arizonae*
58	Saurian	*Gerrhosaurus major*	J	*Salmonella enterica* subsp. *arizonae*
61	Saurian	*Hydrosaurus amboinensis*	J	*Salmonella enterica* subsp. *arizonae*
62	Saurian	*Chlamydosaurus kingii*	J	*Salmonella enterica* subsp. *arizonae*
63	Saurian	*Chlamydosaurus kingii*	J	*Salmonella* spp.
66	Saurian	*Physignatus lesueurii lesueurii*	J	*Salmonella enterica* subsp. *arizonae*
69	Saurian	*Iguana iguana*	J	*Salmonella enterica* subsp. *arizonae*
70	Ophidian	*Pyton regius*	K	*Salmonella enterica* subsp. *arizonae*
73	Ophidian	*Pantherophis guttatus guttatus*	K	*Salmonella* spp.
76	Saurian	*Pogona vitticeps*	L	*Salmonella* spp.

2.2. Antimicrobial Resistance

All of the studied isolates were susceptible to gentamicin (CN) and ciprofloxacin (CIP) (Table 3). High levels of susceptibility to amikacin (AK) (96.87%), sulfamethoxazole/trimethoprim (SXT) (96.87%), nalidixic acid (NA) (93.75%), enrofloxacin (ENR) (90.63%), amoxicillin/clavulanic acid (AMC) (90.63%), ampicillin (AMP) (90.63%), cefotaxime (CTX) (87.50%), tetracycline (TE) (87.50%), and to chloramphenicol (C) (81.25%) were also recorded. On the other hand, 31 of the *Salmonella* isolates (96.87%) were resistant to penicillin (P).

Table 3. Antimicrobial resistance and virulence phenotype results.

Antimicrobial Resistance	Ophidians (%)	Chelonians (%)	Saurians (%)	p Value
AMC	0 (0%)	3 (33.3%)	0 (0%)	0.0286
AMP	0 (0%)	3 (33.3%)	0 (0%)	0.0286
AK	0 (0%)	1 (11.1%)	0 (0%)	N.S.
C	0 (0%)	0 (0%)	1 (5.26%)	N.S.
CN	0 (0%)	0 (0%)	0 (0%)	-
CTX	0 (0%)	0 (0%)	0 (0%)	-
ENR	0 (0%)	0 (0%)	0 (0%)	-
NA	0 (0%)	1 (11.1%)	1 (5.26%)	N.S.
P	4 (100%)	8 (88.89%)	19 (100%)	N.S.
CIP	0 (0%)	0 (0%)	0 (0%)	-
SXT	0 (0%)	0 (0%)	1 (5.26%)	N.S.
TE	0 (0%)	1 (11.1%)	0 (0%)	N.S.
Virulence phenotype				
Hemolytic activity	4 (100%)	9 (100%)	19 (100%)	-
Lipolytic activity	4 (100%)	9 (100%)	19 (100%)	-
DNase activity	4 (100%)	4 (44.44%)	11 (57.89%)	N.S.
Gelatinolytic activity	0 (0%)	0 (0%)	0 (0%)	-

Abbreviations: AMC, amoxicillin/clavulanic acid; AMP, ampicillin; AK, amikacin; C, chloramphenicol; CN, gentamicin; CTX, cefotaxime; ENR, enrofloxacin; NA, nalidixic acid; P, penicillin; CIP, ciprofloxacin; SXT, sulfamethoxazole/trimethoprim; TE, tetracycline; N.S., non-significant.

When comparing groups, resistance to AMC (p = 0.0286) and AMP (p = 0.0286) were associated with Chelonian *Salmonella* spp. isolates, as resistance to both antimicrobials was only detected, and simultaneously, in isolates 26, 36, and 47, all originating from turtles of different owners (Supplementary Table S1). No other statistically significant differences regarding antimicrobial susceptibility were detected.

Only three isolates (9.37%), all from Chelonians, were resistant to three or more of the antimicrobial compounds tested (isolates 26, 36, and 47) (Supplementary Table S1). The multiple resistance patterns were AMC/AMP/P, observed in isolates 26 and 36, and AMC/AMP/P/TE, revealed by isolate 47. None of the isolates was considered to be multidrug resistant, since the detected resistance patterns included antibiotics from the same class.

2.3. Virulence Phenotype

Virulence phenotypic testing revealed that all of the isolates studied expressed both hemolytic and lipolytic behaviors (Table 3). Contrarily, gelatinase activity was not detected in any of the *Salmonella* isolates studied. Overall, DNase activity was observed in more than half (59.37%) of the isolates. No statistically significant differences in phenotypical behavior were identified when comparing isolates from different animal groups.

2.4. Minimum Inhibitory Concentration and Minimum Bactericidal Concentration

The minimum inhibitory concentration (MIC) and minimum bactericidal concentration (MBC) values of chlorhexidine gluconate calculated for each isolate can be found on Supplementary Table S1.

The overall average MIC value was 11.90 mg/L $\pm$ 3.68, ranging from 8.16 mg/L (MIC value observed towards a Chelonian isolate), to 23.81 mg/L (MIC value towards a Chelonian and a Saurian isolates, all from different owners), with a median value of 10.72 mg/L. The majority of the chlorhexidine gluconate MIC values (75%) calculated for each *Salmonella* isolate only ranged between 9.52 mg/L and 14.29 mg/L. When comparing groups, the average MIC values regarding Ophidian, Chelonian, and Saurian isolates were 11.98, 11.25, and 12.19 mg/L, respectively, the differences were not statistically significant ($p = 0.802$) (Table 4).

Table 4. Chlorhexidine gluconate minimum inhibitory concentrations, minimum bactericidal concentrations, minimum biofilm inhibitory concentrations, minimum biofilm eradication concentrations and biofilm formation results.

Heading	Ophidians	Chelonians	Saurians	*p* Value
MIC (mg/L)	11.98 ± 1.46	11.25 ± 4.66	12.19 ± 3.44	N.S.
MBC (mg/L)	86.84 ± 72.75	27.87 ± 11.71	33.87 ± 52.91	N.S.
MBIC (mg/L)	57.15 ± 28.57	64.02 ± 12.32	72.87 ± 39.60	N.S.
MBEC (mg/L)	244.05 ± 131.49 *	333.65 ± 222.2 *	397.39 ± 194.74 *	N.S.
Biofilm formation (days)	5.1 ± 0.49	4.7 ± 1.0	4.2 ± 0.79	N.S.

Abbreviations: MIC, minimum inhibitory concentration; MBC, minimum bactericidal concentration; MBIC, minimum biofilm inhibitory concentration; MBEC, minimum biofilm eradication concentration; N.S., non-significant. * Values above 714.29 mg/L were not included.

Regarding MBC, the overall mean value was 38.8 mg/L ± 50.25, with a minimum value of 9.52 mg/L (observed towards a Chelonian isolate), and a maximum value of 247.62 mg/L (regarding a Saurian isolate), with a median value of 23.22 mg/L. Although a high variability in MBC values was found, towards half of the studied isolates those values ranged between 11.91 mg/L and 23.81 mg/L. When comparing groups, the average MBC values obtained regarding the Ophidian isolates, 86.84 mg/L, the Chelonian isolates, 27.87 mg/L and the Saurian isolates, 33.87 mg/L, were not statistically different ($p = 0.257$).

Chlorhexidine gluconate demonstrated to have a bactericidal effect in the majority of the *Salmonella* isolates (84.37%), since only five isolates (15.63%) had MBC/MIC ratio above 4 (Supplementary Table S1).

2.5. Biofilm Formation in the Air-Liquid Interface

The biofilm formation capability of the *Salmonella* isolates obtained from pet reptiles was studied by observing the development of a pellicle in the air-liquid interface. Of all isolates, only three (9.37%) were not able to form biofilms, thus the vast majority (90.63%) formed a clearly detectable biofilm. The shortest period required for biofilm formation was three days, and the longest was six days. The average number of days until the biofilm was formed was 4.4 days ± 0.90, and the majority of the isolates (75.9%) were able to form the biofilm in five days or less.

The differences on the average number of days until biofilm formation by Ophidian (5.1 days), Chelonian (4.7 days), and Saurian isolates (4.2 days) were considered not to have statistical significance ($p = 0.211$) (Table 4).

2.6. Minimum Biofilm Inhibitory Concentration and Minimum Biofilm Eradication Concentration Determination

The minimum biofilm inhibitory concentration (MBIC) and minimum biofilm eradication concentration (MBEC) values of chlorhexidine gluconate regarding each isolate can be found on Supplementary Table S1.

The MBIC values ranged from 14.29 mg/L to 232.15 mg/L, with an average value of 68.41 mg/L ± 32.68, and a median value of 71.43 mg/L. Despite the broad range of values, 71.43 mg/L of chlorhexidine gluconate was the MBIC value for more than half (59.4%) of the isolates tested. When comparing groups, the recorded average MBIC values regarding Ophidian, 57.15 mg/L, Chelonian, 64.02 mg/L, and Saurian isolates, 72.87 mg/L, did not statistically differ ($p = 0.509$) (Table 4).

Concerning the MBEC values, the average chlorhexidine gluconate biofilm eradication concentration was 360.08 mg/L ± 235.18, with a minimum of 33.34 mg/L and a maximum of 714.29 mg/L, and a median value of 392.86 mg/L. Regarding six isolates, one Ophidian, one Chelonian, and four Saurian related isolates, the MBEC values were considered to be greater than the highest concentration tested, therefore, the results were expressed as >714.29 mg/L.

3. Discussion

Several research groups from multiple countries have reported the isolation of *Salmonella* spp. from pet or captive reptiles, including turtles, lizards, and snakes [26–32]. Although this is not a recent issue, to the author's best knowledge, the present report is the first regarding the isolation of *Salmonella* spp. from healthy pet reptiles in Portugal. Our results point out to an overall *Salmonella* spp. prevalence of 41%, which is similar to studies performed with captive or pet reptiles in Australia (47%) [32], Spain (48%) [14], Norway (43%) [33], or Sweden (49%) [7], but higher than reports from smuggled reptiles in Taiwan (30.9%) [15] or captive animals in Croatia (13%) [29] or in New Zealand (11.4%) [31]. Furthermore, in our study, the prevalence of *Salmonella* spp. was higher in both Ophidians (50%) and Saurians (51.9%), when compared with Chelonians (20.9%) ($p = 0.016$). The lower isolation rates in turtles when compared with other reptiles can be associated with seasonal variations, observed when turtles are preparing for hibernation [28], but also with the diet of these animals [12,15,16]. In fact, the sample collection period occurred before the hibernation stage of Chelonians, during the colder months of the year. Nevertheless, the impact of pet turtles in the reptile-associated salmonellosis scenario should not be underestimated, since exposure to *Salmonella* positive turtles has been linked to disease outbreaks [34–36].

High levels of antimicrobial susceptibility to the majority of the antibiotics tested were found in most the *Salmonella* isolates, and only three isolates (9.37%) were resistant to three or more of the compounds tested. Our results differ from those reported in a recent study carried out in Spain, in which 72% of the isolates were considered to be multidrug resistant [14]. *Salmonella* isolates from reptiles are known to be resistant to several antibiotics frequently used in therapy. This not only implies that reptiles can shed multidrug resistant salmonellae to the environment and to other animals, including humans, but also the genes responsible for those antimicrobial resistances could be transferred to other enteric bacteria [17].

All the isolates studied expressed both hemolytic and lipolytic behaviors on plate tests. These two virulence phenotypes should be further investigated. Hemolysis is not associated with human non-typhoidal salmonellosis cases, and it has not been reported as a virulence trait by other authors, though it was shown that the hemolytic activity in *Salmonella enterica* serovar Typhimurium is dependent of the pathogenicity island 1 type III secretion system [37]. Extracellular lipases have been proposed as potential virulence factors in other pathogenic bacteria, such as *Staphylococcus aureus*, *Staphylococcus epidermis*, or *Pseudomonas aeruginosa* [38], though their role in *Salmonella* spp. virulence does not seem to be fully studied [39]. DNase testing pointed out the presence of extracellular desoxiribonucleases in more than half of the isolates. Gelatinase activity was not detected, even though it is a biochemical characteristic of *Salmonella enterica* subsp. *arizonae* [40]. It is possible that the analyzed isolates harbored the gene responsible for gelatin digestion, even though the isolates under the present study conditions did not express that phenotype. Recently, Salmonellae isolated from ready-to-eat shrimps were also found to express hemolytic, lipolytic, DNA degrading activity and also gelatinase production [41]. Additional studies are necessary in order to understand the extent of the possible role of these phenotypes both in animal and in human *Salmonella* infections. Actually, from the obtained data, the possibility of the same bacterial clone infecting different animals and adapting/evolving within the hosts cannot be excluded. Although a molecular based approach would bring valuable information regarding the identity and the possible genetic relationship between the studied isolates, the present report was designed to clarify the therapeutic potential of chlorhexidine, testing one isolate from each animal. Despite the possible genetic similarities, the foremost important assessed feature of each *Salmonella* isolate was the phenotypical behavior, namely the susceptibility to a commonly used biocide, chlorhexidine gluconate. Thus, the information resulting from this study can be adapted and applied in reptile medicine.

In the present study, the occurrence of both bactericidal and bacteriostatic effects of chlorhexidine gluconate is an example of the duality of the antimicrobial effect that takes place according to the applied concentration. Previous reports revealed chlorhexidine gluconate MIC values ranging from 8 to 16 mg/L when tested towards *Salmonella* Bredeney, Dublin, Gallinarum, Montivideo Virshow and Typhimurium [42]. Another study recorded a range of MIC values for *Salmonella* isolates of animal origin (broilers, cattle and pigs) between 2 and 64 mg/L [43]. More recent studies reported MIC values of 1–8 mg/L in turkey *Salmonella* isolates from commercial processing plants, and MIC values below 4 mg/L to 64 mg/L regarding different *Salmonella* serovars isolated from chicken and in egg production chains [6,44,45]. The overall mean MIC value calculated for the studied *Salmonella* spp. isolates from pet reptiles was 11.90 mg/L, which is coherent with those values. The global mean MBC value is approximately three times the mean MIC. The suggestion that both MIC and MBC values should be included in the monitorization of biocidal susceptibility is consistent with the results obtained in this study considering that both values provide complementary information [46].

Although MIC an MBC values are valuable for evaluating the antimicrobial effect of chlorhexidine gluconate, the previous studies were carried out with planktonic cells. The fact that the *Salmonella* spp. isolates are capable of biofilm formation is worrisome, since *Salmonella* organized in biofilms is less susceptible to disinfectants than planktonic cells, with preliminary studies indicating that disinfectants used at an effective concentration for *Salmonella* biofilm reduction can cause the selection of more virulent cells [47]. The high frequency of the studied reptile *Salmonella* isolates capable of forming biofilms (90.63%) is similar to previously reported data. High frequencies of pellicle formation in the air-liquid interface by *Salmonella* Agona (100%), *Salmonella* Montevideo (100%), and *Salmonella* Senftenberg (88%) were already described [48]. However, in the same study, only 55% of the *Salmonella* Typhimurium isolates tested were biofilm producers [48]. On other studies, the expression of biofilm formation by *Salmonella* Typhimurium isolates varied under the same circumstances, with different strains and morphotypes demonstrating different biofilm capabilities [49,50].

Biofilms are common on liquid-hard surfaces interfaces [51], such as in certain type of reptile cages or in aquariums. In order to simulate a more realistic approach to the effects of chlorhexidine gluconate on *Salmonella* cultures, the antimicrobial action of chlorhexidine gluconate activity was tested on the biofilms formed by the reptile *Salmonella* isolates during a 24 h-period. A chlorhexidine gluconate MBIC value within the concentration limits tested was obtained regarding all the *Salmonella* isolates studied. Regarding the *Salmonella* isolates towards which the MBEC values exceeded 714.29 mg/L, chlorhexidine gluconate was simply not effective in terms of eradicating those biofilms. Overall, chlorhexidine gluconate MBIC and MBEC results show that *Salmonella* biofilms are less susceptible to this biocide, what is consistent with a previous report which stated that three-day old *Salmonella* Typhimurium biofilms were less susceptible to chlorhexidine gluconate when compared to the corresponding planktonic cells [52].

4. Materials and Methods

4.1. Sample Collection and Salmonella spp. Isolation

A total of 78 cloacal swabs were obtained from pet reptiles, specifically 43 Chelonians (commonly referred as turtles), 27 Saurians (commonly named lizards), and eight Ophidians (usually known as snakes). The cloacal swabs were performed using cotton swabs in AMIES transport media (VWR, Amadora, Portugal) during routine health check-ups at the house of the owners or at pet shops, all located in the Lisbon Metropolitan Area, Portugal. All animals were cared for according to the rules given by the current EU (Directive 2010/63/EC) and national (DL 113/2013) legislation and by the competent authority (Direção Geral de Alimentação e Veterinária, DGAV, (www.dgv.min-agricultura.pt/portal/page/portal/DGV, accessed on 20 January 2021) in Portugal. Verbal informed consent was obtained from all the owners. Trained veterinarians performed sample collection of all the samples, following standard routine procedures. After collection, swabs were kept under refrigeration conditions (4 °C) for no longer than 48 h until processing in the Microbiology Laboratory of the Veterinary Medicine Faculty—University of Lisbon for *Salmonella* spp. isolation.

Briefly, each cloacal swab was homogenized and incubated in 5 mL of buffered peptone water (BPW) (Scharlau, Valencia, Spain) for 18 ± 2 h at 37 °C. After the initial incubation, 1 mL of BPW was then added to 10 mL of Muller-Kaufmann Tetrathionate (MKTT) Broth (Oxoid, Hampshire, UK) and incubated for 18–24 h at 37 °C. Simultaneously, 0.1 mL of the BPW solution was added to 10 mL Rappaport–Vassiliadis broth (Oxoid, Hampshire, England) and the resulting suspension was incubated for 18–24 h at 41.5 °C. Afterwards, suspensions were inoculated in Hektoen Agar (Liofilchem, Teramo, Italy) and xylose lysine deoxicholate agar (Scharlau, Valencia, Spain) plates, by streaking, and incubated at 37 °C for 20 ± 2 h. The resulting presumptive *Salmonella* spp. colonies were selected and transferred to triple sugar iron (TSI) Agar (Scharlau, Valencia, Spain) and to urea broth (Oxoid, Dadirlly, France) and incubated for 20 ± 2 h at 37 °C. Presumptive *Salmonella* spp. isolates were identified through the growth pattern in TSI agar and in Urea Broth. The method described is an adaptation of a previously described method [53]. *Salmonella* spp. isolates were identified using biochemical profile system API 20E (BioMérieux, Craponne, France). The biochemical identification was later confirmed by agglutination with Antiserum *Salmonella* OMNIVALENT Omni-O (Bio-Rad Laboratories, Inc., Marnes-la-Coquette, France).

4.2. Antimicrobial Susceptibility Testing

Antimicrobial susceptibility testing was performed by the disk diffusion method, according to Clinical and Laboratory Standards Institute guidelines (CLSI) [54]. The tested antibiotics were amoxicillin/clavulanic acid (AMC, 30 μg), ampicillin (AMP, 10 μg), amikacin (AK, 30 μg), chloramphenicol (C, 30 μg), gentamicin (CN, 10 μg), cefotaxime (CTX, 30 μg), enrofloxacin (ENR, 5 μg), nalidixic acid (NA, 30 μg), penicillin (P, 10 U), ciprofloxacin (CIP, 5 μg), sulfamethoxazole/trimethoprim (SXT, 25 μg), and tetracycline (TE, 30 μg). All antibiotics were purchased from Oxoid, Dadirlly, France. *Escherichia coli* ATCC 25922 was used as the control strain for test performance. Multidrug resistance (MDR) phenotype was considered to be present whenever an isolate revealed resistance to three or more antimicrobial compounds belonging to different classes [55].

4.3. Virulence Phenotype Analysis

In order to assess the virulence phenotype of the *Salmonella* isolates, plate tests were performed for evaluating their DNase, gelatinase, hemolytic and lipase activities.

DNase activity testing was performed by streaking the bacterial isolates on DNase test Agar plates (Liofilchem, Teramo, Italy) supplemented with 0.01% toluidine blue. The plates were incubated for 48 h at 37 °C and positive results showed a transparent halo surrounding the colonies.

Gelatinase activity was tested by streaking the isolates on Gelatinase test Agar plates (Liofilchem, Teramo, Italy), followed by incubation at 37 °C for 48 h. Afterwards, plates were flooded with a mercury chloride solution and the gelatinase positive isolates showed a transparent halo around the colonies.

Production of hemolysins was determined by streaking the isolates on Columbia Agar plates supplemented with 5% sheep blood (BioMérieux, Craponne, France) and incubated for 48 h at 37 °C. The presence of clear halos surrounding the colonies was interpreted as β-hemolysis.

Lipase activity testing was achieved by culturing the isolates in Spirit Blue Agar plates (Difco, Algés, Portugal) supplemented with Tween 80 (30 g/L) and incubating for 48 h at 37 °C. Lipase producing isolates exhibited clear halos around the colonies.

4.4. Chlorhexidine Gluconate Minimum Inhibitory Concentration and Minimum Bactericidal Concentration Determination

The in vitro susceptibility profile of the *Salmonella* isolates to chlorhexidine gluconate was assessed by an adapted protocol based on the microtiter broth dilution method [56,57]. Isolates were grown in a nonselective brain heart infusion (BHI) agar medium (VWR Chemicals, Leuven, Belgium) at 37 °C for 24 h. Bacterial suspensions with 10^8 CFU/mL were prepared directly from plate cultures in sterile normal saline (Merck, Germany) to a 0.5 McFarland suspension. The bacterial suspensions were then diluted in fresh BHI broth (VWR Chemicals, Leuven, Belgium) to a concentration of 10^7 CFU/mL.

Chlorhexidine gluconate dilutions were prepared from a stock solution at a concentration of 4% (*w/v*) (AGA, Lisboa, Portugal). A volume of 25 µL of chlorhexidine gluconate at 0.5, 0.1, 0.05, 0.01, 0.005 and 0.001% were distributed in 96-well flat-bottomed polystyrene microtiter plates (Nunc, Thermo Fisher Scientific, Roskilde, Denmark), apart from the negative and positive controls. All the wells were inoculated with 150 µL of the 10^7 CFU/mL bacterial suspensions, with exception of the negative control wells, which contained only broth medium. Therefore, the final concentration of chlorhexidine gluconate in the wells corresponded to 714.28, 142.86, 71.43, 14.29, 7.14, and 1.43 mg/L. Afterwards, microplates were statically incubated for 24 h at 37 °C. The minimum inhibitory concentration (MIC) was determined as the lowest concentration of chlorhexidine gluconate that visually inhibited microbial growth.

The minimum bactericidal concentration (MBC) value was assessed by inoculating 3 µL of the suspensions from the wells were no growth was observed on BHI agar plates, which were incubated at 37 °C for 24 h. MBC was determined as the lowest chlorhexidine gluconate concentration that did not allow colony development [57,58].

The ratio between MBC and MIC was calculated in order to determine the antimicrobial effect of chlorhexidine gluconate. The effect was considered to be bactericidal when the MBC was no more than fourfold the MIC, or bacteriostatic when the ratio exceeded four [58].

All experiments were conducted in duplicate and independent assays were performed at least three times in different dates.

4.5. Biofilm Formation in the Air-Liquid Interface

Biofilm forming ability was assessed through a biofilm formation assay in the air–liquid interface, by inoculating 0.5 mL of an overnight BHI broth culture, adjusted to a 0.5 McFarland standard, in a 4.5 mL of Luria broth (LB) without NaCl (1:10), prepared using yeast extract (Oxoid, Hampshire, England) and bacto tryptone (BD, Oeiras, Portugal). Isolates were incubated at 28 °C for eight days and each isolate was visually examined for pellicle formation on a daily basis [49]. The isolates capable of forming a pellicle in two distinct occasions were considered to be positive for biofilm formation, and the number of days required until the pellicle was perceivable was used to calculate the mean time for biofilm formation.

All assays were repeated in three independent dates, including 10% replicates.

4.6. Chlorhexidine Gluconate Minimum Biofilm Inhibitory Concentration and Minimum Biofilm Eradication Concentration Determination

The antimicrobial susceptibility of the *Salmonella* isolates when embedded in a 24 h biofilm was evaluated by a modified version of the Calgary Biofilm Pin Lid Device [57,59]. For minimum biofilm inhibitory concentration (MBIC) and minimum biofilm eradication concentration (MBEC) assays, the bacterial isolates were grown in BHI agar medium (VWR Chemicals, Leuven, Belgium) at 37 °C for 24 h. Bacterial suspensions with approximately 10^8 CFU/mL were prepared directly from plate cultures in sterile normal saline

(Merck, Darmstadt, Germany) by comparison with a 0.5 McFarland standard (BioMérieux, Craponne, France). Suspensions were then diluted in fresh BHI broth (VWR Chemicals, Leuven, Belgium) to a concentration of 10^6 CFU/mL. Then, 175 µL of the bacterial suspensions were distributed in 96-well flat-bottomed polystyrene microtiter plates, covered with 96-peg polystyrene lids (Nunc-TSP; Thermo Fisher Scientific, Roskilde, Denmark) and statically incubated for 24 h at 37 °C, allowing biofilm formation on the pegs. Peg lids were then rinsed three times in sterile normal saline to remove planktonic bacteria and placed on new microplates containing the set of chlorhexidine gluconate solutions previously described, corresponding to a final concentration by well of 714.28, 142.86, 71.43, 14.29, 7.14, and 1.43 mg/L.

Microplates were again incubated for 24 h at 37 °C, without shaking. After incubation, peg lids were removed, and the MBIC value was determined as the lowest chlorhexidine gluconate concentration that visually inhibited microbial growth. Subsequently, in order to determine the MBEC value, peg lids were rinsed three times in sterile normal saline, placed in new microplates containing only 175 µL of fresh BHI medium and incubated in an ultrasound bath (Grant MXB14, Essex, England), at 50 Hz during 15 min in order to disperse the biofilm-based bacteria from the peg surface. Afterwards, peg lids were discarded, and microplates were covered with normal lids and incubated for 24 h at 37 °C. The MBEC value was determined through direct observation of bacterial growth in the wells and defined as the lowest chlorhexidine gluconate concentration that visually eliminates the microbial growth [57].

Experiments were conducted in duplicate and independent assays were performed at least two times on different dates.

4.7. Statistical Analysis

For statistical analysis, the associations between frequency of *Salmonella* isolation and reptile group, AMR *Salmonella* and reptile group and virulence phenotype and reptile group were evaluated using the Fisher exact test. Association between different MIC, MEC and MBIC values of chlorhexidine gluconate on *Salmonella* isolates, the number of days until biofilm formation and the reptiles group was assessed recurring to the Brown–Forsythe robustness test based on a one-way ANOVA test. All statistical tests were performed on IBM SPSS Statistical program version 26 for Windows (SPSS Inc., Chicago, IL, USA). Associations were considered to be significant whenever P values were less than 0.05.

5. Conclusions

The present study reports the isolation of *Salmonella* from healthy pet reptiles and stresses their possible role in human non-typhoidal salmonellosis cases. Although presenting high levels of antimicrobial susceptibility, the expression of phenotypical virulence traits and the ability to form biofilms by these isolates are worrisome. Pet reptile owners should always employ good hygiene practices whenever manipulating the animals, but also when in contact with the environment in which the animals are kept. Overall, the use of chlorhexidine gluconate was considered to be effective, both in planktonic cells and biofilms, pointing out the potential of this biocide's use in reptile clinics.

Supplementary Materials: The following are available online at https://www.mdpi.com/2079-638 2/10/3/324/s1, Table S1: Detailed information regarding the studied Salmonella isolates and the results of Antimicrobial Susceptibility analysis, Virulence Phenotype analysis, Minimum Inhibitory Concentration, Minimum Bactericidal Concentration, Minimum Biofilm Inhibitory Concentration and Minimum Biofilm Eradication Concentration Determination and Biofilm Formation.

Author Contributions: Conceptualization: F.B. and M.O.; methodology: A.C.C., I.D. and A.R.; software: J.B.C.; formal analysis: A.C.C., I.D. and J.B.C.; writing—original draft preparation: J.B.C.; writing—review and editing: J.B.C., F.B. and M.O.; supervision: M.O.; funding acquisition: F.B. and M.O. All authors have read and agreed to the published version of the manuscript.

Funding: This research was supported by CIISA—Centro de Investigação Interdisciplinar em Sanidade Animal, Faculdade de Medicina Veterinária, Universidade de Lisboa, Project UIDB/00276/2020 (funded by FCT—Fundação para a Ciência e Tecnologia IP).

Institutional Review Board Statement: Not applicable.

Informed Consent Statement: Informed consent was obtained from all subjects involved in the study.

Data Availability Statement: The data presented in this study are available in Supplementary Table S1.

Conflicts of Interest: The authors declare no conflict of interest.

References

1. European Food Safety Authority and European Centre for Disease Prevention and Control (EFSA and ECDC). The European Union One Health 2018 Zoonoses Report. *EFSA J.* **2019**, *17*, 5926. [CrossRef]
2. Majowicz, S.E.; Musto, J.; Scallan, E.; Angulo, F.J.; Kirk, M.; O'Brien, S.J.; Jones, T.F.; Fazil, A.; Hoekstra, R.M. The global burden of nontyphoidal salmonella gastroenteritis. *Clin. Infect. Dis.* **2010**, *50*, 882–889. [CrossRef]
3. Crum-Cianflone, N.F. Salmonellosis and the gastrointestinal tract: More than just peanut butter. *Curr. Gastroenterol. Rep.* **2008**, *10*, 424–431. [CrossRef]
4. Van Duijkeren, E.; Houwers, D.J. A critical assessment of antimicrobial treatment in uncomplicated Salmonella enteritis. *Vet. Microbiol.* **2000**, *73*, 61–73. [CrossRef]
5. FEDIAF (The European Pet Food Industry). *FEDIAF Annual Report 2019*; FEDIAF: Bruxelles, Belgium, 2020.
6. Hoelzer, K.; Moreno Switt, A.; Wiedmann, M.; Majowicz, S.; Musto, J.; Scallan, E.; Angulo, F.; Kirk, M.; O'Brien, S.; Jones, T.; et al. Animal contact as a source of human non-typhoidal salmonellosis. *Vet. Res.* **2011**, *42*, 34. [CrossRef]
7. Wikström, V.O.; Fernström, L.L.; Melin, L.; Boqvist, S. Salmonella isolated from individual reptiles and environmental samples from terraria in private households in Sweden. *Acta Vet. Scand.* **2014**, *56*, 7. [CrossRef]
8. Brenner, F.W.; Villar, R.G.; Angulo, F.J.; Tauxe, R.; Swaminathan, B. Salmonella nomenclature. *J. Clin. Microbiol.* **2000**, *38*, 2465–2467. [CrossRef] [PubMed]
9. Mermin, J.; Hutwagner, L.; Vugia, D.; Shallow, S.; Daily, P.; Bender, J.; Koehler, J.; Marcus, R.; Angulo, F.J. Reptiles, amphibians, and human Salmonella infection: A population-based, case-control study. *Clin. Infect. Dis.* **2004**, *38*. [CrossRef] [PubMed]
10. Corrente, M.; Sangiorgio, G.; Grandolfo, E.; Bodnar, L.; Catella, C.; Trotta, A.; Martella, V.; Buonavoglia, D. Risk for zoonotic Salmonella transmission from pet reptiles: A survey on knowledge, attitudes and practices of reptile-owners related to reptile husbandry. *Prev. Vet. Med.* **2017**, *146*, 73–78. [CrossRef] [PubMed]
11. Whiley, H.; Gardner, M.G.; Ross, K. A review of salmonella and squamates (Lizards, snakes and amphisbians): Implications for public health. *Pathogens* **2017**, *6*, 38. [CrossRef]
12. Sodagari, H.R.; Habib, I.; Shahabi, M.P.; Dybing, N.A.; Wang, P.; Bruce, M. A review of the public health challenges of salmonella and turtles. *Vet. Sci.* **2020**, *7*, 56. [CrossRef]
13. Bertrand, S.; Rimhanen-Finne, R.; Weill, F.X.; Rabsch, W.; Thornton, L.; Perevoscikovs, J.; van Pelt, W.; Heck, M. Salmonella infections associated with reptiles: The current situation in Europe. *Eur. Surveill.* **2008**, *13*, 18902. [CrossRef]
14. Marin, C.; Lorenzo-Rebenaque, L.; Laso, O.; Villora-Gonzalez, J.; Vega, S. Pet Reptiles: A Potential Source of Transmission of Multidrug-Resistant Salmonella. *Front. Vet. Sci.* **2021**, *7*. [CrossRef] [PubMed]
15. Chen, C.Y.; Chen, W.C.; Chin, S.C.; Lai, Y.H.; Tung, K.C.; Chiou, C.S.; Hsu, Y.M.; Chang, C.C. Prevalence and antimicrobial susceptibility of salmonellae isolates from reptiles in Taiwan. *J. Vet. Diagn. Investig.* **2010**, *22*, 44–50. [CrossRef]
16. Sylvester, W.R.B.; Amadi, V.; Pinckney, R.; Macpherson, C.N.L.; Mckibben, J.S.; Bruhl-Day, R.; Johnson, R.; Hariharan, H. Prevalence, Serovars and Antimicrobial Susceptibility of Salmonella spp. from Wild and Domestic Green Iguanas (Iguana iguana) in Grenada, West Indies. *Zoonoses Public Health* **2014**, *61*, 436–441. [CrossRef]
17. Ebani, V.V. Domestic reptiles as source of zoonotic bacteria: A mini review. *Asian Pac. J. Trop. Med.* **2017**, *10*, 723–728. [CrossRef]
18. Reygaert, W.C. An overview of the antimicrobial resistance mechanisms of bacteria. *AIMS Microbiol.* **2018**, *4*, 482–501. [CrossRef]
19. Bertelloni, F.; Chemaly, M.; Cerri, D.; Le Gall, F.; Ebani, V.V. Salmonella infection in healthy pet reptiles: Bacteriological isolation and study of some pathogenic characters. *Acta Microbiol. Immunol. Hung.* **2016**, *63*, 203–216. [CrossRef]
20. Xia, Y.; Li, H.; Shen, Y. Antimicrobial Drug Resistance in Salmonella enteritidis Isolated from Edible Snakes with Pneumonia and Its Pathogenicity in Chickens. *Front. Vet. Sci.* **2020**, *7*, 463. [CrossRef]
21. Steenackers, H.; Hermans, K.; Vanderleyden, J.; De Keersmaecker, S.C.J. Salmonella biofilms: An overview on occurrence, structure, regulation and eradication. *Food Res. Int.* **2012**, *45*, 502–531. [CrossRef]
22. Trampari, E.; Holden, E.R.; Wickham, G.J.; Ravi, A.; de Martins, L.O.; Savva, G.M.; Webber, M.A. Exposure of Salmonella biofilms to antibiotic concentrations rapidly selects resistance with collateral tradeoffs. *NPJ Biofilms Microbiomes* **2021**, *7*, 1–13. [CrossRef]
23. Mcdonnell, G.; Russell, A.D. Antiseptics and disinfectants: Activity, action, and resistance. *Clin. Microbiol. Rev.* **1999**, *12*, 147–179. [CrossRef]
24. Main, R.C. Should chlorhexidine gluconate be used in wound cleansing? *J. Wound Care* **2008**, *17*, 112–114. [CrossRef]

25. O'Rourke, D.P.; Cox, J.D.; Baumann, D.P. Nontraditional Species. In *Management of Animal Care and Use Programs in Research, Education, and Testing*; CRC Press: Boca Raton, FL, USA, 2020; pp. 579–596.

26. Corrente, M.; Madio, A.; Friedrich, K.G.; Greco, G.; Desario, C.; Tagliabue, S.; D'Incau, M.; Campolo, M.; Buonavoglia, C. Isolation of Salmonella strains from reptile faeces and comparison of different culture media. *J. Appl. Microbiol.* **2004**, *96*, 709–715. [CrossRef]

27. Back, D.-S.; Shin, G.-W.; Wendt, M.; Heo, G.-J. Prevalence of Salmonella spp. in pet turtles and their environment. *Lab. Anim. Res.* **2016**, *32*, 166. [CrossRef]

28. Geue, L.; Löschner, U. Salmonella enterica in reptiles of German and Austrian origin. *Vet. Microbiol.* **2002**, *84*, 79–91. [CrossRef]

29. Lukac, M.; Pedersen, K.; Prukner-Radovcic, E. Prevalence of salmonella in captive reptiles from Croatia. *J. Zoo Wildl. Med.* **2015**, *46*, 234–240. [CrossRef]

30. Pedersen, K.; Lassen-Nielsen, A.-M.; Nordentoft, S.; Hammer, A.S. Serovars of *Salmonella* from captive reptiles. *Zoonoses Public Health* **2009**, *56*, 238–242. [CrossRef] [PubMed]

31. Kikillus, K.; Gartrell, B.; Motion, E. Prevalence of *Salmonella* spp., and serovars isolated from captive exotic reptiles in New Zealand. *N. Z. Vet. J.* **2011**, *59*, 174–178. [CrossRef] [PubMed]

32. Scheelings, T.F.; Lightfoot, D.; Holz, P. Prevalence of salmonella in australian reptiles. *J. Wildl. Dis.* **2011**, *47*, 1–11. [CrossRef]

33. Bjelland, A.M.; Sandvik, L.M.; Skarstein, M.M.; Svendal, L.; Debenham, J.J. Prevalence of Salmonella serovars isolated from reptiles in Norwegian zoos. *Acta Vet. Scand.* **2020**, *62*. [CrossRef]

34. Weltman, A.; Smee, A.; Moll, M.; Deasy, M.; Pringle, J.; Williams, I.; Behravesh, C.B.; Wright, J.; Routh, J.; Longenberger, A. Notes from the field: Outbreak of salmonellosis associated with pet turtle exposures—United States, 2011. *Morb. Mortal. Wkly Rep.* **2012**, *61*, 79.

35. Gambino-Shirley, K.; Stevenson, L.; Wargo, K.; Burnworth, L.; Roberts, J.; Garrett, N.; Van Duyne, S.; McAllister, G.; Nichols, M. Notes from the Field: Four Multistate Outbreaks of Human Salmonella Infections Linked to Small Turtle Exposure—United States, 2015. *MMWR. Morb. Mortal. Wkly. Rep.* **2016**, *65*, 655–656. [CrossRef]

36. Basler, C.; Bottichio, L.; Higa, J.; Prado, B.; Wong, M.; Bosch, S. Multistate Outbreak of Human Salmonella Poona Infections Associated with Pet Turtle Exposure—United States, 2014. *MMWR. Morb. Mortal. Wkly. Rep.* **2015**, *64*, 804. [CrossRef]

37. Miki, T.; Okada, N.; Shimada, Y.; Danbara, H. Characterization of Salmonella pathogenicity island 1 type III secretion-dependent hemolytic activity in Salmonella enterica serovar Typhimurium. *Microb. Pathog.* **2004**, *37*, 65–72. [CrossRef]

38. Pascoal, A.; Estevinho, L.M.; Martins, I.M.; Choupina, A.B. REVIEW: Novel sources and functions of microbial lipases and their role in the infection mechanisms. *Physiol. Mol. Plant Pathol.* **2018**. [CrossRef]

39. Bender, J.; Flieger, A. Lipases as Pathogenicity Factors of Bacterial Pathogens of Humans. In *Handbook of Hydrocarbon and Lipid Microbiology*; Springer: Berlin/Heidelberg, Germany, 2010; pp. 3241–3258.

40. Lamas, A.; Miranda, J.M.; Regal, P.; Vázquez, B.; Franco, C.M.; Cepeda, A. A comprehensive review of non-enterica subspecies of Salmonella enterica. *Microbiol. Res.* **2018**, *206*, 60–73. [CrossRef]

41. Beshiru, A.; Igbinosa, I.H.; Igbinosa, E.O. Biofilm formation and potential virulence factors of Salmonella strains isolated from ready-to-eat shrimps. *PLoS ONE* **2018**, *13*. [CrossRef]

42. Denton, G.W. Chlorhexidine. In *Disinfection, Sterilization and Preservation*; Block, S.S., Ed.; Lippincott Williams & Wilkins: Philadelphia, PA, USA, 2001; pp. 325–333.

43. Aarestrup, F.M.; Hasman, H. Susceptibility of different bacterial species isolated from food animals to copper sulphate, zinc chloride and antimicrobial substances used for disinfection. *Vet. Microbiol.* **2004**, *100*, 83–89. [CrossRef]

44. Beier, R.C.; Anderson, P.N.; Hume, M.E.; Poole, T.L.; Duke, S.E.; Crippen, T.L.; Sheffield, C.L.; Caldwell, D.J.; Byrd, J.A.; Anderson, R.C.; et al. Characterization of salmonella enterica isolates from Turkeys in commercial processing plants for resistance to antibiotics, disinfectants, and a growth promoter. *Foodborne Pathog. Dis.* **2011**, *8*, 593–600. [CrossRef]

45. Long, M.; Lai, H.; Deng, W.; Zhou, K.; Li, B.; Liu, S.; Fan, L.; Wang, H.; Zou, L. Disinfectant susceptibility of different Salmonella serotypes isolated from chicken and egg production chains. *J. Appl. Microbiol.* **2016**, *121*, 672–681. [CrossRef]

46. Maillard, J.Y.; Bloomfield, S.; Coelho, J.R.; Collier, P.; Cookson, B.; Fanning, S.; Hill, A.; Hartemann, P.; McBain, A.J.; Oggioni, M.; et al. Does microbicide use in consumer products promote antimicrobial resistance? A critical review and recommendations for a cohesive approach to risk assessment. *Microb. Drug Resist.* **2013**, *19*, 344–354. [CrossRef] [PubMed]

47. Rodrigues, D.; Cerca, N.; Teixeira, P.; Oliveira, R.; Ceri, H.; Azeredo, J. Listeria monocytogenes and Salmonella enterica Enteritidis biofilms susceptibility to different disinfectants and stress-response and virulence gene expression of surviving cells. *Microb. Drug Resist.* **2011**, *17*, 181–189. [CrossRef]

48. Vestby, L.K.; Møretrø, T.; Langsrud, S.; Heir, E.; Nesse, L.L. Biofilm forming abilities of Salmonella are correlated with persistence in fish meal- and feed factories. *BMC Vet. Res.* **2009**, *5*, 1–6. [CrossRef] [PubMed]

49. Seixas, R.; Machado, J.; Bernardo, F.; Vilela, C.; Oliveira, M. Biofilm formation by salmonella enterica serovar 1,4,[5],12:i:-portuguese isolates: A phenotypic, genotypic, and socio-geographic analysis. *Curr. Microbiol.* **2014**, *68*, 670–677. [CrossRef] [PubMed]

50. Ramachandran, G.; Aheto, K.; Shirtliff, M.E.; Tennant, S.M. Poor biofilm-forming ability and long-term survival of invasive Salmonella Typhimurium ST313. *Pathog. Dis.* **2016**, *74*. [CrossRef]

51. Otter, J.A.; Vickery, K.; Walker, J.T.; deLancey Pulcini, E.; Stoodley, P.; Goldenberg, S.D.; Salkeld, J.A.G.; Chewins, J.; Yezli, S.; Edgeworth, J.D. Surface-attached cells, biofilms and biocide susceptibility: Implications for hospital cleaning and disinfection. *J. Hosp. Infect.* **2015**, *89*, 16–27. [CrossRef]
52. Wong, H.S.; Townsend, K.M.; Fenwick, S.G.; Trengove, R.D.; O'Handley, R.M. Comparative susceptibility of planktonic and 3-day-old Salmonella Typhimurium biofilms to disinfectants. *J. Appl. Microbiol.* **2010**, *108*, 2222–2228. [CrossRef]
53. Hendriksen, R.S. Laboratory Protocols Level 1: Training Course Isolation of Salmonella. In *A Global Salmonella Surveillance and Laboratory. Support Project of the World Health Organization*, 4th ed.; WHO: Geneva, Switzerland, 2003.
54. CLSI. *Performance Standarts for Antimicrobial Susceptibility Testing*; 25th Informational Supplement; CLSI=NCCLS M100-S25; CLSI: Wayne, PA, USA, 2015.
55. Magiorakos, A.-P.; Srinivasan, A.; Carey, R.B.; Carmeli, Y.; Falagas, M.E.; Giske, C.G.; Harbarth, S.; Hindler, J.F.; Kahlmeter, G.; Olsson-Liljequist, B.; et al. Multidrug-resistant, extensively drug-resistant and pandrug-resistant bacteria: An international expert proposal for interim standard definitions for acquired resistance. *Clin. Microbiol. Infect.* **2012**, *18*, 268–281. [CrossRef]
56. Wiegand, I.; Hilpert, K.; Hancock, R.E.W. Agar and broth dilution methods to determine the minimal inhibitory concentration (MIC) of antimicrobial substances. *Nat. Protoc.* **2008**, *3*, 163–175. [CrossRef]
57. Santos, R.; Gomes, D.; Macedo, H.; Barros, D.; Tibério, C.; Veiga, A.S.; Tavares, L.; Castanho, M.; Oliveira, M. Guar gum as a new antimicrobial peptide delivery system against diabetic foot ulcers Staphylococcus aureus isolates. *J. Med. Microbiol.* **2016**, *65*, 1092–1099. [CrossRef] [PubMed]
58. French, G.L. Bactericidal agents in the treatment of MRSA infections—The potential role of daptomycin. *J. Antimicrob. Chemother.* **2006**, *58*, 1107–1117. [CrossRef] [PubMed]
59. Ceri, H.; Olson, M.E.; Stremick, C.; Read, R.R.; Morck, D.; Buret, A. The Calgary Biofilm Device: New technology for rapid determination of antibiotic susceptibilities of bacterial biofilms. *J. Clin. Microbiol.* **1999**, *37*, 1771–1776. [CrossRef] [PubMed]

Article

Spread of Antimicrobial Resistance by *Salmonella enterica* Serovar Choleraesuis between Close Domestic and Wild Environments

María Gil Molino [1,*], **Alfredo García** [2], **Sofía Gabriela Zurita** [1], **Francisco Eduardo Martín-Cano** [3], **Waldo García-Jiménez** [4], **David Risco** [4,5], **Joaquín Rey** [1], **Pedro Fernández-Llario** [4] and **Alberto Quesada** [6,7]

[1] Unidad de Patología Infecciosa, Facultad de Veterinaria, Universidad de Extremadura, 10003 Cáceres, Spain; sofigz32@gmail.com (S.G.Z.); jmrey@unex.es (J.R.)

[2] Área de Producción Animal, CICYTEX-La Orden, 06187 Badajoz, Spain; fredgarsa@gmail.com

[3] Unidad de Reproducción, Facultad de Veterinaria, Universidad de Extremadura, 10003 Cáceres, Spain; femartincano@unex.es

[4] Innovación en Gestión y Conservación de Ingulados S.L. (INGULADOS), 10004 Cáceres, Spain; waldo@ingulados.com (W.G.-J.); david@ingulados.com (D.R.); pedro@ingulados.com (P.F.-L.)

[5] Neobeitar S.L., 10004 Cáceres, Spain

[6] Departamento de Bioquímica, Biología Molecular y Genética, Facultad de Veterinaria, Universidad de Extremadura, 10003 Cáceres, Spain; aquesada@unex.es

[7] INBIO G+C, Universidad de Extremadura, 10003 Cáceres, Spain

* Correspondence: mariamilgm@unex.es

Received: 28 September 2020; Accepted: 28 October 2020; Published: 29 October 2020

Abstract: The *Salmonella enterica* serovar Choleraesuis affects domestic pig and wild boar (WB), causing clinical salmonellosis. Iberian swine production is based on a free-range production system where WB and Iberian pig (IP) share ecosystems. This study focuses on the negative impact on the pork industry of infections due to this serotype, its role in the spread of antibiotic resistance, and its zoonotic potential. Antibiotic resistance (AR) and genetic relationships were analyzed among 20 strains of *S. Choleraesuis* isolated from diseased WB and IP sampled in the southwest region of the Iberian Peninsula. AR was studied using the Kirby–Bauer method with the exception of colistin resistance, which was measured using the broth microdilution reference method. Resistance and Class 1 integrase genes were measured using PCR, and the genetic relationship between isolates and plasmid content by pulsed field gel electrophoresis. The results show a higher incidence of AR in isolates from IP. Phylogenetic analysis revealed seven profiles with two groups containing isolates from IP and WB, which indicates circulation of the same clone between species. Most pulsotypes presented with one plasmid of the same size, indicating vertical transmission. AR determinants *bla*$_{TEM}$ and *tetA* were routinely found in IP and WB, respectively. One isolate from IP expressed colistin resistance and presented the *mcr-1* gene carried by a plasmid. This study suggests that *S. Choleraesuis* circulates between WB and IP living in proximity, and also that the mobilization of AR genes by plasmids is low. Furthermore, the detection of plasmid-mediated colistin resistance in bacteria from IP is alarming and should be monitored.

Keywords: *Salmonella Choleraesuis*; Iberian pig; wild boar; antibiotic resistance; phylogenetic relationship; plasmid replicon typing; colistin

---

1. Introduction

Salmonellosis in swine results in tremendous economic losses in the pork industry [1]. *Salmonella enterica* subsp. *enterica* serovar Choleraesuis (*S. Choleraesuis*) causes clinical salmonellosis in pigs and wild boar (WB) [2], and the identification of epidemiologic groups strongly suggests an exchange of this serovar between WB and domestic pigs [3]. Nowadays, *S. Choleraesuis* is still very common in North America and Asia and, although it is not considered a dominant serovar in pigs from Europe [4,5], different outbreaks have occasionally been reported in recent years [6,7] including in WB [2,3,8–11].

The Iberian pig (IP) is an autochthonous breed that originated in the Iberian Peninsula, for which the production system is mainly associated with extensive management deeply linked to the Mediterranean ecosystem and traditional agroforestry in the southwest of the Iberian Peninsula [12]. This means that WB and IP share the same habitats, leading to subsequent interactions among them [13,14]. WB, as an omnivorous species, is prone to multiple pathogen exposure. They have been shown to carry resistant bacteria [15] and could be a gateway for spread of this resistance from domestic animals or humans to wildlife [16]. Besides, several studies have shown WB as a possible asymptomatic persistent reservoir of *S. Choleraesuis* [17,18].

Although *S. Choleraesuis* is swine-specific and rarely infects other hosts, it is the second most predominant serovar among human isolates in Taiwan and exhibits the highest degree of invasiveness [19,20], which may result in severe disease and death [21]. Most *S. Choleraesuis* isolates from humans and swine exhibit closely related DNA fingerprints, indicating that human infections were acquired from pigs [22], reinforcing the importance of controlling this serotype in Suidae.

Most *S. Choleraesuis* strains that have caused infections in humans, mainly in Asian countries, are multidrug resistant (MDR) [19,23], which has been associated with classical mobile genetic elements (i.e., transposons and plasmids) and integrative elements that can spread antimicrobial resistance genes within the bacterial host genome through gene cassettes by site-specific recombination [24,25]. In addition, plasmids can carry other gene functions such as those involved in virulence by pSCV50 in *S. Choleraesuis* [26]. This 50 kb plasmid does not carry antimicrobial resistance genes, although it can recombine with larger sized plasmids detected in *S. Choleraesuis* where *sul1*, *bla*$_{TEM}$, and extended-spectrum beta-lactamase genes are located [27–29].

In contrast to the limited administration of colistin (polymyxin E) to humans as a last resort antibiotic, it has historically been used for prophylaxis in animal production [30]. Consequently, a dramatic increase of colistin resistance has arisen in naturally sensitive Gram-negative bacteria, with the spread of plasmid carrying *mcr-1* among other resistance determinants [31]. Among different reservoirs, livestock is considered the main source of *mcr* genes worldwide [32], and a global concern exists due to their high mobilization potential by plasmids carrying other resistance determinants [33]. *S. enterica*, one of the most clinically relevant enterobacteria, carries colistin resistance genes in many serovars via different plasmids, including IncHI2 mega-plasmids larger than 200 kb with multiple resistance determinants [34]. In *S. Choleraesuis*, this has been described very recently in one MDR isolate from a human blood infection in Brazil, linked to a 40 kb IncX4 plasmid [35].

The aim of the present investigation was to study the genetic relationship between strains of *S. Choleraesuis* from IP and WB raised in the southwest of the Iberian Peninsula and to address the mechanism of spread of its antimicrobial resistance determinants, including through screening for low-susceptible isolates to colistin in this bacterial pathogen.

2. Results

2.1. Clustering of S. Choleraesuis Isolates by PFGE-XbaI

Pulsed-field gel electrophoresis (PFGE) (*XbaI*) macrorestriction displayed seven different profiles or pulsotypes (PT) grouped into two main clusters: A, with a degree of similarity higher than 75% and B, with more than 80% similarity (Figure 1). Whilst cluster B contains only 4 isolates from 2 estates, all of them from IP, cluster A groups 5 PT that contain 15 isolates from 12 different estates. Within this cluster PT1, PT2 and PT3 showed a degree of similarity higher than 95%. There is remarkable persistency over time for PT1, PT3, and PT5, which were isolated during 5, 3, and 4 year periods, respectively, from the animal populations. Among them, PT1 and PT3 were detected in both IP and WB, indicating bacterial circulation between both suids. The distance between the estates with the same PT was not significantly different than the average distance between all the estates included in the study.

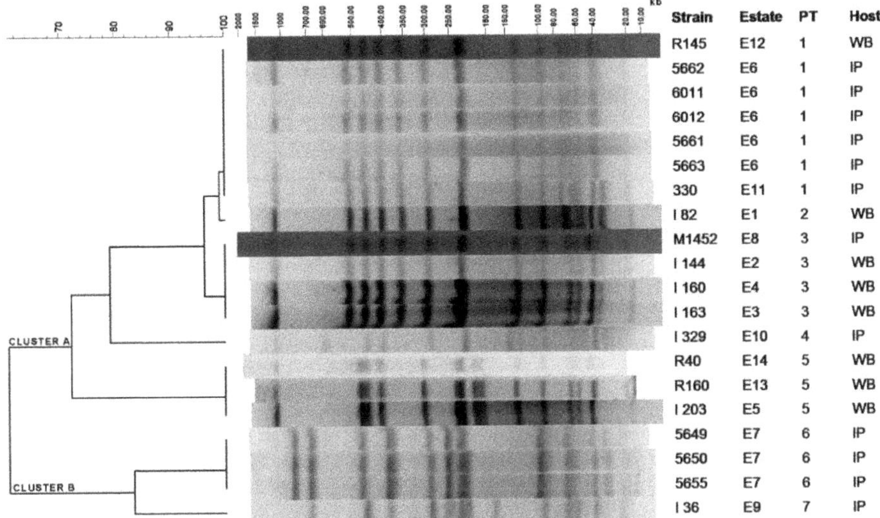

Figure 1. Dendrogram based on PFGE macrorestriction pattern of *S. Choleraesuis* isolates. Dendrogram showing 7 different profiles (PT) further divided into two clusters A and B. Dice coefficients had a 1.5% band position tolerance. The scales at the top indicate the similarity indices (in percentages) and molecular sizes (in kilobases).

2.2. Resistance Determinants against Clinically Relevant Antimicrobials in the S. Choleraesuis Isolates

Resistance against at least one of the 14 tested antibiotics was found in almost all tested strains (19/20; 95%); moreover, 65% (13/20) of the *S. Choleraesuis* isolates were multidrug resistant (MDR) with resistance to 4 or more antibiotics (Table 1). Antimicrobial resistance phenotypes were highly variable, with 14 different patterns existing among the 20 *S. Choleraesuis* isolates (Table 1), even within the same PT, especially if they came from different estates, as observed in PT1 and PT5 (Figure 1). Only three patterns appeared more than once: AMP–TRS–SUL–CHL (3 isolates from PT6), AMP–STR–TRS–SUL (2 isolates from PT1), and NEO (2 isolates from PT3), and none of them were shared between IP and WB. Indeed, the average number of antimicrobials to which isolates presented resistance depended on the host, with 4.9 resistances (or MDR), on average, per isolate in IP and 2.8 in WB. The host effect on MDR of isolates also affects the particular antibiotics found in every spectrum. Among isolates from IP, the most common resistance observed is against ampicillin, followed by sulfonamide, while in those from WB, the lowest susceptibilities were found against aminoglycosides (streptomycin and neomycin) followed by tetracycline and sulfonamide (Table 2). Resistance against colistin, a last resort antibiotic in human health, is found in only one isolate of PT1 from IP. Regardless of their origin, all isolates were susceptible to quinolones or the broad-spectrum cephalosporin cefotaxime.

Table 1. Antibiotic resistance characteristics of *S. Choleraesuis* isolates from Iberian pigs and wild boar in Spain.

PT [1]	Isolate	Origin	Resistance Phenotype	Resistance Genotype	Plasmid Size (kb) [2]
1	R145	WB	AMP–STR–TET–TRS–SUL	bla_{TEM}–*aadA1*–*sul1*–*sul3*–*tetA*	>105
	5662	IP	AMP–DOX–TRS–SUL–CHL	bla_{TEM}–*aadA1*–*sul3*–*Int1*	55
	6011	IP	AMP–STR–TRS–SUL	bla_{TEM}	55
	6012	IP	AMP–STR	bla_{TEM}	ND
	5661	IP	AMP–DOX	bla_{TEM}	55
	5663	IP	AMP–STR–TRS–SUL	bla_{TEM}	ND
	330	IP	AMP–GEN–NEO–STR–TET–DOX–TRS–SUL–COL	*strA*–*strB*–*sul1*–*mcr-1*	55 +> 244 [3]
2	I 82	WB	AMP–NEO–TET–DOX	bla_{TEM}	55
3	M1452	IP	AMP–NEO–STR–TET–DOX–TRS–SUL–CHL	bla_{TEM}–*tetA*–*Int1*(*aadA1*) [4]	55
	I 144	WB	-	-	55
	I 160	WB	NEO	-	55
	I 163	WB	NEO	-	55
4	I 329	IP	AMP–STR–TET–DOX–TRS–SUL	bla_{TEM}–*strA*–*strB*–*sul1*–*Int1*–(*blaPSE1*) [4]	55 + 244
5	R40	WB	STR–TET–DOX–SUL	*aadA1*–*strA*–*strB*–*sul1*–*tetA*	<33 + 55 + 310
	R160	WB	STR–TET–SUL	*strA*–*strB*–*tetA*	<33 + 55 + 240
	I 203	WB	STR–TRS–SUL	*strA*–*strB*–*tetA*	<33 + 55 + 310
6	5649	IP	AMP–TRS–SUL–CHL	bla_{TEM}–*aadA1*–*sul3*–*Int1*	105
	5650	IP	AMP–TRS–SUL–CHL	bla_{TEM}–*aadA1*–*sul3*–*Int1*	105
	5655	IP	AMP–TRS–SUL–CHL	*strA*–*sul3*	105
7	I 36	IP	AMP–NEO–STR–TET–DOX–SUL	*aadA1*–*strA*–*strB*–*sul2*–*tetB*	-

[1] Pulsotype, as deduced from Figure 1. [2] DNA bands detected by PFGE-S1, with size (kb) deduced by proximity to corresponding bands in the *S. braenderup* standard; [3] hybridized to DIG-labeled *mcr-1*; [4] Genes identified in *int1*-linked gene cassettes. None detected. ND, not determined.

Table 2. Prevalence of antimicrobial resistance determinants among *S. Choleraesuis* isolates from Iberian pigs and wild boar in Spain.

Antimicrobials		IP		WB	
		N [1]	Genes [2]	N [1]	Genes [2]
Sulfonamides	Sulfadiazine	10	*sul1* (2), *sul2* (1), *sul3* (4)	4	*sul1* (2), *sul3* (1)
	Cotrimoxazol	9		2	
β-lactams	Ampicillin	12	*bla*$_{TEM}$ (9), *bla*$_{PSE}$ (1)	2	*bla*$_{TEM}$ (2)
Aminoglycosides	Gentamycin	1	-	0	-
	Neomycin	3	-	3	-
	Streptomycin	7	*aadA* (5), *strA* (3), *strB* (3)	4	*aadA* (2), *strA* (3), *strB* (3)
Tetracyclines	Tetracycline	4	*tetA* (1), *tetB* (1)	4	*tetA* (4)
	Doxycycline	6		2	
Phenicols	Chloramphenicol	5		0	-
Polymixins	Colistin	1	*mcr-1* (1)	0	-

[1] Number of isolates sharing resistance to indicated antimicrobial. [2] Number of resistance determinants between parenthesis. None detected.

Antimicrobial resistance determinants were found in all the strains from IP and most (seven out of eight) from WB. The antimicrobial resistance genes detected were highly variable among isolates, with a total of 10 different genotypes, 50% of them with four or more resistance genes (Table 1). Similarly to antimicrobial-resistant phenotypes, genotypes were variable among isolates, even from the same PT, with *bla*$_{TEM}$, *bla*$_{TEM}$–*aadA1*–*sul3*, and *strA*–*strB*–*tetA* found most frequently (Table 1). Considering each resistance gene, the β-lactamase-encoding *bla*$_{TEM}$ was most common with nine strains from IP and two from WB, covering all PT except PT5. However, from WB the most prevalent was *tetA*, found in one isolate from PT1 and three from PT5 in addition to only one isolate from PT3 in IP. The *int1* gene, encoding the class 1 integrase that is frequently linked to antimicrobial resistance gene cassettes, was detected in five isolates, all from IP, although two of them share PT with WB isolates (PT1 and PT3, Figure 1). However, only two isolates presented *int1*-linked gene cassettes of 1000 or 1200 bp length which also coded for *aadA2* or *bla*$_{PSE1}$ genes, respectively (Table 1).

Interestingly, the *mcr-1* (plasmid-mediated colistin resistance) gene was detected in one colistin-resistant isolate from IP belonging to PT1, the most common PT among *S. Choleraesuis* isolates (Table 1). In general, isolates carrying resistance determinants presented low susceptibility to the corresponding antimicrobial(s), although aadA1 and *strA* genes may be expressed weakly or not at all.

2.3. Plasmid Content of S. Choleraesuis Isolates

S1 nuclease treatment and PFGE typing of plasmid content revealed that 19 out the 20 strains carried at least one extrachromosomal molecule of DNA, with five isolates carrying multiple plasmids (Table 1 and Figure 2). The plasmid most frequently found was 50 kb in size, shared by 75% of isolates, including those fully sensitive to antibiotics and lacking resistance genes. Plasmids between 100 and 300 kb were also detected in strains mostly expressing MDR. Due to its clinical relevance, plasmid location was performed for the colistin-resistance *mcr-1* gene identified in this study for the first time in *S. Choleraesuis* isolated from swine (Figure 1). Thus, a plasmid slightly over 240 kb in size was detected that was carrying *mcr-1* from an IP necropsied in 2020, as revealed by specific hybridization with a DIG-labeled probe from a previously characterized sequence [36]. With exceptions, as for the mentioned plasmid carrying *mcr-1* in a PT1 strain, the number and size of plasmids was found to be stable in isolates within every PT.

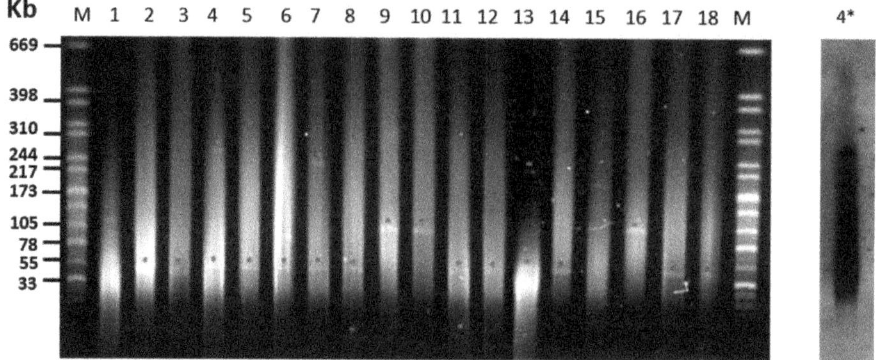

Figure 2. Plasmid analysis of *S. Choleraesuis* isolates from Iberian pigs and wild boar in Spain. PFGE-S1 analysis of isolates. Red asterisks indicate plasmids for which an approximate size has been estimated by comparison with *S. braenderup* molecular weight standards. Analyzed isolates are **M**, *S. braenderup* digested by XbaI; **1**, R145; **2**, 6011; **3**, 5661; **4**, I330; **5**, I160; **6**, I163; **7**, I329; **8**, I203; **9**, 5655; **10**, 5650; **11**, R40; **12**, M1452; **13**, R160 **14**, I82; **15**, 36; **16**, 5649; **17**, I144; **18**, 5662. **4*** Hybridization to a DIG-labeled *mcr-1* probe.

3. Discussion

In this study, isolates of *S. Choleraesuis* from IP and WB have been analyzed in order to trace the spread potential of antimicrobial resistance determinants carried by this serotype in the "dehesa", a traditional agrosystem consisting of grassland with Holm's oaks found in the Iberian Peninsula. The XbaI-PFGE profile of *S. Choleraesuis* isolates revealed different PT, but most of the strains (16/20) belonged to the same cluster with a degree of similarity above 75% (Cluster A), among which PT1, PT3, and PT5 might represent clones with high potential spread both in space and time, in agreement with previous studies in WB [2,3,17] and domestic pigs [5,6]. With regard to phylogeographic analysis, a recent study demonstrated cross-border transmission of *S. Choleraesuis* from pigs between countries that was concordant with the trading network [18]. In our study, genetic relationships were detected not only among bacteria from the same species, but also with the wild ancestor of pigs, the WB, which share the "dehesa" environment with IP [14]. In our study, the geographical link between animals is maximal for WB from estates E4 and E1, the closest to IP farms E6 and E11 (Figure 3) from which *S. Choleraesuis* isolates share PT1, PT2, or PT3 in closely related backgrounds (>95%, Figure 1). On the other hand, it should be noted that there are large distances between these estates; approximately 70 km between E6 and E1 and all of them (E1, E4, E6, and E11) in a radius of 90 km (Figure 3). Apart from the distance, the estates are also separated by several highways (E11 and E4) and a large river (E4). Moreover, one WB isolate from a faraway estate, E12, also shares PT1. When considered together, all these facts suggest that proximity itself is not the main reason for the bacterial relationship and that other factors may be responsible, i.e., human carriers or animal trading, although evidence is lacking. Together with studies showing the spread of *S. Choleraesuis* between WB and domestic pigs [3,18], including asymptomatic WB in Europe [17,37,38], our results show a wildlife reservoir that may spill over to farmed pigs or vice versa.

MDR was detected in 83.3% (10/12) of isolates from IP in this study, higher than the 37.5% (3/8) observed in WB. Similar prevalences of antimicrobial resistance have been reported in *S. Choleraesuis* from domestic pigs in Asia [26,39] but these are higher than previous reports in Europe [5,6]. The data from WB are similar to those previously described by our group [2]. Likewise, the antibiotic groups with higher resistances differ between *S. Choleraesuis* from the analyzed Suidae, showing resistance to ampicillin and sulfonamide for bacteria from IP, and sulfonamide, tetracycline, and streptomycin from WB, although streptomycin resistance had the same ratio in bacteria from both hosts, similarly to previous reports [2,3,6,40,41]. The lack of resistance found against quinolones and cephalosporins is in accordance with most of the studies from Europe [18,42], although outbreaks or sporadic cases of infections caused by *Salmonella* spp. with resistance to these antibiotics are being increasingly reported [39,43].

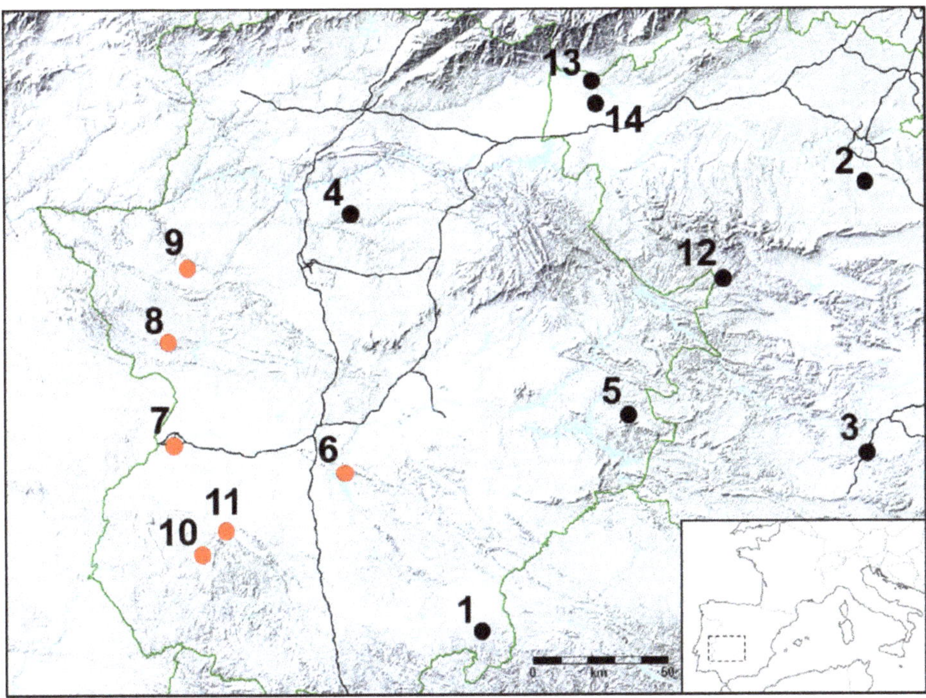

Figure 3. Location of the estates. Geographical map of the southwest Iberian Peninsula displaying the central location of the different estates from where *S. Choleraesuis* suid hosts were sampled. Black dots represent estates where WB were sampled and red dots IP farms. Black lines represent highways and green lines administrative division limits (inset: location of the Iberian Peninsula in southwestern Europe).

Isolates of *S. Choleraesuis* from the two hosts screened in this study, IP and WB, presented quantitative differences in antibiotic resistance found against ampicillin and trimethoprim/sulfamethoxazole, which are higher in the autochthonous pig breed. In contrast, resistance to chloramphenicol, gentamicin or colistin was only detected in IP. This could be due to the fact that many of these antibiotics have been extensively used as growth promoters (beta-lactams) or as prophylactic agents for common diseases such as colibacillosis (colistin) or coccidiosis (sulfonamides) in pig farms for a long time [44], which has been associated with increases in resistant bacteria [45]. Although the IP production system is linked to the dehesa in the last period of fattening, the first stages of breeding mostly take place on farms with semi intensive management systems. It was in these stages where antibiotic abuse has taken place in the past. Considering that frequent use has a stronger association with resistance than sporadic use [46,47], it could explain the lower number of resistances found in WB, as the treatments applied to them, when applied, are scarce and limited to certain short periods of time, which is different to the IP, especially in the early stages of breeding. However, even on estates that did not apply any antibiotic treatment, antibiotic resistances were found in *S. Choleraesuis* from WB. This could be due to the omnivorous behavior of WB, which means they visit communal refuse sites as well as the proximity of farmed animals like IP in free range production systems, where horizontal transmission of bacteria might occur [48,49].

Resistance genes have been previously detected in *S. Choleraesuis* from pigs and humans [18,28,41], but information is scarce in WB [2]. In this study, we described Class 1 integrons in 42% of the *S. Choleraesuis* isolates from IP and none in WB. Around 41% of these integrons carried a resistance gene cassette. These genetic elements play an important role in the development of antibiotic resistance and have a worldwide distribution in Gram-negative bacteria, colonizing both humans and animals [50]. In *S. Choleraesuis* from pigs, finding these elements in a large number of isolates is very common [39,51]. Interestingly, our study shows that the *sul3* gene occurs in 3 out of 5 Salmonella isolates carrying class 1 integrons, although the presence of this integron is more frequently related to the spread of the *sul1* gene [52,53].

Our study reveals that, with exceptions, *S. Choleraesuis* strains from IP or WB carry plasmids which are around 50 kb in size (Figure 1), that isolates lacking antimicrobial resistance did not present additional plasmids, and that bacteria expressing multiple antimicrobial resistance share mega-plasmids, alone or in addition to the 50 kb bands (Table 1). The fact that only closely related isolates share plasmid bands and/or antimicrobial resistance patterns might suggest that clonal spread prevails over horizontal transfer as the common mechanism for dispersion of antimicrobial resistance determinants in the analyzed environment. This study also shows the presence of the colistin-resistant gene *mcr-1* in one of the isolates studied from IP. In this strain, *mcr-1* is carried by a high-molecular weight plasmid (>240 kb), possibly conferring MDR and most probably belonging to the IncHI2-type replicon (different to the recent finding in a human isolate) [35], which could represent a risk for accumulation and/or spread antimicrobial resistance determinants through food chain environments, as for Iberian pigs, and their processed products and humans. Although more studies are needed to determine its prevalence, due to its clinical importance in human health, the presence of these colistin-resistant Salmonella isolates should be monitored in order to control their evolution.

4. Materials and Methods

4.1. Bacterial Strains and Animal Sources

The 20 strains of *S. Choleraesuis* isolated from diseased WB ($n = 8$) and IP ($n = 12$) were analyzed at the Clinical Veterinary Hospital (CVH) at the University of Extremadura. The animals were submitted to the CVH by veterinarians or by a hunting management company (Ingulados S.L.) from Cáceres, Spain, in order to determine the cause of death and control disease on their farms/game estates. After routine necropsy and microbiological analysis, those animals with *S. Choleraesuis* were included in the study. Each isolate from WB was derived from a different outbreak (clinical disease in several animals in a short period of time) and estate (E1–E11), whilst IP belonged to six different estates, among which several animals from the same outbreak were sampled in E6 and E7. All fourteen estates were located in the Central West region of the Iberian Peninsula (Figure 3). The IP estates were either breeding farms connected to a large enclosure of the dehesa ecosystem or just an enclosure where IP underwent a fattening process. The fences or walls that surround those enclosures are strong enough to keep the IP inside, but not enough to prevent the entry of WB and their subsequent interactions with IP. The WB came from game estates where they were occasionally fed and subjected to periodical health inspections, where they are captured, analyzed, and returned to their natural environment.

The clinical isolates came from different organs (liver, kidneys, lungs, and spleen) and were cultured on blood agar, MacConkey agar, and xylose–lysine–deoxycholate agar (XLD) under aerobic conditions for 24 h/37 °C. Colonies compatible with *Salmonella* were confirmed using conventional microbiological methodologies and identified as *Salmonella enterica* serovar Choleraesuis based on *fliC* gene PCR [54].

4.2. Pulsed-Field Gel Electrophoresis (PFGE) Analysis

Determination of the dendrogram of PFGE clusters among isolates of *S. Choleraesuis* was performed by macrorestriction with XbaI followed by PFGE (Chef-DR®III. Bio-Rad; Hercules, CA, USA) according to the PulseNet protocol with pulses oscillating from 2.16 to 63.8 s for 21.5 h [55], and *S. braenderup* was used as the molecular weight standard. The gel was stained with ethidium bromide, and DNA bands were visualized with an UV transilluminator. Images were prepared using Quantity One software (Bio-Rad; Hercules, CA, USA). The different PFGE profiles (PT) were analyzed by InfoQuest FP Software (Version 4.5).

Plasmid size analysis was performed by PFGE under the same conditions described above after incubation of plugs with S1 nuclease (Thermo Fisher, Waltham, MA, USA) according to manufacturer's recommendations. For plasmid hybridization, PFGE was transferred to a nylon membrane and hybridized to a digoxigenin-labeled *mcr-1* probe that was PCR amplified from a previously described *E. coli* strain [36]. Digoxigenin labeling and detection were performed according to manufacturer's instructions (Merck; Darmstadt, Germany).

4.3. Antibiotic Susceptibility Testing

Antibiotic susceptibility was tested by the disc-diffusion method on Mueller–Hinton agar (Kirby–Bauer method) to 13 antimicrobial agents. The following discs (Bio-Rad; Hercules, CA, USA) were used: ampicillin (AMP-10 µg), cefotaxime (CTA-30 µg), ceftiofur (CTF-30 µg) gentamicin (GEN-10 µg), neomycin (NEO-30 µg), streptomycin (STR-10 µg), tetracycline (TET-30 µg), doxycycline (DOX-30 µg), enrofloxacin (ENR-5 µg), nalidixic acid (NAL-30 µg), trimethoprim/sulfamethoxazole (TRS-23.75/1.25 µg), sulfonamide (SUL-200 µg), and chloramphenicol (CHL-30 µg). *E. coli* ATCC 25922 was used as a control strain. Colistin (COL) was not included due to its incompatibility with the disc-diffusion method, but it was tested by MIC determination using the broth microdilution reference method according to ISO 20776–1:2006. Data were interpreted using EUCAST epidemiological cut-off values (www.EUCAST.org).

4.4. Screening for Antibiotic Resistance Genes

After antimicrobial susceptibility testing, resistant strains were screened by PCR for putative determinants with primers and previously described experimental conditions. The following resistance genes were analyzed: *bla*$_{TEM}$ [56], *bla*$_{OXA}$ [57], *tetA* and *tetB* [58], *strA* and *strB* [59], *aadA1* [60], *aph2* [61], *sul1*, *sul2* and *sul3* [62], and *mcr-1* [63]. The Class 1 integrase gene (*int1*) was screened for in all isolates [64] and the presence of a variable region linked to the Class 1 integron was amplified by PCR and sequenced to determine the composition of its gene cassette [65].

5. Conclusions

S. Choleraesuis from IP and WB raised in close environments were found clonally related and transfer antimicrobial resistance determinants mainly by vertical transmission, whereas megaplasmids were detected linked to MDR, including colistin resistance in a single isolate carrying *mcr*-1. The role of *S. Choleraesuis* in the spread of antimicrobial resistance between wild and domestic swine should be carefully surveyed.

Author Contributions: Conceptualization, A.Q. and J.R.; methodology, A.G.; validation, M.G.M., and S.G.Z.; formal analysis, F.E.M.-C. and D.R.; investigation, M.G.M.; resources, W.G.-J. and P.F.-L.; writing—original draft preparation, M.G.M. and A.G.; writing—review and editing, F.E.M.-C. and A.Q.; supervision, A.Q. and J.R.; project administration, A.Q. and J.R.; funding acquisition, A.Q. and J.R. All authors have read and agreed to the published version of the manuscript.

Funding: Work in A.Q. lab is funded by the Spanish Ministry of Economy, Industry and Competitiveness (MINECO, currently MICINN, Grant AGL2016-74882-C3), and the Junta de Extremadura and FEDER (IB16073, IB18047 and GR15075) of Spain. A. G. thanks his current contract (Extremadura government and European Social Fund).

Acknowledgments: The authors thanks Gemma Hannah Louise Gaitskell-Phillips for language editing assistance.

Conflicts of Interest: The authors declare no conflict of interest.

References

1. Khan, M.A.; Cao, L.; Riaz, A.; Li, Y.; Jiao, Q.; Liu, Z.; Wang, H.; Meng, F.; Ma, Z. The Harm of Salmonella to Pig Industry and Its Control Measures. *Int. J. Appl. Agric. Sci.* **2019**, *5*, 24.
2. Gil Molino, M.; Risco Pérez, D.; Gonçalves Blanco, P.; Fernandez Llario, P.; Quesada Molina, A.; García Sánchez, A.; Cuesta Gerveno, J.M.; Gómez Gordo, L.; Martín Cano, F.E.; Pérez Martínez, R. Outbreaks of antimicrobial resistant Salmonella Choleraesuis in wild boars piglets from central-western Spain. *Transbound Emerg. Dis.* **2019**, *66*, 225–233. [CrossRef]
3. Methner, U.; Heller, M.; Bocklisch, H. Salmonella enterica subspecies enterica serovar Choleraesuis in a wild boar population in Germany. *Eur. J. Wildl. Res.* **2010**, *56*, 493–502. [CrossRef]
4. Fedorka-Cray, P.J.; Gray, J.T.; Wray, C. Salmonella infections in pigs. In *Salmonella in Domestic Animals*; Wray, C., Wray, A., Eds.; CABI: London, UK, 2000; pp. 191–207.
5. Asai, T.; Namimatsu, T.; Osumi, T.; Kojima, A.; Harada, K.; Aoki, H.; Sameshima, T.; Takahashi, T. Molecular typing and antimicrobial resistance of Salmonella enterica subspecies enterica serovar Choleraesuis isolates from diseased pigs in Japan. *Comp. Immunol. Microbiol. Infect. Dis.* **2010**, *33*, 109–119. [CrossRef]
6. Pedersen, K.; Sørensen, G.; Löfström, C.; Leekitcharoenphon, P.; Nielsen, B.; Wingstrand, A.; Aarestrup, F.M.; Hendriksen, R.S.; Baggesen, D.L. Reappearance of Salmonella serovar Choleraesuis var. Kunzendorf in Danish pig herds. *Vet. Microbiol.* **2015**, *176*, 282–291. [CrossRef]
7. Baggesen, D.L.; Christensen, J.; Jensen, T.K.; Skov, M.; Sørensen, G.; Sørensen, V. Outbreak of salmonellosis caused by Salmonella enterica subsp. enterica serovar. choleraesuis var. Kunzendorf (*S. choleraesuis*) on a Danish pig farm. *Dan. Veterinærtidsskrift* **2000**, *83*, 6–12.
8. Conedera, G.; Ustulin, M.; Barco, L.; Bregoli, M.; Re, E.; Vio, D. Outbreak of atypical *Salmonella* Choleraesuis in wild boar in North Eastern Italy. In *Trens in Game Meat Hygiene*; Paulsen, P., Bauer, A.F.J.M.S., Eds.; Academic Publishers: Wageningen, The Netherlands, 2014; pp. 151–159.
9. Perez, J.; Astorga, R.; Carrasco, L.; Mendez, A.; Perea, A.; Sierra, M. Outbreak of salmonellosis in farmed European wild boars (Sus scrofa ferus). *Vet. Rec.* **1999**, *145*, 464–465. [CrossRef]
10. Müller, M.; Weber, A.; Tucher, R.; Naumann, L. Case report: Salmonella choleraesuis as a cause of haematogenous osteomyelitis in a wild boar (Sus scrofa). Fallbericht: Osteomyelitis bei einem wildschwein (Sus scrofa) durch salmonella choleraesuis. *Tierarztl. Umsch.* **2004**, *59*, 700–702.
11. Longo, A.; Petrin, S.; Mastrorilli, E.; Tiengo, A.; Lettini, A.A.; Barco, L.; Ricci, A.; Losasso, C.; Cibin, V. Characterizing Salmonella enterica serovar Choleraesuis, var. Kunzendorf: A comparative case study. *Front. Vet. Sci.* **2019**, *6*, 316. [CrossRef]
12. Rodrigáñez, J.; Silió, L.; Rillo, S.M. El cerdo Ibérico y su sistema de producción. *Anim. Genet. Resour. Resour. Génétiques Anim. Recur. Genéticos Anim.* **1993**, *12*, 89–96. [CrossRef]
13. Rodríguez-Prieto, V.; Kukielka, D.; Martínez-López, B.; de las Heras, A.I.; Barasona, J.Á.; Gortázar, C.; Sánchez-Vizcaíno, J.M.; Vicente, J. Porcine reproductive and respiratory syndrome (PRRS) virus in wild boar and Iberian pigs in south-central Spain. *Eur J. Wildl. Res.* **2013**, *59*, 859–867. [CrossRef]
14. Carrasco García de León, R. Factores de Riesgo de Transmisión de Enfermedades en Ungulados Cinegéticos del Centro y sur de España. Available online: https://ruidera.uclm.es/xmlui/handle/10578/9753 (accessed on 30 December 2016).
15. Navarro-Gonzalez, N.; Mentaberre, G.; Porrero, C.M.; Serrano, E.; Mateos, A.; López-Martín, J.M.; Lavín, S.; Domínguez, L. Effect of cattle on Salmonella carriage, diversity and antimicrobial resistance in free-ranging wild boar (Sus scrofa) in northeastern Spain. *PLoS ONE* **2012**, *7*, e51614. [CrossRef] [PubMed]
16. Vittecoq, M.; Godreuil, S.; Prugnolle, F.; Durand, P.; Brazier, L.; Renaud, N.; Arnal, A.; Aberkane, S.; Jean-Pierre, H.; Gauthier-Clerc, M. Antimicrobial resistance in wildlife. *J. Appl. Ecol.* **2016**, *53*, 519–529. [CrossRef]
17. Gil Molino, M.; García Sánchez, A.; Risco Pérez, D.; Gonçalves Blanco, P.; Quesada Molina, A.; Rey Pérez, J.; Martín Cano, F.E.; Cerrato Horrillo, R.; Hermoso-de-Mendoza Salcedo, J.; Fernández Llario, P. Prevalence of Salmonella spp. in tonsils, mandibular lymph nodes and faeces of wild boar from Spain and genetic relationship between isolates. *Transbound. Emerg. Dis.* **2019**, *66*, 1218–1266. [CrossRef]

18. Leekitcharoenphon, P.; Sørensen, G.; Löfström, C.; Battisti, A.; Szabo, I.; Wasyl, D.; Slowey, R.; Zhao, S.; Brisabois, A.; Kornschober, C.; et al. Cross-Border Transmission of Salmonella Choleraesuis var. Kunzendorf in European Pigs and Wild Boar: Infection, Genetics, and Evolution. *Front. Microbiol.* **2019**, *10*, 179. [CrossRef]

19. Chiu, C.H.; Su, L.H.; Chu, C. Salmonella enterica Serotype Choleraesuis: Epidemiology, Pathogenesis, Clinical Disease, and Treatment. *Clin. Microbiol. Rev.* **2004**, *17*, 311–322. [CrossRef]

20. Chen, P.; Wu, C.; Chang, C.; Lee, H.; Lee, N.; Shih, H.; Lee, C.; Ko, N.; Wang, L.; Ko, W. Extraintestinal focal infections in adults with Salmonella enterica serotype Choleraesuis bacteremia. *J. Microbiol. Immunol. Infect.* **2007**, *40*, 240–247.

21. Jean, S.; Wang, J.; Hsueh, P. Bacteremia caused by Salmonella enterica serotype Choleraesuis in Taiwan. *J. Microbiol. Immunol. Infect.* **2006**, *39*, 358.

22. Chiu, C.-H.; Wu, T.-L.; Su, L.-H.; Chu, C.; Chia, J.-H.; Kuo, A.-J.; Chien, M.-S.; Lin, T.-Y. The Emergence in Taiwan of Fluoroquinolone Resistance in Salmonella enterica Serotype Choleraesuis. *N. Engl. J. Med.* **2002**, *346*, 413–419. [CrossRef]

23. Ferstl, P.G.; Reinheimer, C.; Jozsa, K.; Zeuzem, S.; Kempf, V.A.; Waidmann, O.; Grammatikos, G. Severe infection with multidrug-resistant Salmonella choleraesuis in a young patient with primary sclerosing cholangitis. *World J. Gastroenterol.* **2017**, *23*, 2086. [CrossRef]

24. Goldstein, C.; Lee, M.D.; Sanchez, S.; Hudson, C.; Phillips, B.; Register, B.; Grady, M.; Liebert, C.; Summers, A.O.; White, D.G. Incidence of class 1 and 2 integrases in clinical and commensal bacteria from livestock, companion animals, and exotics. *Antimicrob. Agents Chemother.* **2001**, *45*, 723–726. [CrossRef] [PubMed]

25. Bass, L.; Liebert, C.A.; Lee, M.D.; Summers, A.O.; White, D.G.; Thayer, S.G.; Maurer, J.J. Incidence and characterization of integrons, genetic elements mediating multiple-drug resistance, in avianEscherichia coli. *Antimicrob. Agents Chemother.* **1999**, *43*, 2925–2929. [CrossRef]

26. Chu, C.; Chiu, C.-H.; Wu, W.-Y.; Chu, C.-H.; Liu, T.-P.; Ou, J.T. Large Drug Resistance Virulence Plasmids of Clinical Isolates of Salmonella enterica Serovar Choleraesuis. *Antimicrob. Agents Chemother.* **2001**, *45*, 2299–2303. [CrossRef] [PubMed]

27. Tzeng, J.-I.; Chu, C.-H.; Chen, S.-W.; Yeh, C.-M.; Chiu, C.-H.; Chiou, C.-S.; Lin, J.-H.; Chu, C. Reduction of Salmonella enterica serovar Choleraesuis carrying large virulence plasmids after the foot and mouth disease outbreak in swine in southern Taiwan, and their independent evolution in human and pig. *J. Microbiol. Immunol. Infect.* **2012**, *45*, 418–425. [CrossRef]

28. Sirichote, P.; Hasman, H.; Pulsrikarn, C.; Schønheyder, H.C.; Samulioniené, J.; Pornruangmong, S.; Bangtrakulnonth, A.; Aarestrup, F.M.; Hendriksen, R.S. Molecular characterization of extended-spectrum cephalosporinase-producing Salmonella enterica serovar Choleraesuis isolates from patients in Thailand and Denmark. *J. Clin. Microbiol.* **2010**, *48*, 883–888. [CrossRef]

29. Chen, C.-L.; Su, L.-H.; Janapatla, R.P.; Lin, C.-Y.; Chiu, C.-H. Genetic analysis of virulence and antimicrobial-resistant plasmid pOU7519 in Salmonella enterica serovar Choleraesuis. *J. Microbiol. Immunol. Infect.* **2020**, *53*, 49–59. [CrossRef]

30. Catry, B.; Cavaleri, M.; Baptiste, K.; Grave, K.; Grein, K.; Holm, A.; Jukes, H.; Liebana, E.; Navas, A.L.; Mackay, D. Use of colistin-containing products within the European Union and European Economic Area (EU/EEA): Development of resistance in animals and possible impact on human and animal health. *Int. J. Antimicrob. Agents* **2015**, *46*, 297–306. [CrossRef]

31. Darwich, L.; Vidal, A.; Seminati, C.; Albamonte, A.; Casado, A.; López, F.; Molina-López, R.A.; Migura-Garcia, L. High prevalence and diversity of extended-spectrum β-lactamase and emergence of OXA-48 producing Enterobacterales in wildlife in Catalonia. *PLoS ONE* **2019**, *14*, e0210686. [CrossRef]

32. Guenther, S.; Falgenhauer, L.; Semmler, T.; Imirzalioglu, C.; Chakraborty, T.; Roesler, U.; Roschanski, N. Environmental emission of multiresistant Escherichia coli carrying the colistin resistance gene mcr-1 from German swine farms. *J. Antimicrob. Chemother.* **2017**, *72*, 1289–1292.

33. Rhouma, M.; Beaudry, F.; Thériault, W.; Letellier, A. Colistin in Pig Production: Chemistry, Mechanism of Antibacterial Action, Microbial Resistance Emergence, and One Health Perspectives. *Front. Microbiol.* **2016**, *7*, 1789. [CrossRef]

34. Lima, T.; Domingues, S.; Da Silva, G.J. Plasmid-mediated colistin resistance in Salmonella enterica: A review. *Microorganisms* **2019**, *7*, 55. [CrossRef] [PubMed]

35. Santos, C.A.d.; Cunha, M.P.V.; Bertani, A.M.d.J.; de Almeida, E.A.; Gonçalves, C.R.; Sacchi, C.T.; de Paiva, J.B.; Camargo, C.H.; Tiba-Casas, M.R. Detection of multidrug-and colistin-resistant Salmonella Choleraesuis causing bloodstream infection. *J. Antimicrob. Chemother.* **2020**, *75*, 2009–2010. [CrossRef]

36. Sánchez-Benito, R.; Iglesias, M.R.; Quijada, N.M.; Campos, M.J.; Ugarte-Ruiz, M.; Hernández, M.; Pazos, C.; Rodríguez-Lázaro, D.; Garduño, E.; Domínguez, L. Escherichia coli ST167 carrying plasmid mobilisable mcr-1 and blaCTX-M-15 resistance determinants isolated from a human respiratory infection. *Int. J. Antimicrob. Agents* **2017**, *50*, 285. [CrossRef] [PubMed]

37. Zottola, T.; Montagnaro, S.; Magnapera, C.; Sasso, S.; De Martino, L.; Bragagnolo, A.; D'Amici, L.; Condoleo, R.; Pisanelli, G.; Iovane, G. Prevalence and antimicrobial susceptibility of Salmonella in European wild boar (*Sus scrofa*); Latium Region–Italy. *Comp. Immunol. Microbiol. Infect. Dis.* **2013**, *36*, 161–168. [PubMed]

38. Chiari, M.; Zanoni, M.; Tagliabue, S.; Lavazza, A.; Alborali, L.G. Salmonella serotypes in wild boars (Sus scrofa) hunted in northern Italy. *Acta Vet. Scand.* **2013**, *55*, 1–4. [CrossRef]

39. Chang, C.-C.; Lin, Y.-H.; Chang, C.-F.; Yeh, K.-S.; Chiu, C.-H.; Chu, C.; Chien, M.-S.; Hsu, Y.-M.; Tsai, L.-S.; Chiou, C.-S. Epidemiologic relationship between fluoroquinolone-resistant Salmonella enterica serovar Choleraesuis strains isolated from humans and pigs in Taiwan (1997 to 2002). *J. Clin. Microbiol.* **2005**, *43*, 2798–2804. [CrossRef]

40. Donazzolo, C.; Turchetto, S.; Ustulin, M.; Citterio, C.; Conedera, G.; Vio, D.; Cocchi, M. Antimicrobial susceptibility of Salmonella enterica subsp. enterica serovar Choleraesuis strains forum wild boar (Sus scrofa) in Italy. In *Game Meat Hygiene*; Paulsen, P.B.A., Smulders, F.J.M., Eds.; Wageningen Academic Publishers: Wageningen, The Netherlands, 2017; p. 307.

41. Hsu, Y.-M.; Tang, C.-Y.; Lin, H.; Chen, Y.-H.; Chen, Y.-L.; Su, Y.-H.; Chen, D.S.; Lin, J.-H.; Chang, C.-C. Comparative study of class 1 integron, ampicillin, chloramphenicol, streptomycin, sulfamethoxazole, tetracycline (ACSSuT) and fluoroquinolone resistance in various Salmonella serovars from humans and animals. *Comp. Immunol. Microbiol. Infect. Dis.* **2013**, *36*, 9–16. [CrossRef]

42. Chiu, C.-H.; Su, L.-H. Salmonella, Non-Typhoidal Species (*S. choleraesuis, S. enteritidis, S. hadar, S. typhimurium*). Available online: http://www.antimicrobe.org/b258.asp (accessed on 28 October 2020).

43. Su, L.H.; Chiu, C.H.; Chu, C.; Ou, J.T. Antimicrobial resistance in nontyphoid Salmonella serotypes: A global challenge. *Clin. Infect. Dis.* **2004**, *39*, 546–551. [CrossRef]

44. Prescott, J.F. Sulfonamides, Diaminopyrimidines, and Their Combinations. In *Antimicrobial Therapy in Veterinary Medicine*, 5th ed.; Giguère, S., Prescott, J.F., Dowling, P.M., Eds.; John Wiley & Sons, Inc.: Ames, IA, USA, 2013.

45. Nature, E. The antibiotic alarm. *Nature* **2013**, *495*, 141.

46. Sato, T.; Okubo, T.; Usui, M.; Yokota, S.-I.; Izumiyama, S.; Tamura, Y. Association of Veterinary Third-Generation Cephalosporin Use with the Risk of Emergence of Extended-Spectrum-Cephalosporin Resistance in Escherichia coli from Dairy Cattle in Japan. *PLoS ONE* **2014**, *9*, e96101. [CrossRef]

47. Olesen, S.W.; Barnett, M.L.; MacFadden, D.R.; Brownstein, J.S.; Hernández-Díaz, S.; Lipsitch, M.; Grad, Y.H. The distribution of antibiotic use and its association with antibiotic resistance. *Elife* **2018**, *7*, e39435. [CrossRef] [PubMed]

48. Navarro-Gonzalez, N.; Casas-Díaz, E.; Porrero, C.M.; Mateos, A.; Domínguez, L.; Lavín, S.; Serrano, E. Food-borne zoonotic pathogens and antimicrobial resistance of indicator bacteria in urban wild boars in Barcelona, Spain. *Vet. Microbiol.* **2013**, *167*, 686–689. [CrossRef] [PubMed]

49. Navarro-Gonzalez, N.; Castillo-Contreras, R.; Casas-Díaz, E.; Morellet, N.; Porrero, M.C.; Molina-Vacas, G.; Torres, R.T.; Fonseca, C.; Mentaberre, G.; Domínguez, L. Carriage of antibiotic-resistant bacteria in urban versus rural wild boars. *Eur. J. Wildl. Res.* **2018**, *64*, 60. [CrossRef]

50. Fluit, A.; Schmitz, F.J. Resistance integrons and super-integrons. *Clin. Microbiol. Infect.* **2004**, *10*, 272–288. [CrossRef]

51. Lee, M.-F.; Chen, Y.-H.; Peng, C.-F. Molecular characterisation of class 1 integrons in Salmonella enterica serovar Choleraesuis isolates from southern Taiwan. *Int. J. Antimicrob. Agents* **2009**, *33*, 216–222. [CrossRef]

52. Sköld, O. Resistance to trimethoprim and sulfonamides. *Vet. Res.* **2001**, *32*, 261–273. [CrossRef]

53. Antunes, P.; Machado, J.; Sousa, J.C.; Peixe, L. Dissemination of sulfonamide resistance genes (sul1, sul2, and sul3) in Portuguese Salmonella enterica strains and relation with integrons. *Antimicrob. Agents Chemother.* **2005**, *49*, 836–839. [CrossRef]

54. Chiu, T.H.; Pang, J.C.; Hwang, W.Z.; Tsen, H.Y. Development of PCR primers for the detection of Salmonella enterica serovar Choleraesuis based on the fliC gene. *J. Food Prot.* **2005**, *68*, 1575–1580. [CrossRef]

55. Ribot, E.M.; Fair, M.; Gautom, R.; Cameron, D.; Hunter, S.; Swaminathan, B.; Barrett, T.J. Standardization of pulsed-field gel electrophoresis protocols for the subtyping of Escherichia coli O157: H7, Salmonella, and Shigella for PulseNet. *Foodbourne Pathog. Dis.* **2006**, *3*, 59–67. [CrossRef]

56. Briñas, L.; Zarazaga, M.; Sáenz, Y.; Ruiz-Larrea, F.; Torres, C. β-Lactamases in ampicillin-resistant Escherichia coli isolates from foods, humans, and healthy animals. *Antimicrob. Agents Chemother.* **2002**, *46*, 3156–3163. [CrossRef]

57. Chen, X.; Gao, S.; Jiao, X.; Liu, X.F. Prevalence of serogroups and virulence factors of Escherichia coli strains isolated from pigs with postweaning diarrhoea in eastern China. *Vet. Microbiol.* **2004**, *103*, 13–20. [CrossRef]

58. Sengeløv, G.; Agersø, Y.; Halling-Sørensen, B.; Baloda, S.B.; Andersen, J.S.; Jensen, L.B. Bacterial antibiotic resistance levels in Danish farmland as a result of treatment with pig manure slurry. *Environ. Int.* **2003**, *28*, 587–595. [CrossRef]

59. Aarestrup, F.M.; Lertworapreecha, M.; Evans, M.C.; Bangtrakulnonth, A.; Chalermchaikit, T.; Hendriksen, R.S.; Wegener, H.C. Antimicrobial susceptibility and occurrence of resistance genes among Salmonella enterica serovar Weltevreden from different countries. *J. Antimicrob. Chemother.* **2003**, *52*, 715–718. [CrossRef]

60. Hendriksen, R.S.; Bangtrakulnonth, A.; Pulsrikarn, C.; Pornreongwong, S.; Hasman, H.; Song, S.W.; Aarestrup, F.M. Antimicrobial resistance and molecular epidemiology of Salmonella Rissen from animals, food products, and patients in Thailand and Denmark. *Foodborne Pathog. Dis.* **2008**, *5*, 605–619. [CrossRef] [PubMed]

61. Rahmani, M.; Peighambari, S.M.; Svendsen, C.A.; Cavaco, L.M.; Agersø, Y.; Hendriksen, R.S. Molecular clonality and antimicrobial resistance in Salmonella enterica serovars Enteritidis and Infantis from broilers in three Northern regions of Iran. *BMC Vet. Res.* **2013**, *9*, 66. [CrossRef] [PubMed]

62. Kozak, G.K.; Boerlin, P.; Janecko, N.; Reid-Smith, R.J.; Jardine, C. Antimicrobial resistance in Escherichia coli isolates from swine and wild small mammals in the proximity of swine farms and in natural environments in Ontario, Canada. *Appl. Environ. Microbiol.* **2009**, *75*, 559–566. [CrossRef]

63. Liu, Y.-Y.; Wang, Y.; Walsh, T.R.; Yi, L.-X.; Zhang, R.; Spencer, J.; Doi, Y.; Tian, G.; Dong, B.; Huang, X. Emergence of plasmid-mediated colistin resistance mechanism MCR-1 in animals and human beings in China: A microbiological and molecular biological study. *Lancet Infect. Dis.* **2016**, *16*, 161–168. [CrossRef]

64. Leverstein-van Hall, M.; Paauw, A.; Box, A.; Blok, H.; Verhoef, J.; Fluit, A. Presence of integron-associated resistance in the community is widespread and contributes to multidrug resistance in the hospital. *J. Clin. Microbiol.* **2002**, *40*, 3038–3040. [CrossRef]

65. Levesque, C.; Piche, L.; Larose, C.; Roy, P.H. PCR mapping of integrons reveals several novel combinations of resistance genes. *Antimicrob. Agents Chemother.* **1995**, *39*, 185–191. [CrossRef]

Publisher's Note: MDPI stays neutral with regard to jurisdictional claims in published maps and institutional affiliations.

 *antibiotics*

Communication

Emergence of *cfr*-Mediated Linezolid Resistance in *Staphylococcus aureus* Isolated from Pig Carcasses

Hee Young Kang, Dong Chan Moon, Abraham Fikru Mechesso, Ji-Hyun Choi, Su-Jeong Kim, Hyun-Ju Song, Mi Hyun Kim, Soon-Seek Yoon and Suk-Kyung Lim *

Bacterial Disease Division, Animal and Plant Quarantine Agency, 177 Hyeksin 8-ro, Gimcheon-si, Gyeongsangbuk-do 39660, Korea; kanghy7734@korea.kr (H.Y.K.); ansehdcks@korea.kr (D.C.M.); abrahamf@korea.kr (A.F.M.); wlgus01@korea.kr (J.-H.C.); kimsujeong27@gmail.com (S.-J.K.); shj0211@korea.kr (H.-J.S.); kimmh940301@naver.com (M.H.K.); yoonss24@korea.kr (S.-S.Y.)
* Correspondence: imsk0049@korea.kr; Tel.: +82-54-912-0738

Received: 5 October 2020; Accepted: 30 October 2020; Published: 2 November 2020

Abstract: Altogether, 2547 *Staphylococcus aureus* isolated from cattle (n = 382), pig (n = 1077), and chicken carcasses (n = 1088) during 2010–2017 were investigated for linezolid resistance and were further characterized using molecular methods. We identified linezolid resistance in only 2.3% of pig carcass isolates. The linezolid-resistant (LR) isolates presented resistance to multiple antimicrobials, including chloramphenicol, clindamycin, and tiamulin. Molecular investigation exhibited no mutations in the 23S ribosomal RNA. Nevertheless, we found mutations in ribosomal proteins rplC (G121A) and rplD (C353T) in one and seven LR strains, respectively. All the LR isolates carried the multi-resistance gene *cfr*, and six of them co-carried the *mecA* gene. Additionally, all the LR isolates co-carried the phenicol exporter gene, *fexA*, and presented a high level of chloramphenicol resistance. LR *S. aureus* isolates represented 10 genotypes, including major genotypes ST433-t318, ST541-t034, ST5-t002, and ST9-t337. Staphylococcal enterotoxin and leukotoxin-encoding genes, alone or in combination, were detected in 68% of LR isolates. Isolates from different farms presented identical or different pulsed-field gel electrophoresis patterns. Collectively, toxigenic and LR *S. aureus* strains pose a crisis for public health. This study is the first to describe the mechanism of linezolid resistance in *S. aureus* isolated from food animal products in Korea.

Keywords: carcass; *cfr* gene; *fexA* gene; linezolid; mutation; pig; public health; *S. aureus*

1. Introduction

Linezolid belongs to the oxazolidinone antibiotics and is approved for the treatment of severe infections caused by multidrug-resistant Gram-positive pathogens including methicillin-resistant *Staphylococcus aureus* (MRSA) and vancomycin-resistant enterococci [1]. Linezolid interferes with the peptidyltransferase site of the bacterial ribosome. This leads to disruption of protein synthesis and inhibition of bacterial growth [2]. However, the emergence of linezolid-resistant (LR) staphylococci and enterococci poses a significant and interdisciplinary public health challenge [1].

Mutation in the central loop of domain V of 23S ribosomal ribonucleic acid (rRNA) (C2161T) is the primary mechanism of linezolid resistance. However, redundancy of rRNA genes makes it difficult to reach sufficient levels of resistance by a mutation in a single allele [3]. In addition, rRNA mutations often negatively affect ribosome functions and are rapidly reversed in the absence of selection [4]. Therefore, the resistance mechanism based on chemical modification of rRNA such as the acquisition of the multi-resistance gene *cfr* is more common [5]. Linezolid resistance is also associated with mutations in the genes coding for ribosomal proteins (L3 and L4). Moreover, the novel *optrA* and *poxtA* genes have been implicated in transferrable linezolid resistance [1,6].

The *cfr* gene was initially described in a bovine *Staphylococcus sciuri* isolate [7]. It catalyzes the methylation of A2503 in the 23S rRNA of the large ribosomal subunit [8]. The methylation leads to cross-resistance against several antimicrobial classes of drugs (phenicols, lincosamides, pleuromutilins, macrolides, oxazolidinones, and streptogramin A), conferring multidrug resistance [9,10]. Therefore, these antimicrobials can mediate selective pressure in favor of the *cfr* gene. The *cfr* gene was mostly identified on plasmids, often in close proximity to insertion sequences, which play a vital role in the mobility of the *cfr* gene [11]. These mobile structures have been detected among several Gram-positive and Gram-negative bacteria, including bacteria other than staphylococci, *Enterococcus faecalis*, and *Escherichia coli* [9].

The occurrence of LR *S. aureus* in humans and food animals has been increasingly reported in many countries [6,12–15]. Previous studies in South Korea (Korea) demonstrated the occurrence of linezolid resistance in *S. aureus* strains recovered from hospital patients [16,17], pigs, and chicken carcasses [18]. Despite a single report on the occurrence of the *cfr* gene in MRSA recovered from pig carcasses [19], to our knowledge, no attempt has been made on the detailed mechanism of linezolid resistance among *S. aureus* isolates recovered from animal sources in Korea to date. Korea's meat consumption has increased in the past few years, with pork remaining the most popular source. Thus, continuous surveillance on the emergence of antimicrobial-resistant bacterial strains in animal carcasses is essential to safeguard public health. In this study, we aimed to determine the occurrence of linezolid resistance in *S. aureus* isolated from major food animal carcasses from 2010 to 2017, as well as to study the underlying mechanism(s) of resistance.

2. Results and Discussion

2.1. Prevalence and Antimicrobial Susceptibility Profiles of LR S. aureus

Linezolid resistance was detected in 2.3% of *S. aureus* isolated from pigs (Table 1). The low linezolid resistance rate among pig isolates in this study was consistent with those reported in Korea in 2011 (2.9%) [18] and 2015 (0.14%) [19], but lower than a recent report in South Africa (14%) [14]. Similarly, *S. aureus* isolated from medical centers in various countries presented very low linezolid resistance rates ($\approx$1%) [13,20,21]. Agreeing with our recent report [19], resistance was not observed among cattle and chicken isolates. In contrast, previous studies in Korea [18] and South Africa [14] reported that 1.2% and 9% of chicken and cattle carcass isolates, respectively, were resistant to linezolid. Linezolid is not approved for animal use in Korea. Thus, the observed difference in resistance could be associated with the frequent use of phenicols, pleuromutilins, and lincosamides in the Korean pig industry, which might co-select resistance to linezolid [22]. The detection of LR *S. aureus* strains in pig carcasses is worrisome because of the potential transmission to humans through the food supply chain.

Table 1. Prevalence of linezolid resistance in *Staphylococcus aureus* isolated from food animal carcasses in South Korea from 2010 to 2017.

Year	% (No. of Linezolid-Resistant Isolates/No. of Isolates)			
	Cattle	Pig	Chicken	Total
2010	0 (0/39)	0 (0/70)	0 (0/81)	0 (0/190)
2011	0 (0/69)	0 (0/101)	0 (0/137)	0 (0/307)
2012	0 (0/76)	9.8 (12/122)	0 (0/201)	3 (12/399)
2013	0 (0/49)	1.7 (3/178)	0 (0/133)	0.8 (3/360)
2014	0 (0/62)	1.1 (2/182)	0 (0/168)	0.5 (2/412)
2015	0 (0/41)	2.5 (4/160)	0 (0/195)	1 (4/396)
2016	0 (0/29)	1.9 (3/158)	0 (0/77)	1.1 (3/264)
2017	0 (0/17)	0.9 (1/106)	0 (0/96)	0.5 (1/219)
Total	0 (0/382)	2.3 (25/1077)	0 (0/1088)	1.0 (25/2547)

2.2. Mutations and Antimicrobial Resistance Genes

Spontaneous mutation in the multiple copies of 23S rRNA alleles is the primary mechanism of linezolid resistance [23]. None of the identified LR isolates exhibited this type of mutation. Resistance mediated by mutations in the 23S rRNA appears rarely, develops slowly, and is not transferrable between bacterial species [24]. However, all of the identified LR isolates carried the *cfr* gene (Table 2), which is associated with low-level linezolid resistance [6]. Previous studies have also identified *cfr*-harboring *S. aureus* mainly from humans and to a lesser extent from food animals in various countries, including Korea [15,19,25,26]. Notably, all but two of the *cfr*-carrying isolates were from different farms. The extensive dissemination of *cfr*-carrying strains among pig farms could be related to the association of the *cfr* gene to mobile elements [9], which facilitates the mobilization and horizontal transfer [27]. Moreover, the low fitness cost could attribute to the wide dissemination of the *cfr* gene. Previous studies have demonstrated that genes that come at low cost can stably persist in the cells, even when pathogens were not exposed to antibiotics [27–30].

Linezolid resistance mediated by the *cfr* gene has also been shown to coexist with other resistance mechanisms [17]. We identified mutations in ribosomal proteins rplC (G121A) and rplD (C353T) in one and seven LR strains, respectively (Table 2). These types of mutations have been linked with resistance or decreased susceptibility to linezolid [31]. The difference in linezolid resistance mechanisms between human isolates, mutations in the 23S rRNA gene [17], and pig isolates in this study indicates the presence of unique clones in the pig industry.

All the identified *cfr*-carrying isolates were resistant to multiple antimicrobials including chloramphenicol, clindamycin, and tiamulin, and co-carried phenicol exporter gene *fexA* (Table 2). The *cfr* gene has been reported to confer resistance to antimicrobials that are widely used in veterinary medicine, such as macrolides, tetracyclines, phenicols, and lincosamides [5]. Previous studies have also shown the co-existence of the *cfr* gene and other resistance genes, which facilitates its co-selection and spread [26,32]. Moreover, six of the LR strains co-carried the *mecA* gene. The co-existence of the *mecA* and *cfr* genes is an unwelcome development because linezolid is among the last resort of antimicrobial agents against MRSA in humans.

Table 2. Characteristics of linezolid-resistant *S. aureus* isolated from pig carcasses.

Isolate	Year	Provinces	Farm ID	MIC (µg/mL)					Other Resistance Phenotype	Genetic Resistance Marker								MLST	Spa Type	SCCmec Type	Virulence Patterns	Pulso Type
				LNZ	CHL	CLI	TIA	SYN		mecA	cfr	fexA	optrA	poxtA	23S rRNA	rplC	rplD					
V02-12-023	2012	Gyeonggi	GG-1	8	>64	>4	>4	>4	ERY, GEN, KAN, PEN, TMP	-	+	+	-	-	WT	WT	WT	5	t002	-	*seg-sei-sem-sen-seo*	A
V02-12-027	2012	Chungnam	CN-1	8	>64	>4	>4	4	FOX, PEN, TET	+	+	+	-	-	WT	WT	WT	398	t034	V		ND
V04-12-005	2012	Chungnam	CN-2	16	>64	>4	>4	2	GEN, KAN, PEN, TET	-	+	+	-		WT	WT	WT	5	t002	-	*seg-sei-sem-sen-seo-lukED*	A
V08-12-002	2012	Gyeongbuk	GB-1	8	>64	>4	>4	>4	FOX, CIP, ERY, GEN, KAN, PEN, TET	+	+	+	-	-	WT	WT	WT	541	t034	V		ND
V13-12-013	2012	Gyeongbuk	GB-2	16	>64	>4	>4	4	GEN, KAN, PEN, TET	-	+	+	-	-	WT	WT	C353T	433	t318	-	*seg*	B
V14-12-001	2012	Chungnam	CN-3	8	>64	>4	>4	4	TET	-	+	+	-	-	WT	WT	C353T	433	t318	-	*seg*	B
V14-12-008	2012	Chungnam	CN-3	16	>64	>4	>4	4	FOX, ERY, PEN, TET	+	+	+	-	-	WT	WT	WT	541	t034	V		ND
V14-12-011	2012	Gyeonggi	GG-2	16	>64	>4	>4	2	FOX, ERY, PEN, TET	+	+	+	-	-	WT	WT	WT	541	t034	V		ND
V14-12-012	2012	Incheon	IC-1	8	>64	>4	>4	>4	FOX, ERY, PEN, TET	+	+	+	-	-	WT	WT	WT	541	t034	V		ND
V14-12-015	2012	Chungnam	CN-4	8	>64	>4	>4	>4	CIP, ERY, GEN, KAN, PEN, TET, TMP	-	+	+	-	-	WT	WT	WT	541	t034	-		ND
V14-12-016	2012	Chungnam	CN-5	16	>64	>4	>4	4	-	-	+	+	-	-	WT	WT	C353T	433	t318	-	*seg*	B
V14-12-017	2012	Gyeonggi	GG-3	16	>64	>4	>4	4	-	-	+	+	-	-	WT	WT	C353T	433	t318	-	*seg*	B
V04-13-019	2013	Chungbuk	CB-1	16	>64	>4	>4	4	PEN	-	+	+	-	-	WT	WT	WT	9	t337	-	*seg-sei-sem-sen-seo*	C
V04-13-032	2013	Chungnam	CN-6	16	>64	>4	>4	4	PEN	-	+	+	-	-	WT	WT	WT	9	t337	-	*seg-sei-sem-sen-seo*	C
V08-13-003	2013	Gyeongbuk	GB-3	8	>64	>4	>4	4	PEN	-	+	+	-	-	WT	WT	WT	5	t548	-	*seg-sei-sem-sen-seo-lukED*	A
V04-14-023	2014	Chungbuk	CB-2	8	>64	>4	>4	2	PEN	-	+	+	-	-	WT	G121A	WT	5	t002	-	*seg-sei-sem-sen-seo-lukED*	A-1
V14-14-006	2014	Chungnam	CN-7	8	>64	>4	>4	4	CIP, GEN, KAN, PEN	-	+	+	-	-	WT	WT	C353T	433	t318	-	*seg*	B
V02-15-007	2015	Gyeonggi	GG-4	8	>64	>4	>4	2	GEN, KAN, PEN	-	+	+	-	-	WT	WT	WT	2007	t8314	-	*seg-sei-sem-sen-seo*	D
V14-15-002	2015	Incheon	IC-2	8	>64	>4	>4	2	TET	-	+	+	-	-	WT	WT	C353T	433	t318	-	*seg*	B
V14-15-016	2015	Incheon	IC-3	8	>64	>4	>4	>4	FOX, ERY, PEN, TET	+	+	+	-	-	WT	WT	WT	541	t034	V		ND
V15-15-012	2015	Jeonnam	JN-1	8	>64	>4	>4	4	PEN	-	+	+	-	-	WT	WT	WT	9	t337	-	*seg-sei-sem-sen-seo*	C
V03-16-003	2016	Gangwon	GW-1	8	>64	>4	>4	4	GEN, KAN, PEN	-	+	+	-	-	WT	WT	WT	5	t002	-	*seg-sei-sem-sen-seo-lukED*	A
V06-16-007	2016	Jeonbuk	JB-1	8	>64	>4	>4	2	PEN, TET	-	+	+	-	-	WT	WT	WT	9	t899	-	*seg-sei-sem-sen-seo*	C-1
V14-16-004	2016	Gyeonggi	GG-5	8	>64	>4	>4	4	CIP, ERY, PEN, TET, TMP	-	+	+	-	-	WT	WT	WT	398	t1170	-		ND
V13-17-011	2017	Gyeongbuk	GB-4	8	64	>4	>4	4	-	-	+	+	-	-	WT	WT	C353T	433	t021	-	*seg*	B

Abbreviations: LNZ, linezolid; CHL, chloramphenicol; CLI, clindamycin; TIA, tiamulin; SYN, quinupristin/dalfopristin; FOX, cefoxitin; CIP, ciprofloxacin; ERY, erythromycin; GEN, gentamicin; KAN, kanamycin; PEN, penicillin; TET, tetracycline; TMP, trimethoprim; WT, wild type; MLST, multi-locus sequence type. SCCmec typing was performed for methicillin-resistant *Staphylococcus aureus* (MRSA) strains only.

2.3. Molecular Characteristics of LR S. aureus Isolates

The potential risk of transmission of *cfr*-carrying *S. aureus* between pigs and humans is a growing health concern. In this study, the 25 LR-isolates belonged to ST433-t318 (n = 6); ST541-t034 (n = 6); ST5-t002 (n = 4); ST9-t337 (n = 3); and each of ST5-t548, ST9-t899, ST398-t034, ST398-t1170, ST433-t021, and ST2007-t8314. Five of these lineage types (ST9, ST398, ST433, ST541, and ST2007) were livestock-associated (LA) strains, while ST5 *S. aureus* was the only human-associated (HA) strain. Except for ST2007, all the LA and HA strains were reported in pigs and farmers in Korea [19,33,34], indicating the possibility of transmission between pigs and humans. Korea is one of the markets with the fastest growing consumption of pork in the world. Hence, the emergence of *cfr*-carrying *S. aureus* with unique molecular characteristics in pig carcasses is concerning.

We identified LR *S. aureus* strains with sequence type (ST2007) and spa types (ST5-t548 and ST433-t318) that had not been reported in Korea, suggesting the emergence of new clones that carried the *cfr* gene and/or have mutations in ribosomal protein rplD. Although the linezolid resistance profiles are unknown, the ST5-t548 [35], ST433-t318 [36], and ST2007-t8314 [37] strains were detected in humans and/or pigs in China, Poland, and the United States, respectively. Moreover, we observed LR-ST398 *S. aureus* carrying a novel spa type (t1170) in farm GG-5, suggesting an evolutionary change in *S. aureus*.

Staphylococcal enterotoxin and leukotoxin-encoding genes, alone or in combination, were detected in 68% of LR isolates: *seg* (28%, 7/25), *seg-sei-sem-sen-seo* (24%, 6/25), and *seg-sei-sem-sen-seo-lukED* (16%, 4/25) (Table 2). Eight (32%) isolates, including the five MRSA strains, were negative for any of the tested virulence factor genes. In agreement with Price et al. [38], the HA-ST5 strains appeared to be more virulent than the LA strains. Additionally, multiple virulence factor genes were detected in one of the LA strains, ST9. *S. aureus* harboring the classical enterotoxins and leucotoxins can spread to humans either through contact or via the food chain and are capable of causing food-related illnesses in humans [39].

Analysis using pulsed-field gel electrophoresis (PFGE) revealed four distinct PFGE types, with identical PFGE types in isolates belonging to the same sequence types (Figure S1). Isolates from different farms in the same or different provinces presented identical or different PFGE patterns. These results might suggest cross-contamination in the slaughterhouse, or clonal dissemination and/or persistence of specific clones among farms, not only within a province but also in different provinces.

3. Materials and Methods

3.1. Sample Collection and Isolation of S. aureus

A total of 2547 *S. aureus* isolates (382 cattle, 1077 pig, and 1088 chicken carcass isolates) were obtained from 16 laboratories/centers participating in the Korean Veterinary Antimicrobial Resistance Monitoring System. Sample collection and isolation of *S. aureus* were performed as described previously [19]. Briefly, the back and chest of cattle and pig carcasses were swabbed with sterile gauze pads wetted with buffered peptone water (BPW) (Becton Dickinson, Sparks, MD, USA), while the whole carcasses of chickens were rinsed in Phosphate Buffered Water (PBW). Homogenized samples were inoculated into tryptic soy broth (Becton Dickinson) containing 6.5% sodium chloride and incubated at 37 °C for 16 h. Following incubation, one or two loops from each enrichment broth were streaked onto mannitol salt agar (Difco, Detroit, MI, USA). Suspected colonies were then identified by matrix-assisted laser desorption/ionization time-of-flight mass spectrometry (Biomerieux, Marcy L'Etoile, France). *S. aureus* and MRSA isolates were further confirmed by a multiplex polymerase chain reaction (PCR) assay specific for the *16S rRNA*, *clfA*, and *mecA* genes [40].

3.2. Antimicrobial Susceptibility Testing and Detection of Resistance Genes

Linezolid susceptibility was determined by the broth dilution method [41], using linezolid-containing plates (1–8 µg/mL) (EUST, TREK Diagnostics Systems, Cleveland, OH). The LR isolates were screened for the presence of *cfr*, *fexA*, *optrA*, and *poxtA* genes using PCR [6,42]. The susceptibility profiles of the identified LR isolates were further evaluated for the following 19 antimicrobial agents using antibiotic-containing plates (EUST, TREK Diagnostics Systems, Cleveland, OH): cefoxitin (0.5–16 µg/mL), chloramphenicol (4–64 µg/mL), ciprofloxacin (0.25–8 µg/mL), clindamycin (0.12–4 µg/mL), erythromycin (0.25–8 µg/mL), fusidic acid (0.5–4 µg/mL), gentamicin (1–32 µg/mL), kanamycin (4–64 µg/mL), mupirocin (0.5–256 µg/mL), penicillin (0.12–2 µg/mL), quinupristin/dalfopristin (0.5–4 µg/mL), rifampin (0.02–0.5 µg/mL), streptomycin (1–16 µg/mL), sulfamethoxazole (64–512 µg/mL), tetracycline (0.5–16 µg/mL), tiamulin (0.5–4 µg/mL), trimethoprim (2–32 µg/mL), and vancomycin (1–16 µg/mL). Briefly, approximately 5×10^5 colony forming unit (cfu)/mL inoculums, prepared from overnight cultures, were inoculated on the minimum inhibitory concentration (MIC) panels and incubated at 35 °C for 20–24 h. *S. aureus* ATCC 25,923 was used as a reference strain. The MIC values were interpreted according to the Clinical and Laboratory Standards Institute [41] and the European Committee on Antimicrobial Susceptibility Testing [43] guidelines.

3.3. Detection of Mutations

The central loop of domain V of the 23S rRNA and the genes encoding ribosomal proteins L3 (*rplC*) and L4 (*rplD*) were amplified using primers, as described previously [17,32]. The nucleotide and amino acid sequences of *rplC*, *rplD*, and domain V of the 23S rRNA gene, for each of the isolates tested, were compared with those of the wild-type linezolid-susceptible *S. aureus* ATCC29213 strain (GenBank accession no. NZ_MOPB01000038.1). Analysis and comparison were performed using the basic local alignment search tool (BLAST) program (http://www.ncbi.nlm.nih.gov/BLAST) and ExPASY proteomics tools (http://www.expasy.ch/tools/#similarity).

3.4. Molecular Typing of LR S. aureus

The LR isolates were further characterized by multilocus sequence typing (MLST). Sequences of the PCR products were compared with sequences available on the MLST website for *S. aureus* [44]. *S. aureus* protein A (spa) typing was performed using the method described by Enright et al. [45], and the spa types were assigned using the Ridom Staph Type server (Ridom GmbH, Wurzburg, Germany) (www.spaserver.ridom.de). Additionally, the staphylococcal cassette chromosome mec (*SCCmec*) typing was carried out in all LR isolates that harbored the *mecA* gene using PCR [46]. The detection of genes encoding the virulence determinants such as Panton–Valentine leucocidin (PVL), leukotoxins (lukED), exfoliatins (*eta* and *etb*), toxic shock syndrome toxin 1 (*tsst1*), and staphylococcal enterotoxins (*sea, seb, sec, sed, see, seg, seh, sei, selj, sek, sell, sem, sen, seo, sep, seq,* and ser) was performed by PCR [47]. The isolates were also investigated for three genes (*scn, chp,* and *sak*) that represent components of the immune evasion cluster [48].

Pulsed-field gel electrophoresis (PFGE) analysis of *SmaI*-digested chromosomal DNA was performed to investigate clonality [49]. Briefly, chromosomal DNA sample plugs were digested with 50 U of *SmaI* (Takara Bio, Otsu, Japan) and separated by electrophoresis on 1.0% SeaKem Gold agarose (Lonza, Allendale, NJ, USA) in 0.5× Tris–borate–Ethylenediaminetetraacetic acid EDTA buffer at 14 °C for 20 h using a CHEF-Mapper (Bio-Rad, Hercules, CA, USA) with the following parameters: initial switch time, 5.3 s; final switch time, 34.9 s; angle, 120°; gradient, 6.0 V/cm; ramping factor, linear. Results were analyzed using Bionumerics software, version 4.0 (Applied Maths, Sint-Martens-Latem, Belgium), and relatedness was calculated using the unweighted pair-group method with arithmetic averages (UPGMA) algorithm, on the basis of the Dice similarity index.

4. Conclusions

The occurrence of linezolid resistance is still rare among *S. aureus* isolates from animal carcasses. Nevertheless, we detected the multi-resistance gene *cfr* and the novel phenicol exporter gene *fexA* among all the LR *S. aureus* isolated from pigs. Mutations in ribosomal proteins rplC and rplD were also detected in some of the strains. Resistant strains could be transmitted to humans through the food supply chain, subsequently limiting the treatment options for multidrug-resistant *S. aureus*. Therefore, frequent screening of pig carcasses, farmers, and slaughterhouse environments, as well as thorough cooking of pig meat, should be implemented to detect the emergence and persistence of toxigenic and LR *S. aureus* strains in order to prevent dissemination to humans.

Supplementary Materials: The following are available online at http://www.mdpi.com/2079-6382/9/11/769/s1, Figure S1: *Sma I*-digested pulse-field gel electrophoresis patterns of linezolid-resistant *S. aureus* isolated from pig carcasses in Korea. Genomic DNA of ST398 and ST541 are not digested by *SmaI*, and hence pulsed-field gel electrophoresis (PFGE) patterns were not determined.

Author Contributions: Conceptualization, S.-K.L., and D.C.M.; methodology, H.Y.K., A.F.M., and D.C.M.; software, A.F.M., H.Y.K., J.-H.C., and S.-J.K.; validation, A.F.M., M.H.K., and H.-J.S.; formal analysis, H.-J.S., and M.H.K.; investigation, A.F.M., H.Y.K., H.-J.S., M.H.K., J.-H.C., and S.-J.K.; data curation, D.C.M., and H-J.S.; writing—original draft preparation, H.Y.K., and A.F.M.; writing—review and editing, A.F.M., S.-S.Y., S.-K.L., and D.C.M.; supervision, S.-S.Y., S.-K.L., and D.C.M.; project administration, D.C.M. and H.Y.K.; funding acquisition; S.K.L. All authors have read and agreed to the published version of the manuscript.

Funding: This research was funded from the Animal and Plant Quarantine Agency, Ministry of Agriculture, Food, and Rural Affairs, Korea, grant number N-1543081-2017-24-01.

Conflicts of Interest: The authors declare no conflict of interest. The funders had no role in the design of the study; in the collection, analyses, or interpretation of data; in the writing of the manuscript; or in the decision to publish the results.

References

1. Mendes, R.E.; Hogan, P.A.; Jones, R.N.; Sader, H.S.; Flamm, R.K. Surveillance for linezolid resistance via the Zyvoxw Annual Appraisal of Potency and Spectrum (ZAAPS) programme (2014): Evolving resistance mechanisms with stable susceptibility rates. *J. Antimicrob. Chemother.* **2016**, *71*, 1860–1865. [CrossRef] [PubMed]

2. Wang, Y.; Lv, Y.; Cai, J.; Schwarz, S.; Cui, L.; Hu, Z.; Zhang, R.; Li, J.; Zhao, Q.; He, T.; et al. A novel gene, *optrA*, that confers transferable resistance to oxazolidinones and phenicols and its presence in *Enterococcus faecalis* and *Enterococcus faecium* of human and animal origin. *J. Antimicrob. Chemother.* **2015**, *70*, 2182–2190. [CrossRef] [PubMed]

3. Lobritz, M.; Hutton-Thomas, R.; Marshall, S.; Rice, L.B. Recombination proficiency influences frequency and locus of mutational resistance to linezolid in *Enterococcus faecalis*. *Antimicrob. Agents Chemother.* **2003**, *47*, 3318–3320. [CrossRef] [PubMed]

4. Wolter, N.; Smith, A.M.; Farrell, D.J.; Klugman, K.P. Heterogeneous macrolide resistance and gene conversion in the pneumococcus. *Antimicrob. Agents Chemother.* **2006**, *50*, 359–361. [CrossRef] [PubMed]

5. Toh, S.M.; Xiong, L.; Arias, C.A.; Villegas, M.V.; Lolans, K.; Quinn, J.; Mankin, A.S. Acquisition of a natural resistance gene renders a clinical strain of methicillin-resistant *Staphylococcus* aureus resistant to the synthetic antibiotic linezolid. *Mol. Microbiol.* **2007**, *64*, 1506–1514. [CrossRef] [PubMed]

6. Wang, J.; Lin, D.C.; Guo, X.M.; Wei, H.K.; Liu, X.Q.; Chen, X.J.; Guo, J.Y.; Zeng, Z.L.; Liu, J.H. Distribution of the multidrug resistance gene *cfr* in *Staphylococcus* isolates from pigs, workers, and the environment of a hog market and a slaughterhouse in Guangzhou, China. *Foodborne Pathog. Dis.* **2015**, *12*, 598–605. [CrossRef]

7. Schwarz, S.; Werckenthin, C.; Kehrenberg, C. Identification of a plasmid borne chloramphenicol-florfenicol resistance gene in *Staphylococcus sciuri*. *Antimicrob. Agents Chemother.* **2000**, *44*, 2530–2533. [CrossRef] [PubMed]

8. Mendes, R.E.; Deshpande, L.M.; Jones, R.N. Linezolid update: Stable in vitro activity following more than a decade of clinical use and summary of associated resistance mechanisms. *Drug Resist. Updates* **2014**, *17*, 1–12. [CrossRef] [PubMed]

9. Shen, J.; Wang, Y.; Schwarz, S. Presence and dissemination of the multiresistance gene *cfr* in Gram-positive and Gram-negative bacteria. *J. Antimicrob. Chemother.* **2013**, *68*, 1697–1706. [CrossRef]

10. Long, K.S.; Poehlsgaard, J.; Kehrenberg, C.; Schwarz, S.; Vester, B. The *cfr* rRNA methyltransferase confers resistance to phenicols, lincosamides, oxazolidinones, pleuromutilins, and streptogramin A antibiotics. *Antimicrob. Agents Chemother.* **2006**, *50*, 2500–2505. [CrossRef]

11. Wendlandt, S.; Shen, J.; Kadlec, K.; Wang, Y.; Li, B.; Zhang, W.J.; Feßler, A.T.; Wu, C.; Schwarz, S. Multidrug resistance genes in staphylococci from animals that confer resistance to critically and highly important antimicrobial agents in human medicine. *Trends Microbiol.* **2015**, *23*, 44–54. [CrossRef]

12. Kosowska-Shick, K.; Julian, K.G.; McGhee, P.L.; Appelbaum, P.C.; Whitener, C.J. Molecular and epidemiologic characteristics of linezolid-resistant coagulase-negative *Staphylococci* at a tertiary care hospital. *Diagn. Microbiol. Infect. Dis.* **2010**, *68*, 34–39. [CrossRef]

13. Ross, J.E.; Farrell, D.J.; Mendes, R.E.; Sader, H.S.; Jones, R.N. Eight-year (2002–2009) summary of the Linezolid (Zyvox®annual appraisal of potency and spectrum; ZAAPS) program in European countries. *J. Chemother.* **2011**, *23*, 71–76. [CrossRef] [PubMed]

14. Pekana, A.; Green, E. Antimicrobial resistance profiles of *Staphylococcus aureus* isolated from meat carcasses and bovine milk in abattoirs and dairy farms of the Eastern Cape, South Africa. *Int. J. Environ. Res. Public Health.* **2018**, *15*, 2223. [CrossRef]

15. Li, D.; Wu, C.; Wang, Y.; Fan, R.; Schwarz, S.; Zhang, S. Identification of multiresistance gene *cfr* in methicillin-resistant *Staphylococcus aureus* from pigs: Plasmid location and integration into a staphylococcal cassette chromosome mec complex. *Antimicrob. Agents Chemother.* **2015**, *59*, 3641–3644. [CrossRef] [PubMed]

16. An, D.; Lee, M.Y.; Jeong, T.D.; Sung, H.; Kim, M.N.; Hong, S.B. Co-emergence of linezolid-resistant *Staphylococcus aureus* and *Enterococcus faecium* in a patient with methicillin-resistant *S. aureus* pneumonic sepsis. *Diagn. Microbiol. Infect. Dis.* **2011**, *69*, 232–233. [CrossRef]

17. Yoo, I.Y.; Kang, O.K.; Shim, H.J.; Huh, H.J.; Lee, N.Y. Linezolid resistance in methicillin-resistant *Staphylococcus aureus* in Korea: High rate of false resistance to linezolid by the VITEK 2 system. *Ann. Lab. Med.* **2020**, *40*, 57–62. [CrossRef] [PubMed]

18. Lim, S.K.; Nam, H.M.; Jang, G.C.; Kim, S.R.; Chae, M.H.; Jung, S.C.; Kang, D.; Kim, J. Antimicrobial resistance in *Staphylococcus aureus* isolated from raw meats in slaughterhouse in Korea during 2010. *Korean J. Vet. Publ. Health* **2011**, *35*, 231–238.

19. Moon, D.C.; Tamang, M.D.; Nam, H.M.; Jeong, J.H.; Jang, G.C.; Jung, S.C.; Park, Y.H.; Lim, S.K. Identification of livestock-associated methicillin-resistant *Staphylococcus aureus* isolates in Korea and molecular comparison between isolates from animal carcasses and slaughterhouse workers. *Foodborne Pathog. Dis.* **2015**, *12*, 327–334. [CrossRef]

20. Farrell, D.J.; Mendes, R.E.; Ross, J.E.; Jones, R.N. Linezolid surveillance program results for 2008 (LEADER Program for 2008). *Diagn. Microbiol. Infect. Dis.* **2009**, *65*, 392–403. [CrossRef] [PubMed]

21. Jones, R.N.; Kohno, S.; Ono, Y.; Ross, J.E.; Yanagihara, K. ZAAPS International Surveillance Program (2007) for linezolid resistance: Results from 5591 Gram-positive clinical isolates in 23 countries. *Diagn. Microbiol. Infect. Dis.* **2009**, *64*, 191–201. [CrossRef]

22. Tamang, M.D.; Moon, D.C.; Kim, S.R.; Kang, H.Y.; Lee, K.; Nam, H.M.; Jang, G.C.; Lee, H.S.; Jung, S.C.; Lim, S.K. Detection of novel oxazolidinone and phenicol resistance gene *optrA* in *Enterococcal* isolates from food animals and animal carcasses. *Vet. Microbiol.* **2017**, *201*, 252–256. [CrossRef]

23. Patel, S.N.; Memari, N.; Shahinas, D.; Toye, B.; Jamieson, F.B.; Farrell, D.J. Linezolid resistance in *Enterococcus faecium* isolated in Ontario, Canada. *Diagn. Microbiol. Infect. Dis.* **2013**, *77*, 350–353. [CrossRef] [PubMed]

24. Stefani, S.; Bongiorno, D.; Mongelli, G.; Campanile, F. Linezolid resistance in staphylococci. *Pharmaceuticals* **2010**, *3*, 1988–2006. [CrossRef] [PubMed]

25. Argudín, M.A.; Tenhagen, B.A.; Fetsch, A.; Sachsenröder, J.; Käsbohrer, A.; Schroeter, A.; Hammer, J.A.; Hertwig, S.; Helmuth, R.; Bräunig, J.; et al. Virulence and resistance determinants of German *Staphylococcus aureus* ST398 isolates from nonhuman sources. *Appl. Environ. Microbiol.* **2011**, *77*, 3052–3060. [CrossRef] [PubMed]

26. Kehrenberg, C.; Cuny, C.; Strommenger, B.; Schwarz, S.; Witte, W. Methicillin-resistant and -susceptible *Staphylococcus aureus* strains of clonal lineages ST398 and ST9 from swine carry the multidrug resistance gene *cfr*. *Antimicrob. Agents Chemother.* **2009**, *53*, 779–781. [CrossRef] [PubMed]

27. LaMarre, J.M.; Locke, J.B.; Shaw, K.J.; Mankin, A.S. Low fitness cost of the multidrug resistance gene *cfr*. *Antimicrob. Agents Chemother.* **2011**, *55*, 3714–3719. [CrossRef]

28. Andersson, D.I.; Hughes, D. Antibiotic resistance and its cost: Is it possible to reverse resistance? *Nat. Rev. Microbiol.* **2010**, *8*, 260–271. [CrossRef]

29. Foucault, M.L.; Depardieu, F.; Courvalin, P.; Grillot-Courvalin, C. Inducible expression eliminates the fitness cost of vancomycin resistance in enterococci. *Proc. Natl. Acad. Sci. USA* **2010**, *107*, 16964–16969. [CrossRef]

30. Rodriguez-Rojas, A.; Macia, M.D.; Couce, A.; Gomez, C.; Castaneda-Garcia, A.; Oliver, A.; Blazquez, J. Assessing the emergence of resistance: The absence of biological cost in vivo may compromise fosfomycin treatments for *P. aeruginosa* infections. *PLoS ONE* **2010**, *5*, e10193. [CrossRef]

31. Locke, J.B.; Hilgers, M.; Shaw, K.J. Mutations in ribosomal protein L3 are associated with oxazolidinone resistance in Staphylococci of clinical origin. *Antimicrob. Agents Chemother.* **2009**, *53*, 5275–5278. [CrossRef]

32. Gales, A.C.; Deshpande, L.M.; De Souza, A.G.; Pignatari, A.C.C.; Mendes, R.E. MSSA ST398/t034 carrying a plasmid-mediated *Cfr* and *Erm(B)* in Brazil. *J. Antimicrob. Chemother.* **2015**, *70*, 303–305. [CrossRef] [PubMed]

33. Lim, S.K.; Nam, H.M.; Jang, G.C.; Lee, H.S.; Jung, S.C.; Kwak, H.S. The first detection of methicillin-resistant *Staphylococcus aureus* ST398 in pigs in Korea. *Vet. Microbiol.* **2012**, *155*, 88–92. [CrossRef] [PubMed]

34. Moon, D.C.; Jeong, S.K.; Hyun, B.H.; Lim, S.K. Prevalence and characteristics of methicillin-resistant *Staphylococcus aureus* isolates in pigs and pig farmers in Korea. *Foodborne Pathog. Dis.* **2019**, *16*, 256–261. [CrossRef]

35. Li, X.; Fang, F.; Zhao, J.; Lou, N.; Li, C.; Huang, T.; Li, Y. Molecular characteristics and virulence gene profiles of *Staphylococcus aureus* causing bloodstream infection. *Braz. J. Infect. Dis.* **2018**, *22*, 487–494. [CrossRef] [PubMed]

36. Mroczkowska, A.; Zmudzki, J.; Marszaøek, N.; Orczykowska-Kotyna, M.; Komorowska, I.; Nowak, A.; Grzesiak, A.; Czyzewska-Dors, E.; Dors, A.; Pejsak, Z.; et al. Livestock-Associated *Staphylococcus aureus* on Polish pig farms. *PLoS ONE* **2017**, *12*, e0170745. [CrossRef] [PubMed]

37. Sun, J.; Yang, M.; Sreevatsan, S.; Bender, J.B.; Singer, R.S.; Knutson, T.P.; Marthaler, D.G.; Davies, P.R. Longitudinal study of *Staphylococcus aureus* colonization and infection in a cohort of swine veterinarians in the United States. *BMC Infect. Dis.* **2017**, *17*, 690. [CrossRef]

38. Price, L.B.; Stegger, M.; Hasman, H.; Aziz, M.; Larsen, J.; Andersen, P.S.; Pearson, T.; Waters, A.E.; Foster, J.T.; Schupp, J.; et al. *Staphylococcus aureus* CC398: Host adaptation and emergence of methicillin resistance in livestock. *MBio* **2012**, *3*, 1–6. [CrossRef]

39. Kérouanton, A.; Hennekinne, J.A.; Letertre, C.; Petit, L.; Chesneau, O.; Brisabois, A.; De Buyser, M.L. Characterization of *Staphylococcus aureus* strains associated with food poisoning outbreaks in France. *Int. J. Food Microbiol.* **2007**, *115*, 369–375. [CrossRef]

40. Mason, W.J.; Blevins, J.S.; Beenken, K.; Wibowo, N.; Ojha, N.; Smeltzer, M.S. Multiplex PCR protocol for the diagnosis of staphylococcal infection. *J. Clin. Microbiol.* **2001**, *39*, 3332–3338. [CrossRef]

41. Clinical and Laboratory Standards Institute. *Performance Standards for Antimicrobial Susceptibility Testing; Twentieth Informational Supplement*; CLSI Document; Clinical and Laboratory Standards Institute: Wayne, PA, USA, 2018; p. M100.

42. Kehrenberg, C.; Schwarz, S. Distribution of florfenicol resistance genes *fexA* and *cfr* among chloramphenicol-resistant *Staphylococcus* isolates. *Antimicrob. Agents Chemother.* **2006**, *50*, 1156–1163. [CrossRef]

43. European Committee on Antimicrobial Susceptibility Testing. Breakpoint Tables for Interpretation of MICs and Zone Diameters. 2018, Version 8.1. Available online: http://www.eucast.org (accessed on 7 July 2020).

44. *Staphylococcus aureus* MLST Database. Available online: https://pubmlst.org/saureus/ (accessed on 9 August 2020).

45. Enright, M.C.; Day, N.P.J.; Davies, C.E.; Peacock, S.J.; Spratt, B.G. Multilocus sequence typing for characterization of methicillin-resistant and methicillin-susceptible clones of *Staphylococcus aureus*. *J. Clin. Microbiol.* **2000**, *38*, 1008–1015. [CrossRef] [PubMed]

46. Boyle-Vavra, S.; Daum, R.S. Community-acquired methicillin-resistant *Staphylococcus aureus*: The role of Panton-Valentine leukocidin. *Lab. Investig.* **2007**, *87*, 3–9. [CrossRef]

47. Van Duijkeren, E.; Ikawaty, R.; Broekhuizen-Stins, M.J.; Jansen, M.D.; Spalburg, E.C.; de Neeling, A.J.; Allaart, J.G.; van Nes, A.; Wagenaar, J.A.; Fluit, A.C. Transmission of methicillin-resistant *Staphylococcus aureus* strains between different kinds of pig farms. *Vet. Microbiol.* **2008**, *126*, 383–389. [CrossRef]

48. Sung, J.M.L.; Lloyd, D.H.; Lindsay, J.A. *Staphylococcus aureus* host specificity: Comparative genomics of human versus animal isolates by multi-strain microarray. *Microbiology* **2008**, *154*, 1949–1959. [CrossRef] [PubMed]

49. McDougal, L.K.; Steward, C.D.; Killgore, G.E.; Chaitram, J.M.; McAllister, S.K.; Tenover, F.C. Pulsed-field gel electrophoresis typing of oxacillin-resistant *Staphylococcus aureus* isolates from the United States: Establishing a national database. *J. Clin. Microbiol.* **2003**, *41*, 5113–5120. [CrossRef] [PubMed]

Publisher's Note: MDPI stays neutral with regard to jurisdictional claims in published maps and institutional affiliations.

Article

Isolation and Characterization of Multidrug-Resistant *Escherichia coli* and *Salmonella* spp. from Healthy and Diseased Turkeys

Md. Tawyabur [1], Md. Saiful Islam [1], Md. Abdus Sobur [1], Md. Jannat Hossain [1,2], Md. Muket Mahmud [1], Sumon Paul [1], Muhammad Tofazzal Hossain [1], Hossam M. Ashour [3,4,*] and Md. Tanvir Rahman [1,*]

[1]   Department of Microbiology and Hygiene, Faculty of Veterinary Science, Bangladesh Agricultural University, Mymensingh 2202, Bangladesh; soukhin1993@gmail.com (M.T.); dvm41257@bau.edu.bd (M.S.I.); dvm38397@bau.edu.bd (M.A.S.); dvm38553@bau.edu.bd (M.J.H.); dvm41187@bau.edu.bd (M.M.M.); sumonmbjust@gmail.com (S.P.); tofazzalmh@bau.edu.bd (M.T.H.)
[2]   Department of Microbiology and Public Health, Khulna Agricultural University, Khulna 9208, Bangladesh
[3]   Department of Integrative Biology, College of Arts and Sciences, University of South Florida, St. Petersburg, FL 33701, USA
[4]   Department of Microbiology and Immunology, Faculty of Pharmacy, Cairo University, Cairo 11562, Egypt
*   Correspondence: hossamking@mailcity.com (H.M.A.); tanvirahman@bau.edu.bd (M.T.R.)

Received: 23 August 2020; Accepted: 25 October 2020; Published: 2 November 2020

Abstract: Diseases caused by *Escherichia coli* (*E. coli*) and *Salmonella* spp. can negatively impact turkey farming. The aim of this study was to isolate and characterize multidrug-resistant (MDR) *E. coli* and *Salmonella* spp. in healthy and diseased turkeys. A total of 30 fecal samples from healthy turkeys and 25 intestinal samples from diseased turkeys that died of enteritis were collected. Bacterial isolation and identification were based on biochemical properties and polymerase chain reaction (PCR). Antibiogram profiles were determined by disk diffusion. The tetracycline-resistance gene *tetA* was detected by PCR. All samples were positive for *E. coli*. Only 11 samples (11/30; 36.67%) were positive for *Salmonella* spp. from healthy turkeys, whereas 16 (16/25; 64%) samples were positive for *Salmonella* spp. from diseased turkeys. *E. coli* isolated from diseased turkeys showed higher resistance to levofloxacin, gentamicin, chloramphenicol, ciprofloxacin, streptomycin, and tetracycline. *Salmonella* spp. isolated from healthy turkeys exhibited higher resistance to gentamicin, chloramphenicol, ciprofloxacin, streptomycin, imipenem, and meropenem. All *E. coli* and *Salmonella* spp. from both healthy and diseased turkeys were resistant to erythromycin. *Salmonella* spp. from both healthy and diseased turkeys were resistant to tetracycline. Multidrug resistance was observed in both *E. coli* and *Salmonella* spp. from diseased turkeys. Finally, the *tetA* gene was detected in 93.1% of the *E. coli* isolates and in 92.59% of the *Salmonella* spp. isolates. To the best of our knowledge, this is the first study to isolate and characterize *tetA*-gene-containing MDR *E. coli* and *Salmonella* spp. from healthy and diseased turkeys in Bangladesh. Both microorganisms are of zoonotic significance and represent a significant public health challenge.

Keywords: avian colibacillosis; salmonellosis; antibiotic resistance; MDR; *tetA*; public health

---

1. Introduction

Turkey (*Meleagris gallopavo*) farming is a profitable business in many countries. In Bangladesh, turkey farming generates a higher profit than broiler and layer farming due to lower feeding cost, higher market price, and high demand from consumers. In addition, turkey is generally more adaptable under different weather conditions and less prone to disease than other poultry birds [1,2].

In Bangladesh, there are more than 600 small- and medium-sized commercial turkey farms [3]. With strong support of the Bangladesh government, the number of farms is increasing [3]. According to the Household Income and Expenditure Survey 2016 in Bangladesh [4], the average daily protein intake per capita was 63.50 g, of which meat, poultry, and eggs contributed 12.65% of the total proteins. Furthermore, poultry contributed 37% of the overall meat production in Bangladesh [5]. In rural areas, rearing poultry is a common additional source of income [6]. The challenges of turkey farming include potential outbreaks of infectious and non-infectious diseases, which have been shown to impact more than a third of turkey farmers in Bangladesh [7]. Infections caused by *Escherichia coli* and *Salmonella* spp. have negative impacts on turkey farming as they lower egg production, reduce hatchability, and increase mortality rates [8]. Thus, the control of *E. coli* and *Salmonella* infections in turkey farms is crucial.

E. coli is a zoonotic commensal pathogen that is capable of causing infections in the gastrointestinal tract (GIT), respiratory tract, and bloodstream in both humans and animals [9,10]. Avian colibacillosis caused by *E. coli* is responsible for turkey cellulitis, colisepticemia, swollen head syndrome, synovitis, salpingitis, coligranuloma, osteomyelitis, omphalitis, peritonitis, panophthalmitis, and is often deadly for turkeys [11,12]. It also causes urinary tract infections (UTIs), abdominal sepsis, and meningitis. It is important to note that *E. coli* is responsible for about 80% of UTIs in humans [13,14].

Salmonella spp. can cause salmonellosis (especially pullorum disease and fowl typhoid) in turkeys [15,16]. *Salmonella* infections reduce hatchability, fertility, growth, and increase mortality rates in poultry [17]. Due to their zoonotic nature, *Salmonella* spp. can be transmitted to humans through the food chain. This can lead to the development of salmonellosis, gastroenteritis, enteric fever [18,19], and can sometimes cause life-threatening consequences [20].

The excessive use of antibiotics in farms led to the emergence of antibiotic-resistant bacteria such as *E. coli*, *Salmonella* spp., and *Campylobacter* spp. in poultry [21,22]. High levels of antibiotic-resistant or multidrug-resistant (MDR) *E. coli* and *Salmonella* spp. can constitute a more significant problem in turkeys than in other livestock species [21,23]. Mutations in *E. coli* and *Salmonella* spp. could result in the acquisition of antibiotic resistance [24]. Mobile genetic elements allowed bacteria to acquire and disseminate antibiotic resistance [25]. The implications of this acquired antibiotic resistance for public health necessitates attention from both clinical and economic experts [26].

Antimicrobial resistance (AMR) poses a significant threat to human health [27]. AMR is responsible for approximately 700,000 human deaths every year throughout the world [28]. This figure could significantly increase in the near future if we do not discover novel and effective antibiotics [29]. The antibiotic resistance in farm animals is clearly intertwined with the presence of this problem in humans [30,31]. In addition, the indiscriminate use of antibiotics in livestock is one of the main causes of AMR [25,26]. The overuse of antibiotics by farm owners in poultry farms, a common practice in developing countries, is a major reason for the development of MDR bacteria [32,33]. This overuse typically occurs without consulting any veterinarians and without any previous testing of the animals. The development of MDR bacteria in poultry has been previously reported in previous studies [22,33–35]. Poultry farmers have been using different types of poultry in recent years including broilers, layers, and turkeys. These animals are hosted close to each other, which can lead to the horizontal transmission of MDR bacteria to turkeys. The dissemination of MDR bacteria to humans exposes the population to risk, especially the immunocompromised individuals, and exacerbates healthcare costs, and ultimately increases the usage of antibiotics [36].

The present study was designed to isolate and characterize MDR *E. coli* and *Salmonella* spp. from both healthy and diseased turkeys. There is an urgent need to design proper surveillance and control programs for the detection and control of antibiotic-resistant bacteria in turkey farms.

2. Results

2.1. Prevalence of E. coli and Salmonella spp.

All 55 samples were positive for *E. coli* (using PCR targeting the *malB* gene), whereas 27 samples (27/55; 49.09%) were positive for *Salmonella* spp. (using PCR targeting the *invA* gene). The prevalence of *E. coli* in turkeys was significantly higher than *Salmonella* spp. (chi-square test, 95% CI, $p < 0.001$). The prevalence of *Salmonella* spp. was significantly higher in diseased (64%; 16/25) than in healthy turkeys (36.67; 11/30) (chi-square test, 95% CI, $p < 0.05$). No significant difference between healthy and diseased turkeys was observed in the case of *E. coli* (Table 1).

Table 1. Prevalence and resistance profiles of *E. coli* and *Salmonella* spp. isolated from turkeys.

Microorganism	Categories	Prevalence	Antibiotic Resistance Pattern (%)								
			LEV	E	GEN	C	CIP	S	IMP	MEM	TE
E. coli	Healthy	30 (100)	4 (13.33)	30 (100)	0 (0)	0 (0)	17 (56.67)	4 (13.33)	0 (0)	30 (100)	4 (13.33)
	Diseased	25 (100)	11 (44)	25 (100)	9 (36)	11 (44)	20 (80)	5 (20)	0 (0)	10 (40)	25 (100)
	p-value (Healthy vs. Diseased)	N/C	0.011	N/C	<0.001	<0.001	0.066	0.716	N/C	<0.001	<0.001
Salmonella spp.	Healthy	11 (36.67)	2 (18.18)	11 (100)	5 (45.45)	6 (54.54)	6 (54.54)	4 (36.36)	4 (36.36%)	7 (63.63)	11 (100)
	Diseased	16 (64)	4 (25)	16 (100)	0 (0)	2 (12.5)	6 (37.5)	2 (12.5)	4 (25%)	4 (25)	16 (100)
	p-value (Healthy vs. Diseased)	0.043	1.000	N/C	0.006	0.033	0.438	0.187	0.675	0.061	N/C

A *p*-value less than 0.05 was deemed to be statistically significant; N/C, not computed; *E. coli*, *Escherichia coli*; LEV, Levofloxacin; E, Erythromycin; GEN, Gentamicin; C, Chloramphenicol; CIP, Ciprofloxacin; S, Streptomycin; IMP, Imipenem; MEM, Meropenem; TE, Tetracycline.

2.2. Antibiotic Profiles of Isolated E. coli and Salmonella spp.

Antibiotic sensitivity tests revealed that all *E. coli* isolates were resistant to erythromycin; whereas all *Salmonella* isolates were resistant to erythromycin and tetracycline. Additionally, *E. coli* isolates were resistant to ciprofloxacin (67.27%), meropenem (72.73%), and tetracycline (52.73%). *Salmonella* spp. were resistant to ciprofloxacin (44.44%) and meropenem (40.74%). *E. coli* isolates were highly sensitive to imipenem (92.73%)

E. coli isolated from diseased turkeys showed higher resistance to levofloxacin (chi-square test, 95% CI, $p = 0.011$), gentamicin ($p < 0.001$), chloramphenicol ($p < 0.001$), and tetracycline ($p < 0.001$); whereas isolates from healthy turkeys showed higher resistance to meropenem ($p < 0.001$). Interestingly, *Salmonella* spp. isolated from healthy turkeys exhibited higher resistance to gentamicin, chloramphenicol, ciprofloxacin, streptomycin, imipenem, and meropenem than *Salmonella* spp. isolated from diseased turkeys. However, only a few cases were statistically significant (Table 1).

2.3. Detection of tetA Gene

Of the 29 *E. coli* isolates phenotypically resistant to tetracycline, *tetA* was detected in 27 (27/29; 93.1%). In the case of *Salmonella* spp., *tetA* was detected in 25 of the 27 isolates (25/27; 92.59%). The prevalence of *tetA* was similar in healthy and diseased turkeys for both *E. coli* and *Salmonella* spp. (Figure 1).

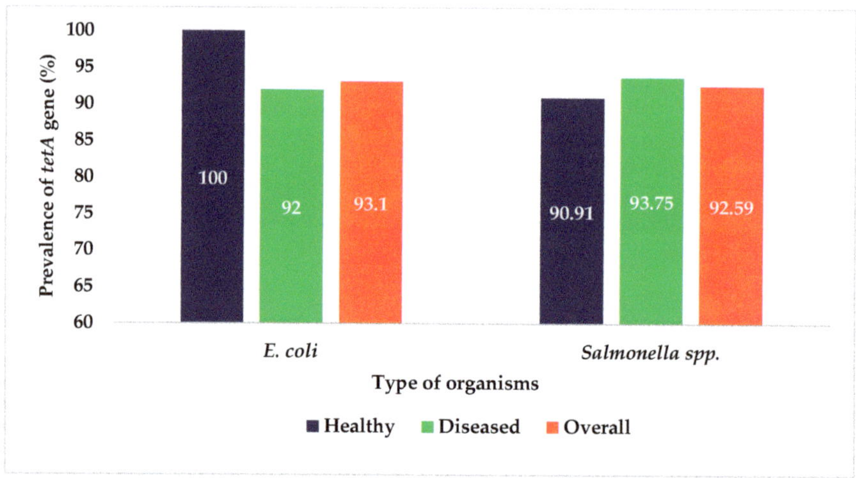

Figure 1. Prevalence of *tetA* gene in *E. coli* and *Salmonella* spp. isolated from turkeys.

2.4. Detection of MDR E. coli and Salmonella spp.

As shown in Table 2, antibiogram typing revealed that most *E. coli* isolates (48/55; 87.27%) and most *Salmonella* isolates (24/27; 88.89%) exhibited multi-drug resistance. For *E. coli*, the percentage of MDR isolates was higher from diseased turkeys (24/25; 96%) than from healthy turkeys (24/30; 80%). For *Salmonella*, the percentage of MDR isolates was also higher in diseased turkeys (16/16; 100%) than in healthy turkeys (11/16; 72.72%). However, the differences were not statistically significant in either case (chi-square test, 95% CI, $p > 0.05$).

E. coli isolated from healthy turkeys showed eight resistance patterns, while *E. coli* isolated from diseased turkeys showed ten resistance patterns. *Salmonella* isolated from healthy and diseased turkeys showed four and seven resistance patterns, respectively (Table 2). Among the antibiogram types, pattern E-MEM-CIP showed the highest prevalence in *E. coli* (14 isolates). On the other hand, the E-CIP-TE pattern showed the highest prevalence in *Salmonella* (five isolates) (Table 2).

Table 2. Multidrug resistance profiles of *E. coli* and *Salmonella* spp. isolated from healthy and diseased turkeys.

Microorganism	Source	Pattern No.	Antibiotic Resistance Patterns	No. of Antibiotics (Classes)	No. of MDR Isolates (%)	Total (%)	p-Value (Healthy vs. Diseased)
E. coli ($n = 55$)	Healthy Turkeys ($n = 30$)	1	E, MEM, CIP	3 (3)	14	24 (80%)	0.112
		2	E, MEM, TE	3 (3)	1		
		3	E, MEM, LEV	3 (3)	2		
		4	E, MEM, S	3 (3)	3		
		5	E, MEM, CIP, LEV	4 (3)	1		
		6	E, MEM, LEV, TE	4 (4)	1		
		7	E, MEM, CIP, TE	4 (4)	1		
		8	E, MEM, S, CIP, TE	5 (5)	1		
	Diseased Turkeys ($n = 25$)	1	E, CIP, TE	3 (3)	4	24 (96%)	
		2	E, MEM, TE	3 (3)	3		
		3	E, CIP, LEV, TE	4 (3)	3		
		4	E, GEN, S, CIP, TE	5 (4)	3		
		5	E, MEM, C, CIP, TE	5 (5)	2		
		6	E, MEM, C, S, TE	5 (5)	1		
		7	E, C, GEN, CIP, LEV, TE	6 (5)	4		
		8	E, MEM, C, CIP, LEV, TE	6 (5)	2		
		9	E, MEM, C, GEN, CIP, LEV, TE	7 (6)	1		
		10	E, MEM, C, GEN, S, CIP, LEV, TE	8 (6)	1		

Table 2. *Cont.*

Microorganism	Source	Pattern No.	Antibiotic Resistance Patterns	No. of Antibiotics (Classes)	No. of MDR Isolates (%)	Total (%)	*p*-Value (Healthy vs. Diseased)
Salmonella spp. (*n* = 27)	Healthy Turkeys (*n* = 11)	1	E, MEM, C, CIP, TE	5 (5)	3	8 (72.73%)	0.056
		2	E, C, GEN, CIP, TE	5 (5)	1		
		3	E, MEM, IMP, C, GEN, S, TE	7 (5)	2		
		4	E, MEM, IMP, GEN, S, CIP, LEV, TE	8 (6)	2		
	Diseased Turkeys (*n* = 16)	1	E, MEM, TE	3 (3)	3	16 (100%)	
		2	E, IMP, TE	3 (3)	3		
		3	E, CIP, TE	3 (3)	5		
		4	E, LEV, TE	3 (3)	2		
		5	E, IMP, C, TE	4 (4)	1		
		6	E, C, S, LEV, TE	5 (5)	1		
		7	E, MEM, S, CIP, LEV, TE	6 (5)	1		

A *p*-value less than 0.05 was deemed to be statistically significant; *E. coli*, *Escherichia coli*; TE, Tetracycline; E, Erythromycin; C, Chloramphenicol; LEV, Levofloxacin; GEN, Gentamicin; MEM, Meropenem; IMP, Imipenem; S, Streptomycin; CIP, Ciprofloxacin.

3. Discussion

In this study, we report the detection of MDR *E. coli* and *Salmonella* spp. from healthy and diseased turkeys. This is significant to human health due to the zoonotic nature of these pathogens. Moreover, most *E. coli* and *Salmonella* spp. isolates were found to be MDR, which makes it difficult to treat the infected turkeys [37–42]. Antibiograms can guide the choice of therapies for colibacillosis and salmonellosis in turkeys. The incorrect choice of antibiotics is not only associated with the development of AMR but can also have significant negative economic impacts.

Whereas all samples were positive for *E. coli*, only 49.09% (27/55) of the samples were positive for *Salmonella* spp., which were significantly more prevalent in diseased than in healthy turkeys. The isolation and characterization of *E. coli* and *Salmonella* spp. from turkeys revealed the presence of the *tetA* gene. The gut microflora of poultry typically includes *E. coli* and *Salmonella* spp. [43]. Detection of *Salmonella* spp. in diseased turkeys that died of enteritis suggests that *Salmonella* was the causative factor of enteritis. Previously, Kar et al. [8] reported the detection of *E. coli* and *Salmonella* spp. from cloacal swabs of turkeys but did not use any molecular techniques, such as the PCR technology used in this study. PCR is a robust and rapid detection method with increased sensitivity and specificity for detecting *Salmonella* in food, environmental, and clinical samples [44]. The *invA* gene has been the target for many PCR protocols, as it is found in almost all known serovars of *Salmonella* [45]. This gene encodes an inner membrane protein necessary for invasion of epithelial cells by *Salmonella* [46]. We were able to observe higher rates of *E. coli* and *Salmonella* spp. compared to the study of Kar et al. [8], which may be attributed to the highly sensitive nature of the molecular techniques used in this study.

The detection of *E. coli* and *Salmonella* spp. from fecal materials and intestinal contents of healthy turkeys indicates intestinal colonization [47]. The findings also indicate that fecal materials may be a source of transmission of *E. coli* and *Salmonella* spp. to other birds. The detection of the virulence gene *invA* in the isolated *Salmonella* spp. indicates the potential pathogenic nature of these isolates. It is also possible for these pathogens to be introduced into the food chain causing food-borne diseases [48].

Antibiotic resistance is a major public health problem. The misuse and abuse of antimicrobial agents contributed to the emergence and dissemination of antibiotic-resistant pathogens in animals and humans [49]. Location-specific information on antibiotic resistance patterns in different geographical areas is important for the successful treatment of outbreaks and infections. The isolated *E. coli* and *Salmonella* spp. were found to be resistant to levofloxacin, erythromycin, ciprofloxacin, meropenem, and tetracycline. This antibiotic resistance profile can be due to the frequent use of antibiotics in poultry for therapeutic and growth promotion purposes [32,33]. The presence of antibiotic-resistant *E. coli* and *Salmonella* spp. in fecal materials of healthy turkeys indicates the role of these birds as spreaders of resistant microorganisms in farm environments.

Several studies detected the *tetA* gene in *E. coli* and *Salmonella* spp. from dairy farms, boiler farms, house flies, and aquatic environments [31,33,50–52]. However, there were no studies on the detection of the *tetA* gene in *E. coli* and *Salmonella* from turkeys. Among the isolates phenotypically resistant to tetracycline, 93.1% of the *E. coli* isolates and 92.59% of *Salmonella* spp. isolates were positive for the *tetA* gene. The *tetA* has been shown to be the most common genetic component in tetracycline-resistant *E. coli* and *Salmonella* spp. [9,53–55]. Generally remaining in mobile genetic components (integrons, transposons, and plasmids), *tetA* can be easily transferred to different bacteria.

Resistance to carbapenems (imipenem and meropenem) may be due to the transmission of bacteria from human sources, especially that carbapenems are not approved for use in livestock [56]. Future detailed studies at the genetic level are needed to test this hypothesis. According to the WHO, carbapenem-resistant *E. coli* and *Salmonella* spp. are considered to be among the most critical pathogens [57]. The detection of carbapenem-resistant *E. coli* and *Salmonella* spp. in turkeys has to be treated as an urgent public health problem.

Antibiotic treatment failures in poultry has been highly attributed to the MDR nature of the pathogens [58]. In the present study, the majority of the isolated *E. coli* (48/55; 87.27%) and *Salmonella* spp. (24/27; 88.89%) were MDR. More MDR *E. coli* and *Salmonella* spp. were retrieved from diseased turkeys than from healthy turkeys. The higher MDR in diseased turkeys may have been caused by the selection pressure resulting from the excessive use of several classes of antibiotics. However, the differences were statistically insignificant as in Table 2 ($p = 0.112$ and $p = 0.056$ for *E. coli* and *Salmonella* spp., respectively). The statistical insignificance indicates that the bacteria were MDR regardless of whether the source was healthy or diseased turkeys. To avoid the development of MDR, the use of antibiotics should be more strategic and selective.

4. Materials and Methods

4.1. Ethics Statement

No ethical permission was required for the study. During sample collection, verbal permission was taken from farm owners.

4.2. Study Design

A pilot survey was conducted prior to the start of the current study to identify the different turkey farming areas in Bangladesh, disease outbreaks in these farms, and antibiotic treatment regimens. Based on the survey results, seven antibiotics were selected. In addition, two carbapenem antibiotics were included based on reports that indicated that *E. coli* could be resistant to carbapenems in poultry [31,50,59]. Guided by bird mortality rates and antibiotic use reports from the survey, five farms from two districts were selected for sample collection. The birds were categorized into healthy and diseased birds. Six healthy and five diseased bird samples were randomly collected from each farm resulting in a total of 55 samples from the five farms. Freshly dropped feces from healthy birds and intestinal contents from diseased birds that had avian colibacillosis and/or Salmonellosis were collected for analysis.

4.3. Study Areas and Collection of Samples

The study was conducted in two districts of Bangladesh namely Mymensingh (24.7539° N, 90.4073° E) and Tangail (24.2513° N, 89.9167° E) during the period from June 2018 to November 2019. The study areas are represented in Figure 2.

Freshly dropped fecal samples ($n = 30$) were aseptically collected using sterile cotton buds from healthy turkeys. During the postmortem examination, 5 g of intestinal contents ($n = 25$) was collected from each turkey that died of enteritis and had lesions of avian colibacillosis and/or salmonellosis.

Immediately after collection, samples were transferred to sterile zip-lock bags. Samples were transported to the laboratory maintaining cold chain. Collected samples were transferred into sterile test tubes containing freshly prepared nutrient broth (5 mL) and were incubated aerobically at 37 °C overnight for the growth of bacteria.

4.4. Isolation of E. coli and Salmonella spp.

Isolation of *E. coli* and *Salmonella* spp. was based on culture on Eosin Methylene Blue (EMB) and Xylose Lysine Deoxycholate (XLD) agar (HiMedia, India) plates, respectively. Initially, freshly grown broth cultures were streaked on EMB and XLD agar media using sterile inoculating loops. This was followed by aerobic incubation of the inoculated agar plates at 37 °C overnight to obtain pure colonies. Single green-colored metallic-sheen colonies on EMB agar media and black-centered colonies on XLD agar media represented the growth of *E. coli* and *Salmonella* spp., respectively. For further confirmation, selected colonies were subjected to morphological study by Gram staining and biochemical tests such as the methyl red test, sugar fermentation test, Voges–Proskauer test, motility test, urease test, and indole test [22,31].

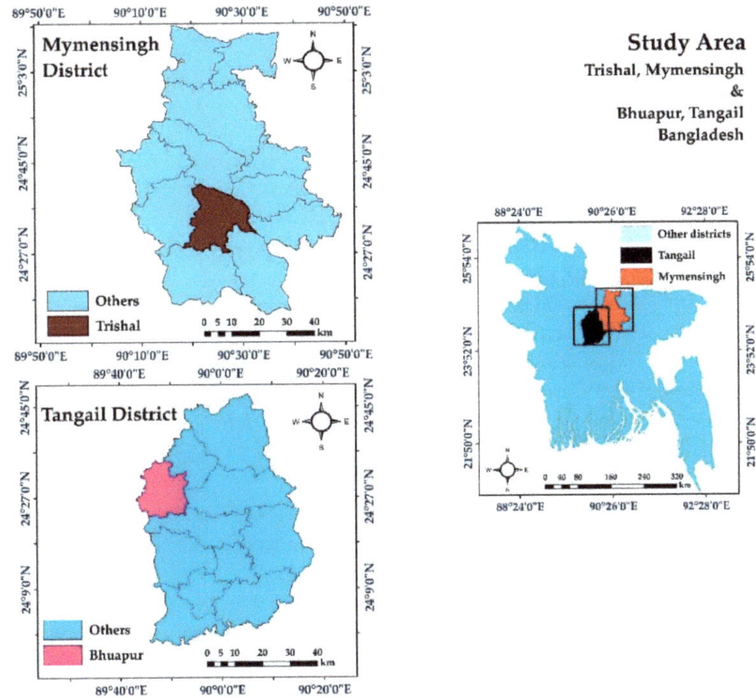

Figure 2. Map of the study area. Images were extracted from DIVA-GIS using Geographical Information System (GIS). The map was developed using ArcMap version 10.7.

4.5. Molecular Detection of E. coli and Salmonella spp.

Isolation of *E. coli* and *Salmonella* spp. were confirmed by polymerase chain reaction (PCR) targeting *E. coli* 16S rRNA gene and *Salmonella* genus specific *invA* genes respectively (Table 3).

Table 3. List of primers used for detecting *E. coli*, *Salmonella* spp., and tetracycline-resistance gene.

Target Gene	Primer Sequence (5′–3′)	Amplicon Size (bp)	Annealing Temperature (°C)	References
malB	F: GACCTCGGTTTAGTTCACAGA R: CACACGCTGACGCTGACCA	585	55	[60]
invA	F: ATCAGTACCAGTCGTCTTATCTTGAT R: TCTGTTTACCGGGCATACCAT	211	58	[61]
tetA	F: GGTTCACTCGAACGACGTCA R: CTGTCCGACAAGTTGCATGA	577	57	[62]

For PCR, genomic DNA of *E. coli* and *Salmonella* spp. was extracted by the boiling method as described by Sobur et al. [50]. Briefly, a pure colony collected from freshly grown culture was initially taken into an Eppendorf tube containing molecular-grade water (100 μL) followed by mixing gently through vortexing. Subsequently, the mixture was boiled for 10 min, cooled for 10 min, and centrifuged for 10 min at 1400 rpm. Finally, the supernatant was collected as the source for the genomic DNA for PCR and stored at −20 °C until further use.

PCR tests were carried out in a final volume of 25 μL with 12.5 μL of the master mix (2X) (Promega, Madison, WI, USA), 4 μL of genomic DNA (50 ng/μL), 1 μL of each primer, and 6.5 μL of nuclease-free water. After amplification, PCR products were subjected to gel electrophoresis in 1.5% agarose, followed by staining and visualizing by 0.25% ethidium bromide solution and ultraviolet trans-illuminator (Biometra, Göttingen, Germany). A DNA ladder (100 bp; Promega, Madison, WI, USA) was used to assess the sizes of PCR amplicons.

4.6. Antibiotic Sensitivity Test

Antibiotic sensitivity testing of isolated *E. coli* and *Salmonella* spp. was carried out using the disk diffusion assay as previously described [63]. Antibiotic classes included fluoroquinolones (levofloxacin, LEV—5 µg; ciprofloxacin, CIP—5 µg), aminoglycosides (gentamicin, GEN—10 µg; streptomycin, S—10 µg), carbapenems (Meropenem, MEM—10 µg; imipenem, IMP—10 µg), amphenicols (chloramphenicol, C—10 µg), macrolides (erythromycin, E—15 µg), and tetracyclines (tetracycline, TE—30 µg) purchased from Hi Media (India). Sensitivity tests were performed on freshly grown isolates having a concentration equivalent to 0.5 McFarland standard using Mueller-Hinton agar media (Hi Media, India). All results were interpreted according to the guidelines provided by Clinical and Laboratory Standards Institute [64]. Furthermore, isolates showing resistance against three or more different classes of antibiotics were defined as MDR [65].

4.7. Molecular Detection of Tetracycline Resistance tetA Gene

E. coli and *Salmonella* isolates resistant to tetracycline were screened by PCR for the detection of the tetracycline-resistance *tetA* gene using the primer and protocol described by Randall et al. [62].

4.8. Statistical Analysis

Chi-square tests were performed using the SPSS software (IBM SPSS version 25.0, IBM, Chicago, IL, USA). p-values less than 0.05 ($p < 0.05$) were considered to be statistically significant.

5. Conclusions

The isolation and characterization of *tetA*-gene-containing-MDR *E. coli* and *Salmonella* spp. from turkeys are concerning. The potential ability of these MDR bacteria to enter into the food chain can expose humans to serious health risks. Bacterial surveillance programs should be implemented in order to control the emergence of bacterial resistance in turkey farms in Bangladesh and elsewhere in the world. This should be a concerted effort that is best carried out via bacterial surveillance networks across different countries. Additionally, holistic and multi-sectoral approaches, such as the one health approach, need to be implemented [66]. Guided by top health professionals and scientists, these strategies can provide effective solutions to the complex, multifaceted global challenge of AMR.

Author Contributions: Conceptualization and design of the study, H.M.A. and M.T.R.; Sampling, M.T. and M.M.M.; Methodology, M.T., M.S.I., M.A.S. and S.P.; data analysis and interpretation, M.S.I., M.A.S., H.M.A. and M.T.R.; writing—original draft preparation, M.T., M.S.I., M.A.S., M.J.H., H.M.A. and M.T.R.; supervision, M.T.H., H.M.A., and M.T.R.; critical revisions and writing of the revised manuscript, H.M.A. and M.T.R. All authors have read and agreed to the final published version of the manuscript.

Funding: This research received no external funding.

Acknowledgments: We are grateful to farm owners for giving us access to samples during the course of the study.

Conflicts of Interest: The authors declare no conflict of interest.

References

1. Anandh, M.A.; Jagatheesan, P.N.R.; Kumar, P.S.; Rajarajan, G.; Paramasivam, A. Effect of egg weight on egg traits and hatching performance of turkey (*Meleagris gallopavo*) eggs. *Iran. J. Appl. Anim. Sci.* **2012**, *2*, 391–395.
2. Jahan, B.; Ashraf, A.; Rahman, M.A.; Molla, M.H.R.; Chowdhury, S.H.; Megwalu, F.O. Rearing of high yielding turkey poults: Problems and future prospects in Bangladesh: A review. *SF J. Biotechnol. Biomed. Eng.* **2018**, *1*, 1008.

3. The Independent. Turkey Rearing Gains Popularity. Available online: http://www.theindependentbd.com/printversion/details/185891 (accessed on 12 September 2020).

4. Bangladesh Bureau of Statistics (BBS). Report of the Household Income and Expenditure Survey 2016. Available online: https://drive.google.com/file/d/1TmUmC-0M3wC5IN6_tUxZUvTW2rmUx-Mce/view?usp=sharing (accessed on 12 September 2020).

5. Begum, I.A.; Alam, M.J.; Buysse, J.; Frija, A.; Van Huylenbroeck, G. A comparative efficiency analysis of poultry farming systems in Bangladesh: A Data Envelopment Analysis approach. *Appl. Econ.* **2011**, *44*, 3737–3747. [CrossRef]

6. Islam, M.S.; Sabuj, A.A.M.; Haque, Z.F.; Pondit, A.; Hossain, M.G.; Saha, S. Seroprevalence of avian reovirus in backyard chickens in different areas of Mymensingh district in Bangladesh. *J. Adv. Vet. Anim. Res.* **2020**, *7*, 546–553. [CrossRef] [PubMed]

7. Asaduzzaman, M.; Salma, U.; Ali, H.S.; Hamid, M.A.; Miah, A.G. Problems and prospects of turkey (*Meleagris gallopavo*) production in Bangladesh. *Res. Agric. Livest. Fish.* **2017**, *4*, 77–90. [CrossRef]

8. Kar, J.; Barman, T.R.; Sen, A.; Nath, S.K. Isolation and identification of *Escherichia coli* and *Salmonella* sp. from apparently healthy Turkey. *Int. J. Adv. Res. Biol. Sci.* **2017**, *4*, 72–78.

9. Hopkins, K.L.; Davies, R.H.; Threlfall, E.J. Mechanisms of quinolone resistance in *Escherichia coli* and *Salmonella*: Recent developments. *Int. J. Antimicrob. Agents* **2005**, *25*, 358–373. [CrossRef] [PubMed]

10. Rahman, M.T.; Sobur, M.A.; Islam, M.S.; Ievy, S.; Hossain, M.J.; El Zowalaty, M.E.; Rahman, A.T.; Ashour, H.M. Zoonotic Diseases: Etiology, Impact, and Control. *Microorganisms* **2020**, *8*, 1405. [CrossRef] [PubMed]

11. Barnes, H.J.; Gross, W.B. Colibacillosis. In *Diseases of Poultry*, 10th ed.; Mosby-Wolfe Medical Publication Ltd: London, UK, 1997; pp. 131–139.

12. De Oliveira, A.L.; Newman, D.M.; Sato, Y.; Noel, A.; Rauk, B.; Nolan, L.K.; Barbieri, N.L.; Logue, C.M. Characterization of Avian Pathogenic *Escherichia coli* (APEC) Associated With Turkey Cellulitis in Iowa. *Front. Vet. Sci.* **2020**, *7*, 380. [CrossRef]

13. Foxman, B. The epidemiology of urinary tract infection. *Nat. Rev. Urol.* **2010**, *7*, 653–660. [CrossRef]

14. Mellata, M. Human and avian extraintestinal pathogenic *Escherichia coli*: Infections, zoonotic risks, and antibiotic resistance trends. *Foodborne Pathog. Dis.* **2013**, *10*, 916–932. [CrossRef] [PubMed]

15. Aury, K.; Chemaly, M.; Petetin, I.; Rouxel, S.; Picherot, M.; Michel, V.; Le Bouquin, S. Prevalence and risk factors for *Salmonella enterica* subsp. enterica contamination in French breeding and fattening turkey flocks at the end of the rearing period. *Prev. Vet. Med.* **2010**, *94*, 84–93. [CrossRef] [PubMed]

16. Abdukhalilova, G.; Kaftyreva, L.; Wagenaar, J.A.; Tangyarikov, B.; Bektimirov, A.; Akhmedov, I.; Khodjaev, Z.; Kruse, H. World Health Organization. Occurrence and antimicrobial resistance of *Salmonella* and *Campylobacter* in humans and broiler chicken in Uzbekistan. *Public Health Panor.* **2016**, *2*, 340–347.

17. Andino, A.; Hanning, I. *Salmonella enterica*: Survival, Colonization, and Virulence Differences among Serovars. *Sci. World J.* **2015**, *2015*, 520179. [CrossRef] [PubMed]

18. Zhao, S.; Qaiyumi, S.; Friedman, S.; Singh, R.; Foley, S.L.; White, D.G.; McDermott, P.F.; Donkar, T.; Bolin, C.; Munro, S.; et al. Characterization of *Salmonella enterica* serotype Newport isolated from humans and food animals. *J. Clin. Microbiol.* **2003**, *41*, 5366–5371. [CrossRef]

19. Varga, C.; Guerin, M.T.; Brash, M.L.; Slavic, D.; Boerlin, P.; Susta, L. Antimicrobial resistance in fecal *Escherichia coli* and *Salmonella enterica* isolates: A two-year prospective study of small poultry flocks in Ontario, Canada. *BMC Vet. Res.* **2019**, *15*, 1–10. [CrossRef]

20. Helms, M.; Vastrup, P.; Gerner-Smidt, P.; Mølbak, K. Excess mortality associated with antimicrobial drug-resistant *Salmonella Typhimurium*. *Emerg. Infect. Dis.* **2002**, *8*, 490–495. [CrossRef]

21. Yeh, J.C.; Chen, C.L.; Chiou, C.S.; Lo, D.Y.; Cheng, J.C.; Kuo, H.C. Comparison of prevalence, phenotype, and antimicrobial resistance of *Salmonella* serovars isolated from turkeys in Taiwan. *Poult. Sci.* **2018**, *97*, 279–288. [CrossRef]

22. Ievy, S.; Islam, M.S.; Sobur, M.A.; Talukder, M.; Rahman, M.B.; Khan, M.F.R.; Rahman, M.T. Molecular Detection of Avian Pathogenic *Escherichia coli* (APEC) for the First Time in Layer Farms in Bangladesh and Their Antibiotic Resistance Patterns. *Microorganisms* **2020**, *8*, 1021. [CrossRef]

23. Poppe, C.; Martin, L.C.; Gyles, C.L.; Reid-Smith, R.; Boerlin, P.; McEwen, S.A.; Prescott, J.F.; Forward, K.R. Acquisition of resistance to extended-spectrum cephalosporins by *Salmonella enterica* subsp. enterica serovar Newport and Escherichia coli in the turkey poult intestinal tract. *Appl. Environ. Microbiol.* **2005**, *71*, 1184–1192. [CrossRef]

24. Blair, J.M.; Webber, M.A.; Baylay, A.J.; Ogbolu, D.O.; Piddock, L.J. Molecular mechanisms of antibiotic resistance. *Nat. Rev. Microbiol.* **2015**, *13*, 42–51. [CrossRef] [PubMed]
25. Fluit, A.; Schmitz, F.J. Class 1 integrons, gene cassettes, mobility, and epidemiology. *Eur. J. Clin. Microbiol. Infect. Dis.* **1999**, *18*, 761–770. [CrossRef] [PubMed]
26. Tirumalai, M.R.; Karouia, F.; Tran, Q.; Stepanov, V.G.; Bruce, R.J.; Ott, C.M.; Pierson, D.L.; Fox, G.E. Evaluation of acquired antibiotic resistance in *Escherichia coli* exposed to long-term low-shear modeled microgravity and background antibiotic exposure. *Mbio* **2019**, *10*, e02637-18. [CrossRef] [PubMed]
27. Alexander, H.K.; MacLean, R.C. Stochastic bacterial population dynamics restrict the establishment of antibiotic resistance from single cells. *Proc. Natl. Acad. Sci. USA* **2020**, *117*, 19455–19464. [CrossRef]
28. Clifford, K.; Darash, D.; da Costa, C.P.; Meyer, H.; Islam, M.T.; Meyer, H.; Klohe, K.; Winklerc, A.; Rahman, M.T.; Islam, M.T.; et al. Antimicrobial resistance in livestock and poor quality veterinary medicines. *Bull. World Health Organ.* **2018**, *96*, 662–664. [CrossRef]
29. O'Neill, J. Tackling Drug-Resistant Infections Globally: Final Report and Recommendations. 2016. Available online: https://amr-review.org/sites/default/files/160518_Final%20paper_with%20cover.pdf (accessed on 18 August 2020).
30. Woolhouse, M.; Ward, M.; van Bunnik, B.; Farrar, J. Antimicrobial resistance in humans, livestock and the wider environment. *Philos. Trans. R. Soc. B* **2015**, *370*, 20140083. [CrossRef]
31. Sobur, M.A.; Sabuj, A.A.M.; Sarker, R.; Rahman, A.M.M.T.; Kabir, S.M.L.; Rahman, M.T. Antibiotic-resistant *Escherichia coli* and *Salmonella* spp. associated with dairy cattle and farm environment having public health significance. *Vet. World* **2019**, *12*, 984–993. [CrossRef]
32. Wright, G.D. The antibiotic resistome: The nexus of chemical and genetic diversity. *Nat. Rev. Microbiol.* **2007**, *5*, 175–186. [CrossRef]
33. Alam, S.B.; Mahmud, M.; Akter, R.; Hasan, M.; Sobur, A.; Nazir, K.H.M.; Noreddin, A.; Rahman, T.; El Zowalaty, M.E.; Rahman, M. Molecular detection of multidrug resistant *Salmonella* species isolated from broiler farm in Bangladesh. *Pathogens* **2020**, *9*, 201. [CrossRef]
34. Nandi, S.P.; Sultana, M.; Hossain, M.A. Prevalence and characterization of multidrug-resistant zoonotic *Enterobacter* spp. in poultry of Bangladesh. *Foodborne Pathog. Dis.* **2013**, *10*, 420–427. [CrossRef]
35. Azad, M.A.R.A.; Amin, R.; Begum, M.I.A.; Fries, R.; Lampang, K.N.; Hafez, H.M. Prevalence of antimicrobial resistance of *Escherichia coli* isolated from broiler at Rajshahi region, Bangladesh. *Br. J. Biomed. Multidisc. Res.* **2017**, *1*, 6–12.
36. Van Duin, D.; Paterson, D.L. Multidrug-resistant bacteria in the community: Trends and lessons learned. *Infect. Dis. Clin.* **2016**, *30*, 377–390. [CrossRef] [PubMed]
37. Lu, Y.; Zhao, H.; Sun, J.; Liu, Y.; Zhou, X.; Beier, R.C.; Wu, G.; Hou, X. Characterization of multidrug-resistant *Salmonella enterica* serovars Indiana and Enteritidis from chickens in Eastern China. *PLoS ONE* **2014**, *9*, e96050. [CrossRef]
38. Routh, J.A.; Pringle, J.; Mohr, M.; Bidol, S.; Arends, K.; Adams-Cameron, M.; Hancock, W.T.; Kissler, B.; Rickert, R.; Folster, J.; et al. Nationwide outbreak of multidrug-resistant *Salmonella Heidelberg* infections associated with ground turkey: United States, 2011. *Epidemiol. Infect.* **2015**, *143*, 3227–3234. [CrossRef]
39. Sinwat, N.; Angkittitrakul, S.; Coulson, K.F.; Pilapil, F.M.I.R.; Meunsene, D.; Chuanchuen, R. High prevalence and molecular characteristics of multidrug-resistant *Salmonella* in pigs, pork and humans in Thailand and Laos provinces. *J. Med. Microbiol.* **2016**, *65*, 1182–1193. [CrossRef] [PubMed]
40. Shecho, M.; Thomas, N.; Kemal, J.; Muktar, Y. Cloacael carriage and multidrug resistance *Escherichia coli* O157: H7 from poultry farms, eastern Ethiopia. *J. Vet. Med.* **2017**, *2017*, 8264583. [CrossRef]
41. Davis, G.S.; Waits, K.; Nordstrom, L.; Grande, H.; Weaver, B.; Papp, K.; Horwinski, J.; Koch, B.; Hungate, B.A.; Liu, C.M.; et al. Antibiotic-resistant *Escherichia coli* from retail poultry meat with different antibiotic use claims. *BMC Microbiol.* **2018**, *18*, 174. [CrossRef]
42. Díaz-Jiménez, D.; García-Meniño, I.; Fernández, J.; García, V.; Mora, A. Chicken and turkey meat: Consumer exposure to multidrug-resistant *Enterobacteriaceae* including *mcr*-carriers, uropathogenic E. coli and high-risk lineages such as ST131. *Int. J. Food Microbiol.* **2020**, *331*, 108750. [CrossRef]
43. Winsor, D.K.; Bloebaum, A.P.; Mathewson, J.J. Gram-negative, aerobic, enteric pathogens among intestinal microflora of wild turkey vultures (*Cathartes aura*) in west central Texas. *Appl. Environ. Microbiol.* **1981**, *42*, 1123–1124. [CrossRef]

44. Toze, S. PCR and the detection of microbial pathogens in water and wastewater. *Water Res.* **1999**, *33*, 3545–3556. [CrossRef]

45. Chiu, C.H.; Ou, J.T. Rapid identification of Salmonella serovars in feces by specific detection of virulence genes, *invA* and *spvC*, by an enrichment broth culture-multiplex PCR combination assay. *J. Clin. Microbiol.* **1996**, *34*, 2619–2622. [CrossRef] [PubMed]

46. Darwin, K.H.; Miller, V.L. Molecular Basis of the Interaction of *Salmonella* with the Intestinal Mucosa. *Clin. Microbiol. Rev.* **1999**, *12*, 405–428. [CrossRef] [PubMed]

47. Saelinger, C.A.; Lewbart, G.A.; Christian, L.S.; Lemons, C.L. Prevalence of *Salmonella* spp. in cloacal, fecal, and gastrointestinal mucosal samples from wild North American turtles. *J. Am. Vet. Med. Assoc.* **2006**, *229*, 266–268. [CrossRef] [PubMed]

48. Havelaar, A.H.; Kirk, M.D.; Torgerson, P.R.; Gibb, H.J.; Hald, T.; Lake, R.J.; Praet, N.; Bellinger, D.C.; de Silva, N.R.; Gargouri, N.; et al. World Health Organization global estimates and regional comparisons of the burden of foodborne disease in 2010. *PLoS Med.* **2015**, *12*, e1001923. [CrossRef]

49. Simonsen, G.S.; Tapsall, J.W.; Allegranzi, B.; Talbot, E.A.; Lazzari, S. The antimicrobial resistance containment and surveillance approach-a public health tool. *Bull. World Health* **2004**, *82*, 928–934.

50. Sobur, A.; Haque, Z.F.; Sabuj, A.A.; Ievy, S.; Rahman, A.T.; El Zowalaty, M.E.; Rahman, T. Molecular detection of multidrug and colistin-resistant *Escherichia coli* isolated from house flies in various environmental settings. *Future Microbiol.* **2019**, *14*, 847–858. [CrossRef]

51. Sobur, A.; Hasan, M.; Haque, E.; Mridul, A.I.; Noreddin, A.; El Zowalaty, M.E.; Rahman, T. Molecular Detection and Antibiotyping of Multidrug-Resistant *Salmonella* Isolated from Houseflies in a Fish Market. *Pathogens* **2019**, *8*, 191. [CrossRef]

52. Nahar, A.; Islam, M.A.; Sobur, M.A.; Hossain, M.J.; Binte, S.; Zaman, M.; Rahman, B.; Kabir, S.L.; Rahman, M.T. Detection of tetracycline resistant *E. coli* and *Salmonella* spp. in sewage, river, pond and swimming pool in Mymensingh, Bangladesh. *Afr. J. Microbiol. Res.* **2018**, *13*, 382–387.

53. Bryan, A.; Shapir, N.; Sadowsky, M.J. Frequency and distribution of tetracycline resistance genes in genetically diverse, nonselected, and nonclinical *Escherichia coli* strains isolated from diverse human and animal sources. *Appl. Environ. Microbiol.* **2004**, *70*, 2503–2507. [CrossRef]

54. Strahilevitz, J.; Jacoby, G.A.; Hooper, D.C.; Robicsek, A. Plasmid-mediated quinolone resistance: A multifaceted threat. *Clin. Microbiol. Rev.* **2009**, *22*, 664–689. [CrossRef]

55. Møller, T.S.; Overgaard, M.; Nielsen, S.S.; Bortolaia, V.; Sommer, M.O.; Guardabassi, L.; Olsen, J.E. Relation between *tetR* and *tetA* expression in tetracycline resistant *Escherichia coli*. *BMC Microbiol.* **2016**, *16*, 1–8. [CrossRef] [PubMed]

56. Poirel, L.; Stephan, R.; Perreten, V.; Nordmann, P. The carbapenemase threat in the animal world: The wrong culprit. *J. Antimicrob. Chemother.* **2014**, *69*, 2007–2008. [CrossRef]

57. Tacconelli, E.; Carrara, E.; Savoldi, A.; Harbarth, S.; Mendelson, M.; Monnet, D.L.; Pulcini, C.; Kahlmeter, G.; Kluytmans, J.; Carmeli, Y.; et al. Discovery, research, and development of new antibiotics: The WHO priority list of antibiotic-resistant bacteria and tuberculosis. *Lancet Infect. Dis.* **2018**, *18*, 318–327. [CrossRef]

58. Jajere, S.M. A review of *Salmonella enterica* with particular focus on the pathogenicity and virulence factors, host specificity and antimicrobial resistance including multidrug resistance. *Vet. World* **2019**, *12*, 504–521. [CrossRef]

59. Sobur, A.M.; Ievy, S.; Haque, Z.F.; Nahar, A.; Zaman, S.B.; Rahman, M.T. Emergence of colistin-resistant *Escherichia coli* in poultry, house flies, and pond water in Mymensingh, Bangladesh. *J. Adv. Vet. Anim. Res.* **2019**, *6*, 50–53. [PubMed]

60. Wang, R.; Cao, W.; Cerniglia, C.E. PCR detection and quantitation of predominant anaerobic bacteria in human and animal fecal samples. *Appl. Environ. Microbiol.* **1996**, *62*, 1242–1247. [CrossRef]

61. Fratamico, P.M. Comparison of culture, polymerase chain reaction (PCR), TaqMan *Salmonella*, and Transia Card *Salmonella* assays for detection of *Salmonella* spp. in naturally-contaminated ground chicken, ground turkey, and ground beef. *Mol. Cell. Probes* **2003**, *17*, 215–221. [CrossRef]

62. Randall, L.P.; Cooles, S.W.; Osborn, M.K.; Piddock, L.J.; Woodward, M.J. Antibiotic resistance genes, integrons and multiple antibiotic resistance in thirty-five serotypes of *Salmonella enterica* isolated from humans and animals in the UK. *J. Antimicrob. Chemother.* **2004**, *53*, 208–216. [CrossRef]

63. Bayer, A.W.; Kirby, W.M.; Sherris, J.C.; Turck, M. Antibiotic susceptibility testing by a standardized single disc method. *Am. J. Clin. Pathol.* **1966**, *45*, 493–496. [CrossRef]

64. CLSI. *Performance Standards for Antimicrobial Susceptibility Testing*, 26th ed.; CLSI Supplement M100s; Clinical and Laboratory Standards Institute: Wayne, PA, USA, 2016.
65. Sweeney, M.T.; Lubbers, B.V.; Schwarz, S.; Watts, J.L. Applying definitions for multidrug resistance, extensive drug resistance and pandrug resistance to clinically significant livestock and companion animal bacterial pathogens. *J. Antimicrob. Chemother.* **2018**, *73*, 1460–1463. [CrossRef]
66. Ashour, H.M. One Health—People, Animals, and the Environment. *Clin. Infect. Dis.* **2014**, *59*, 1510. [CrossRef]

Publisher's Note: MDPI stays neutral with regard to jurisdictional claims in published maps and institutional affiliations.

 *antibiotics*

Article

Genetic Subtyping, Biofilm-Forming Ability and Biocide Susceptibility of *Listeria monocytogenes* Strains Isolated from a Ready-to-Eat Food Industry

Joana Catarina Andrade [1], António Lopes João [2], Carlos de Sousa Alonso [2], António Salvador Barreto [1] and Ana Rita Henriques [1,*]

[1] CIISA–Centre for Interdisciplinary Research in Animal Health, Faculty of Veterinary Medicine, University of Lisbon, 1300-477 Lisbon, Portugal; ajoanacatarina@gmail.com (J.C.A.); asbarreto@fmv.ulisboa.pt (A.S.B.)

[2] Laboratório de Bromatologia e Defesa Biológica do Exército, 1849-012 Lisbon, Portugal; joao.aebl@exercito.pt (A.L.J.); alonso.ces@exercito.pt (C.d.S.A.)

* Correspondence: anaritah@fmv.ulisboa.pt; Tel.: +351-213-652-834

Received: 25 June 2020; Accepted: 15 July 2020; Published: 16 July 2020

Abstract: *Listeria monocytogenes* is a foodborne pathogen of special concern for ready-to-eat food producers. The control of its presence is a critical step in which food-grade sanitizers play an essential role. *L. monocytogenes* is believed to persist in food processing environments in biofilms, exhibiting less susceptibility to sanitizers than planktonic cells. This study aimed to test the susceptibility of *L. monocytogenes* in planktonic culture and biofilm to three commercial food-grade sanitizers and to benzalkonium chloride; together with the genetic subtyping of the isolates. *L. monocytogenes* isolates were collected from raw materials, final products and food-contact surfaces during a 6-year period from a ready-to-eat meat-producing food industry and genetically characterized. Serogrouping and pulsed-field gel electrophoresis (PFGE) revealed genetic variability and differentiated *L. monocytogenes* isolates in three clusters. The biofilm-forming ability assay revealed that the isolates were weak biofilm producers. *L. monocytogenes* strains were susceptible both in the planktonic and biofilm form to oxidizing and ethanol-based compounds and to benzalkonium chloride, but not to quaternary ammonium compound. A positive association of biofilm-forming ability and LD_{90} values for quaternary ammonium compound and benzalkonium chloride was found. This study highlights the need for preventive measures improvement and for a conscious selection and use of sanitizers in food-related environments to control *Listeria monocytogenes*.

Keywords: biocide; *Listeria monocytogenes*; biofilm; planktonic culture; pulsed-field gel electrophoresis

1. Introduction

Listeria monocytogenes is an ubiquitous small Gram-positive bacterium widespread in the natural environment [1]. It is also an opportunistic pathogen responsible for human listeriosis, a severe disease with high hospitalization and case fatality rates [2,3]. Its psychrotrophic nature and the ability to survive and multiply under extreme physicochemical conditions [4] may explain the difficulty of controlling its presence in refrigerated environments [5].

This pathogen is often associated to ready-to-eat (RTE) food products, with contamination occurring during food processing production [6,7]. Incoming raw materials, food handlers, and even processed ingredients and products are frequent sources of *L. monocytogenes* contamination [8]. After entering a food producing facility, *L. monocytogenes* can become a long-term resident, being able to persist for months or years within the premises, including food contact equipments [9]. Once established,

L. monocytogenes biofilms can persist, resulting in the potential continuous contamination of the food products [10].

L. monocytogenes has the ability to adhere to different surfaces within the food industry, such as plastic, rubber, stainless steel, glass, and produce biofilms [5,11]. Biofilm formation is affected by many factors, such as strain-specific properties, composition of the attachment surface, and environmental conditions [12]. Previous works relating *L. monocytogenes* serotypes and biofilm formation remained inconclusive, although several authors have addressed it [13–16].

In the biofilm, bacteria are embedded by an extracellular matrix able to function as a structural scaffold and protective barrier against various stresses and antimicrobials, like those encountered in the food processing environment [13,17]. Biofilms are associated to increased resistance to sanitizing compounds, due to bacterial exposure to sublethal biocide concentrations, acquiring resistance to antimicrobials over time [17,18].

The validation of sanitizers is essential to avoid the misuse of biocides that may end-up promoting resistance of *L. monocytogenes* virulent strains. Still, the effectiveness of commercial food-grade sanitizers is tested on planktonic microorganisms, but the biofilm environment may change the response of every strain involved [19]. Among food-grade sanitizers used in RTE food processing premises, oxidizing disinfectants and quaternary ammonium compounds are the most popular, due to their broad-spectrum activity against bacteria, high efficacy and low cost [20,21]. Nevertheless, *L. monocytogenes* resistance to these compounds has been described, whether in planktonic cultures or in biofilms [10,22]. The same was reported for benzalkonium chloride, a quaternary ammonium compound [23–25].

In this work, the susceptibility of *L. monocytogenes* in planktonic culture and biofilm to three commercial food-grade sanitizers and to benzalkonium chloride was assessed. For that, *L. monocytogenes* isolates collected from a RTE food-producing industry during a 6-year period were genetically characterized and their biofilm-forming ability was assessed, prior to biocide susceptibility testing.

2. Results and Discussion

2.1. Characterization of L. monocytogenes Isolates Collection

The overall proportion of positive samples (food and food related environment) contaminated by *L. monocytogenes* was 26.3% (20/76) (Table 1). This high percentage is in line with other studies in Portugal [26] that reported 25% of positive samples in ham, 11.1% in blood sausage and 2.3% in dry cured ham collected from producers and retailers.

2.2. L. monocytogenes Confirmation and Serogrouping by PCR

All of the *L. monocytogenes* presumptive isolates (*n* = 20) obtained by conventional microbiological methods belonged to the *Listeria* genus, but only 17 were confirmed as *L. monocytogenes* by PCR [27]. Among these 17 isolates, four different molecular serogroups were identified (Table 2).

Most of the isolates belonged to serogroup IIc (52.9%), followed by serogroup IIa (35.3%), IIb and IVb (each with 5.9%). In line with our results, other authors have reported similar findings. Lotfollahi et al. [28] found serogroup IIc to be the most prevalent in *L. monocytogenes* isolates from several foods retailed in Iranian markets. In another study, Montero et al. [29] found serogroup IIa to be the most common one in RTE meat-based products collected from different retail stores and industrial processing plants in Santiago, Chile, although serogroup IIb, IIc and IVb strains were also present. In an investigation assessing serogroup diversity of *L. monocytogenes* isolates in food from central and northern regions in Italy, 67.5% of isolates belonged to serogroup IIa [30]. Rodríguez-López et al. [31] reported similar results in samples collected from different food-related premises in Northwest Spain during 2010 and 2011, of which only 5.9% of isolates belonged to serogroup IVb. Torresi et al. [32] reported a predominance of serogroup IIa and IIc strains in several different cheeses in Italy.

Molecular serotyping is a rapid and useful method for first-level characterization of *L. monocytogenes* [16]. Still, to allow for a more reliable characterization of strains and contamination routes investigation, other molecular subtyping methods, such as pulsed-field gel electrophoresis (PFGE) should be used [33].

Table 1. Samples collected in the assessed industry with positive *L. monocytogenes* detection by conventional microbiological methods.

Date of Collection	Sample Description/Type	Presumptive *L. monocytogenes* Isolate Code
July 2010	*Chourição*/Final product	FP1
February 2013	*Chouriço*/Final product	FP2
March 2013	Seasoned ham meat/Intermediate product	IP1
March 2013	In-use meat mincing machine/Equipment	E1
March 2013	Meat sausage/Final product	FP3
April 2013	Unseasoned ham meat/Intermediate product	IP2
April 2013	Seasoned ham meat/Intermediate product	IP3
April 2013	Pork meat/Raw material	RM1
April 2013	In-use meat mincing machine/Equipment	E2
April 2013	Raw meat transport box/Equipment	E3
July 2013	Pork meat/Raw material	RM2
October 2013	*Chouriço*/Final product	FP4
February 2014	Lard for *chouriço*/Raw material	RM3
February 2014	*Chouriço*/Final product	FP5
April 2014	Boneless pork shoulder/Raw material	RM4
May 2014	*Chouriço*/Final product	FP6
May 2014	*Alheira*/Final product	FP7
January 2015	Boneless pork shoulder/Raw material	RM5
February 2015	*Farinheira*/Final product	FP8
April 2015	*Chouriço*/Final product	FP9

Table 2. Description of the obtained serogroups among *L. monocytogenes* confirmed isolates (*n* = 17).

Serogroup	Proportion	Isolate Code [1]
IIa	6 (35.3%)	FP1, RM2, RM3, FP7, FP8, FP9
IIb	1 (5.9%)	FP5
IIc	9 (52.9%)	FP2, IP1, E1, FP3, IP2, RM1, E2, E3, RM4
IVb	1 (5.9%)	RM5

[1] *L. monocytogenes* isolates share the same code with the sample from which they were recovered.

2.3. Pulsed-Field Gel Electrophoresis Typing

Figure 1 presents the resulting dendrogram of 17 *L. monocytogenes* strains considering *Apa*I and *Asc*I restriction patterns and serogroups. Pulsotypes were considered to be clones when they had at least 90% of similarity.

The different food and environment samples presented six PFGE types. Three clusters were identified (indicated as A, B and C in Figure 1), while FP5, RM5 and FP1 pulsotypes had a distinct PFGE profile.

The first cluster (Figure 1, cluster A) includes 9 strains, corresponding to 52.9% of all the analyzed isolates. These strains with identical restriction patterns and exhibiting the same serogroup (serogroup IIc) were collected from raw materials, intermediate products, finished products, and food-contact surfaces in a time frame of 14 months (from February 2013 to April 2014). When comparing cluster A food-contact surfaces and finished products strains' profiles, results suggest the possibility of a common source. It is noteworthy that *L. monocytogenes* strain RM4 collected in 2014 has 91.6% similarity with strains collected in 2013. This is suggestive of a potential persistent contamination within the food industry, although more studies should be considered in order to establish source attribution. Kurpas et al. [34] linked *L. monocytogenes* presence in food processing environments, such as abattoirs, RTE meat-processing industries and retail establishments to cross-contamination.

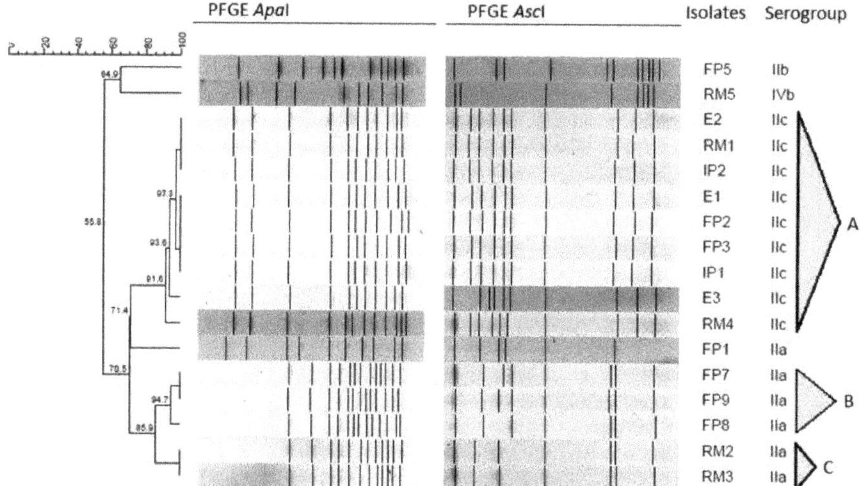

Figure 1. Dendrogram of the *Apa*I-*Asc*I PFGE profiles and corresponding serogroup for 17 *L. monocytogenes* selected isolates.

Cluster B includes three strains (FP7, FP8 and FP9) collected from different finished products between May 2014 and April 2015, all belonging to serogroup IIa. *L. monocytogenes* strain FP7 shares an indistinguishable profile with strain FP9, collected 1 year later. Apart from being suggestive of persistence over time, which might be due to *L. monocytogenes* survival and growth in niches within the food environment, these strains belong to serogroup IIa, which is the one most commonly associated to food-related environments [1,30].

Cluster C includes 2 strains (RM2 and RM3) collected between 2013 and 2014, from different raw materials. As seen before in cluster A, *L. monocytogenes* pulsotypes identified in raw materials exhibit high similarity with pulsotypes from equipment and finished products. These pulsotypes may persist due to the repeated re-introduction of strains from the external environment into food processing facilities over time [35]. Suppliers should be addressed to understand the origin of some strains, although results underline cross-contamination as a possible way of disseminating *L. monocytogenes* in the assessed food industry. A strict selection and control of suppliers seems to be a preventive measure of upmost importance [36]. Three distinct pulsotypes can also be seen in the resulting dendrogram. FP5 and RM5 strains were collected 1 year apart from each other and presented distinct pulsotypes (64.9% of similarity), belonging to serogroups IIb and IVb, respectively. FP1 isolate exhibits a different PFGE profile from other serogroup IIa strains (71.4% of similarity), which might be due to the fact that serogroup IIa includes atypical strains [27,37].

The presence of serogroups IIa, IIb and IVb isolates suggests a potential public health hazard associated with these RTE meat-based products consumption, since these are the serogroups more commonly associated to human infection [38,39].

2.4. Biofilm Formation Assay

After serogrouping and PFGE typing, 10 *L. monocytogenes* strains were selected for the biofilm formation assay in order to have representatives with different profiles (serogroups and pulsotypes). *L. monocytogenes* CECT 4031, CECT 911, CECT 935, and CECT 937 were also included in order to investigate differences between strains of different serogroups.

The assessed strains in biofilms revealed cvOD values ranging from 0.068 ± 0.001 to 0.1240 ± 0.006 and viable cell counts of 6.0 ± 0.4 log cfu/mL to 7.6 ± 0.4 log cfu/mL after 5 days of growth in polystyrene microtiter wells (Figure 2).

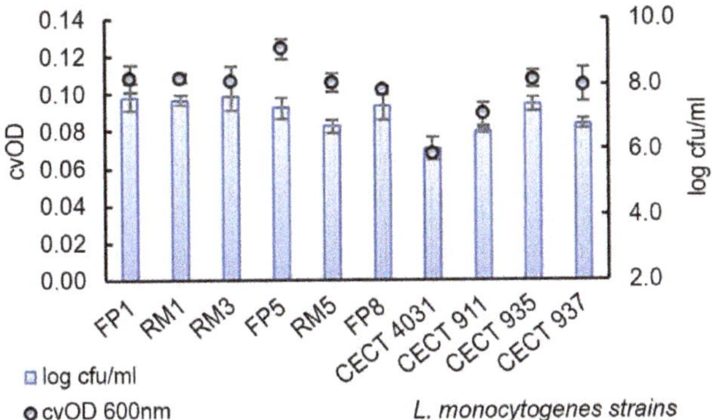

Figure 2. Average and standard deviation of log cfu/mL and cvOD of 5-day *L. monocytogenes* biofilms.

According to Stepanović et al. [40] classification, all the strains (n = 10) revealed a weak biofilm-forming ability. Similar results were obtained by Meloni et al. [41] when studying *L. monocytogenes* isolates from fermented sausage processing plants: 65% of all isolates were weak biofilm producers. However, in our work, the assessed strains exhibited significantly different degrees of biofilm-forming ability based on cvOD values (p = 0.0066), and VCC results did not reflect the same biofilm-forming ability classes as those obtained using cvOD values. Considering VCC values, all the strains, except *L. monocytogenes* FP1 and RM3 isolates, revealed lower values than *L. monocytogenes* CECT 935, which exhibited 7.4 ± 0.2 log cfu/mL. On the other hand, when considering cvOD values, *L. monocytogenes* FP1, RM1, and FP6 isolates exhibited higher cvOD values than reference *L. monocytogenes* CECT 935 (0.1078 ± 0.005). Considering cvOD and VCC values, *L. monocytogenes* CECT 4031 revealed the lowest values for both parameters at 30 °C. The obtained difference between these two parameters is due to the nature of each method of determination. While cvOD measures the turbidity of a suspension and quantifies total biomass (viable and non-viable cells, but also extracellular matrix components), VCC only considers live cells [42]. Taking into account the selected methods to analyze biofilm formation-VCC (log cfu/mL) and cvOD, Pearson's correlation analysis was performed. According to Pearson's correlation coefficient (ρ = 0.7749, p = 0.009), there is a positive and strong correlation between both parameters, which indicates that both methods present a good relationship, being reliable to quantifying *L. monocytogenes* biofilm formation, complementing each other.

When relating the biofilm-forming ability using cvOD values with the assessed *L. monocytogenes* serogroups, no significant differences were found (p = 0.526) and the same happened for VCC values (p = 0.929) (Table 3).

Table 3. Biofilm-forming ability of *L. monocytogenes* strains according to the respective serogroups.

L. monocytogenes Serogroup	n	Log cfu/mL (Mean ± SD)	cvOD (Mean ± SD)
IIa	4	7.2 ± 0.8	0.096 ± 0.019
IIb	2	7.0 ± 0.4	0.114 ± 0.002
IIc	2	7.0 ± 0.7	0.099 ± 0.003
IVb	2	7.1 ± 0.5	0.107 ± 0.001

Similar results were obtained by Di Bonaventura et al. [43] when studying the association of phylogeny and biofilm production. Nevertheless, this study's results counteract the ones obtained by Meloni et al. [41], in which serotypes 1/2a, 1/2b and 4b isolates presented a higher adherence when compared to serotype 1/2c isolates.

Other authors have shown that *L. monocytogenes* strains from different sources and serogroups are able to produce biofilms on a variety of surfaces, depending on the strain, surface and culture conditions [13,44]. Previous works reported that *L. monocytogenes* strains varied significantly in their ability to produce biofilm, but no trends could be observed when isolates' serotype and source were compared [3,40]. It is important to highlight that since there is a link between virulence and *L. monocytogenes* serotype, a continuous discussion relating biofilm formation and serotypes goes on, in order to determine whether biofilm formation is related to disease incidence [1,14].

For further testing, five *L. monocytogenes* strains (RM1, RM3, RM5, CECT 4031, and CECT 935) were selected based on serogrouping and biofilm formation parameters data analyses.

2.5. Biocides Activity Testing Assay

2.5.1. Activity towards *L. monocytogenes* Planktonic Suspension

The effect of food-grade commercial sanitizers, including an oxidizing compound (OxC), a quaternary ammonium compound (QaC) and an ethanol-based compound (EthC) on the selected five *L. monocytogenes* strains was assessed. Tested concentrations were selected based on the manufacturer's recommendation for use in food contact surfaces. The manufacturer's recommended concentrations for OxC and EthC were found to be equally effective in inactivating the five tested strains in planktonic suspension, although this was not observed for QaC.

L. monocytogenes planktonic cells were inactivated by 50 ppm or more (100 and 150 ppm) of OxC. Norwood and Gilmour [45] reported that a 30 sec exposure to 10 ppm free chlorine was enough to completely eliminate planktonic *L. monocytogenes* culture.

L. monocytogenes strains were exposed to increasing concentrations of EthC (50%, 70%, and 100%) that seemed to be effective in inactivating planktonic cells. Similar results were obtained by Aarnisalo et al. [46].

L. monocytogenes planktonic forms enumeration after QaC treatment was not possible to perform within the tested concentration range. Some authors have reported resistance to QaCs in *L. monocytogenes* strains [47–49] and active efflux pumps are considered the main mechanism for *L. monocytogenes* tolerance to QaCs [50]. Because it was not possible to determine *L. monocytogenes* susceptibility to QaC, benzalkonium chloride (BaC) was used to evaluate *L. monocytogenes* planktonic cells susceptibility. Figure 3 presents the effects of BaC treatment on the five selected *L. monocytogenes* strains planktonic suspensions. As shown, all strains in the planktonic form presented different susceptibilities to BaC, being affected by different concentrations. Reference strains *L. monocytogenes* CECT 4031 and CECT 935 were the most susceptible, presenting more than 4-log cfu/mL reduction when exposed to 0.8 ppm of BaC. *L. monocytogenes* RM1, RM3 and RM5 strains were less susceptible, presenting 4-log cfu/mL reduction only for concentrations higher than 12.5 ppm for RM1 and RM5 and 20 ppm for RM3. To have an 8-log cfu/mL reduction, *L. monocytogenes* CECT 4031 and CECT 935 planktonic cells were exposed to 2 ppm of BaC. The same was observed when RM1 and RM5 and RM3 were subjected to 25 ppm and 150 ppm, respectively. In line with our results, Nocker et al. [51] reported that the exposure of *L. monocytogenes* strains to BaC concentrations higher than 30 ppm for 30 min was able to reduce bacterial colonies as measured by plate counts.

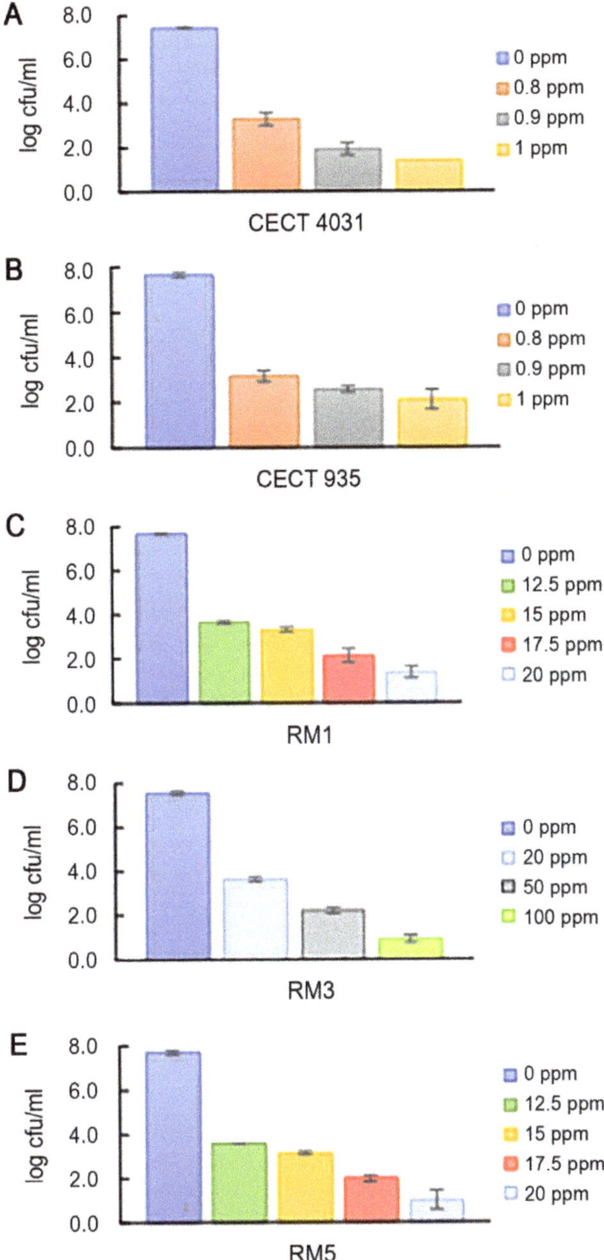

Figure 3. Viable cell counts average and standard deviation (error bars) of the tested planktonic *L. monocytogenes* strains after treatment with BaC. (**A**) *L. monocytogenes* CECT 4031; (**B**) *L. monocytogenes* CECT 935; (**C**) *L. monocytogenes* RM1; (**D**) *L. monocytogenes* RM3; (**E**) *L. monocytogenes* RM5.

2.5.2. Activity towards *L. monocytogenes* 5-day-old Biofilms

The biocide activity testing assay on biofilms was based on the enumeration of viable cells. The three commercial compounds were tested on 5-day-old biofilms according to the manufacturer's recommended concentrations. As was observed for *L. monocytogenes* planktonic cells, both OxC and EthC tested concentrations, which were within the manufacturer's recommended concentrations, were able to eliminate biofilms of all the tested isolates in 5 min at 20 °C. In fact, it was reported that 200 ppm of sodium hypochlorite, an OxC, is enough to eliminate at least 20% of *L. monocytogenes* biofilms [3,45]. In contrast, after QaC's treatment, no susceptibility to this biocide was observed. Figure 4 presents the effect on VCC after treatment with QaC on the selected *L. monocytogenes* 5-day-old biofilms.

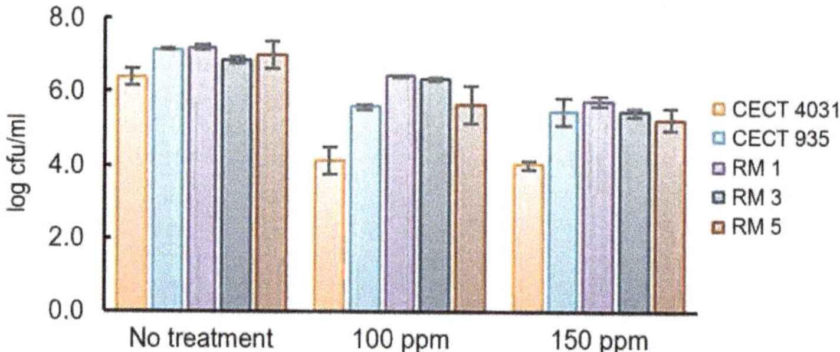

Figure 4. Average and standard deviation (error bars) of tested *L. monocytogenes* 5-day-old biofilms' VCC after QaC treatment.

In general, QaC was not effective in removing *L. monocytogenes* 5-day-old biofilms. As shown in Figure 4, when exposed to 150 ppm of QaC, *L. monocytogenes* CECT 4031 presented the highest reduction (from 6.4 log cfu/mL to 4.0 log cfu/mL). The remaining *L. monocytogenes* strains presented approximately 1-log cfu/mL reduction in VCC values. QaC resistance in *L. monocytogenes* biofilms has been reported [52,53]. Taking into account that this biocide is commonly used in food-related environments, these results are worrisome, as *L. monocytogenes* biofilms present a potential risk in food safety [54].

Figure 5 presents the tested concentration range of BaC's in *L. monocytogenes* 5-day-old biofilms. In general, *L. monocytogenes* 5-day-old biofilms' VCC were affected by different BaC concentrations, as happened for planktonic suspensions. While *L. monocytogenes* CECT 4031 was the most susceptible to BaC's treatment and also presented the lowest biofilm-forming ability, *L. monocytogenes* RM3 strain was the less susceptible, but presented the highest biofilm-forming ability based on VCC values. A 3-log cfu/mL reduction was observed for *L. monocytogenes* CECT 4031 after 5 min of exposure to 10 ppm of BaC. On the other hand, for a similar reduction on *L. monocytogenes* RM3 biofilm, 250 ppm of BaC were necessary. Comparing these results to those obtained for planktonic cells, it seems that *L. monocytogenes* biofilms are less susceptible to BaC's tested concentrations, since a higher BaC's concentration is needed to have an equivalent log cfu/mL reduction.

One example is *L. monocytogenes* RM3 isolate that in biofilm presented a 2-log cfu/mL reduction when exposed to 100 ppm of BaC and a 3-log cfu/mL reduction when exposed to 250 ppm, while the exposure to 150 ppm of BaC in the planktonic form was enough to cause a 8-log cfu/mL reduction. It has been previously discussed that in biofilm form, *L. monocytogenes* is more resistant to stress and sanitizing agents than planktonic cells [41,55]. Nakamura et al. [54], when assessing the sanitizing effect of BaC in *L. monocytogenes* planktonic cells and biofilms, reported that biofilm formation and tolerance to BaC might be related. Tolerance to BaC has also been reported by Piercey et al. [23] after testing BaC resistance and susceptibility based on the minimum inhibitory concentration, and by Xu et al. [24] after investigating phenotypic and genotypic tolerance to BaC based on susceptibility testing and molecular methods. Although in the last years several studies have focused on biofilm elimination, possible facilitating strategies are still unclear.

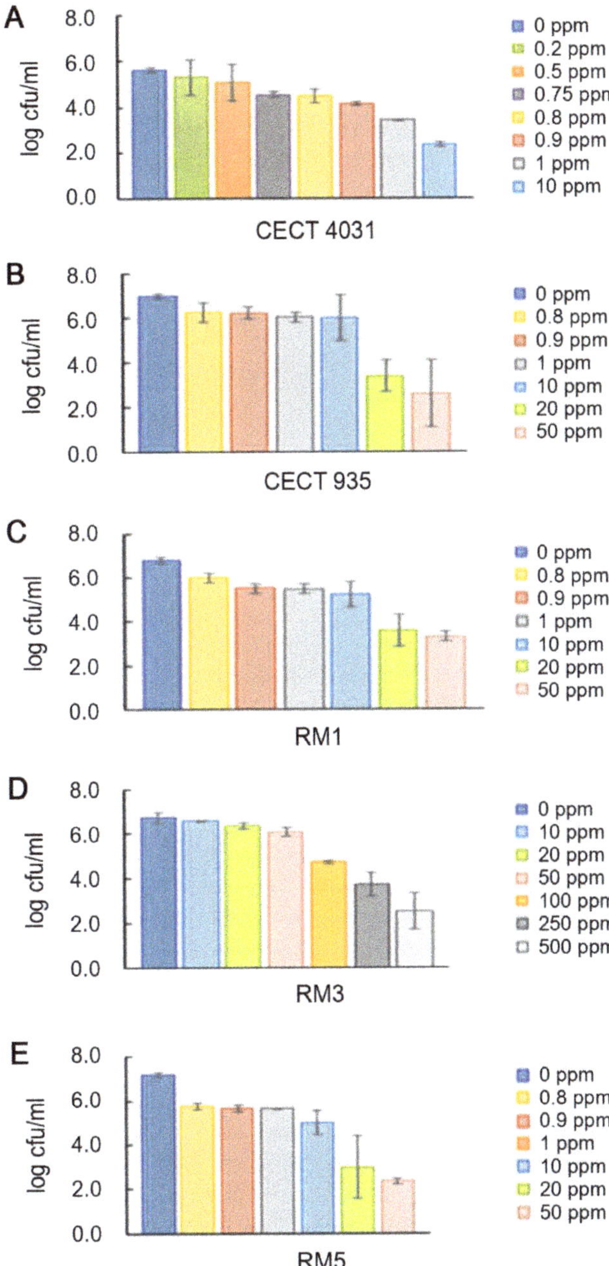

Figure 5. Viable cell counts average and standard deviation (error bars) of the tested *L. monocytogenes* 5-day-old biofilms after BaC treatment. (**A**) *L. monocytogenes* CECT 4031; (**B**) *L. monocytogenes* CECT 935; (**C**) *L. monocytogenes* RM1; (**D**) *L. monocytogenes* RM3; (**E**) *L. monocytogenes* RM5.

To assess *L. monocytogenes* susceptibility to QaC and BaC, LD_{90} values were calculated. Figure 6 presents QaC LD_{90} values. These values ranged from 298.0 to 532.2 ppm, and were higher than the manufacturer's recommended concentrations to be used in food-related surfaces (maximum recommended concentration: 150 ppm).

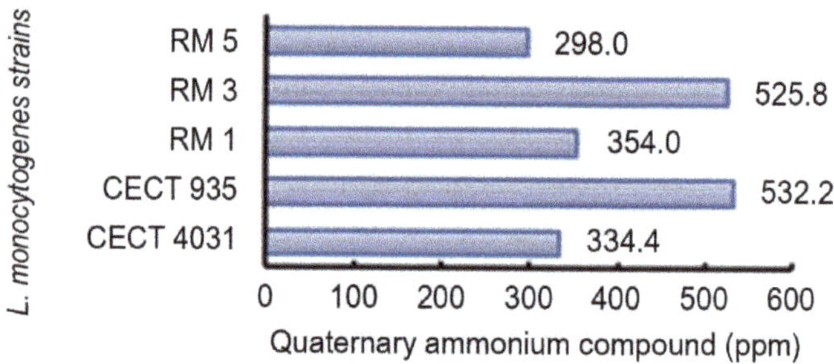

Figure 6. LD_{90} estimated values of *L. monocytogenes* tested strains in biofilm exposed to QaC.

This fact is relevant, because QaC is a commercial biocide that might be used in sublethal concentrations, which might induce *L. monocytogenes* resistance [46,56]. *L. monocytogenes* QaC resistance has been previously described, both for planktonic cells and biofilms [10,22,35].

BaC estimated LD_{90} values for *L. monocytogenes* tested strains (Figure 7) that ranged from 1.0 to 102.0 ppm in the planktonic form and from 17.8 to 675.2 ppm in biofilms, presenting significant differences ($p < 0.0001$).

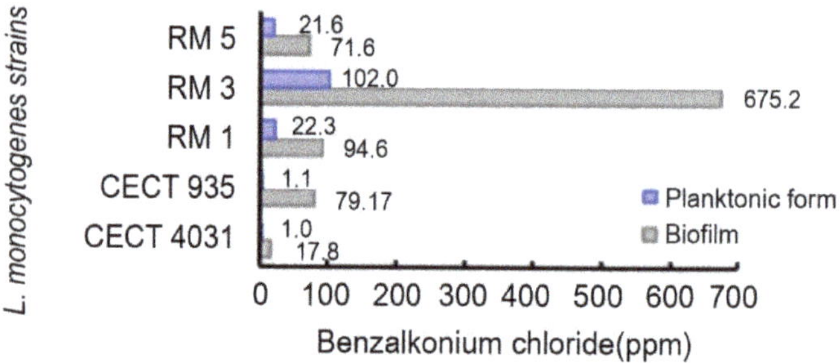

Figure 7. LD_{90} estimated values of *L. monocytogenes* tested strains in biofilm and in the planktonic form exposed to BaC.

L. monocytogenes biofilms exhibited a reduced susceptibility to BaC, compared to the planktonic forms. The biofilm structure may play an important role as the extracellular matrix acts like a barrier, preventing contact with antimicrobial agents [57,58]. In this study, the higher the biofilm-forming ability, the higher were the LD_{90} values for QaC and BaC. This positive association of biofilm-forming ability and LD_{90} values was moderate, both for QaC and for BaC (Table 4).

Table 4. Pearson's correlation coefficients for biofilm-forming ability parameters and LD_{90} values for QaC and BaC.

LD_{90} Values	Log cfu/mL	QaC LD_{90}
QaC	0.652	1
BaC	0.554	0.607

These results emphasize the importance of the cautious selection and use of sanitizers in food-producing premises. In fact, the equipment's sanitizing method should be re-assessed and validated in order to control *L. monocytogenes* contamination, as it might select isolates that are able to survive and adapt to the food processing environment [59], acting as potential contamination sources for RTE food produced in those surfaces. Taken together, biofilm-forming ability and LD_{90} values underline the need to select different sanitizers, using rotating schemes, in order to prevent biocide resistance over time. Also, different strategies should be considered, other than the use of chemical biocides, as novel technologies, to control *L. monocytogenes* in the food production environment [60,61].

3. Materials and Methods

3.1. Characterization of L. monocytogenes Isolates Collection

A collection of presumptive *L. monocytogenes* isolates ($n = 20$) was gathered from raw materials, intermediate, and final products, as well as industrial environment samples (food contact surfaces) of a RTE meat-based food producing industry (Table 1). This industry was located in Évora, Alentejo and produced pork meat delicatessens. *L. monocytogenes* isolates were collected during a 6-year period (2010–2015) as a result of routine microbiological sampling for industrial hygiene and food safety verification purposes, according to ISO 11290:1996 [62]. From a total of 76 collected samples, five raw materials, three intermediate products, nine finished RTE meat products and three food-contact surfaces were positive for *L. monocytogenes*. The strains were preserved in brain hearth infusion (BHI) broth (Scharlab, S.B., Barcelona, Spain) with 15% glycerol (Merck KGaA, Darmstadt, Germany) at −80 °C and revivified before use.

3.2. L. monocytogenes Confirmation and Serogrouping by PCR

Presumptive *L. monocytogenes* isolates ($n = 20$) were confirmed by PCR and serogrouped using a multiplex PCR and an additional PCR based on the amplification of the *fla*A gene [27]. *L. monocytogenes* confirmed isolates ($n = 17$) were selected for further genetic characterization.

3.3. Pulsed-Field Electrophoresis Typing

PFGE typing of the selected isolates was performed according to the PulseNet standardized procedure for *L. monocytogenes* [63]. Briefly, bacterial genomic DNA in 1.5% agarose (SeaKem Gold Agarose, Cambrex, East Rutherford, NJ, USA) plugs was digested in separate reactions with 10U *Asc*I (NZYTech, Lisbon, Portugal) for 2h at 37 °C, and with 50U *Apa*I (NZYTech) for 2h at 25 °C. Electrophoresis of the resulting DNA fragments was performed in 1% agarose gel (SeaKem Gold), with a lambda PFG ladder standard (New England Biolabs, Massachusetts, USA) in 0.5 X solution of Tris–borate–EDTA buffer (NZYTech) at 14 °C, with 6 V/cm, initial pulsed time of 4.0 s and final pulsed time of 40 s, included angle of 120° over 19 h using a CHEF-Dr III System (Bio-Rad Laboratories, Hercules, CA, USA). Gels were stained with ethidium bromide (Sigma, St. Louis, MO, USA) and photographed under UV transillumination.

3.4. Biofilm-Forming Ability Assay

To assess biofilm formation, six *L. monocytogenes* strains were selected (RM1, RM3, RM5, FP1, FP5, and FP8) to have representatives from different serogroups and PFGE types. Also, four *L. monocytogenes* reference strains from the Spanish Type Culture Collection (CECT) were used: CECT 4031 (serogroup IIa), CECT 937 (serogroup IIb), CECT 911 (serogroup IIc), and CECT 935 (serogroup IVb). These strains present the same serogroups as the ones detected in this study isolates (Table 2), allowing for the comparison with existing studies.

The protocol proposed by Romanova et al. [51] was used with some modifications to obtain a 5-day *L. monocytogenes* mono-cultural biofilm. A single colony of each selected strain was inoculated in buffered peptone water (BPW) (Scharlab, S.B) and incubated for 16–18 h at 30 °C. Optical density at 600 nm (OD_{600nm}) was assessed in Ultrospec 2000 (Pharmacia Biotech, Washington, WA, USA) to obtain a concentration of 8 log cfu/mL. For each strain, 4 μL were transferred into three separate wells of polystyrene flat-bottomed microtiter plates (Normax, Marinha Grande, Portugal) filled with 200 μL of BPW. Three wells were used as negative controls, with BPW alone. The plates were lidded and statically incubated at 30 °C for 5 days. After this, the solution was removed from the wells that were rinsed once with sterile distilled water to remove loosely associated bacteria and the attached biofilms were assessed by viable cells enumeration and crystal violet staining.

Considering both evaluation methods, this assay was performed in triplicate, with three replicates for each strain.

3.4.1. Enumeration of Viable Cells in Biofilms

The biofilm was detached from the well surface mechanically into 100 μL of BPW using a mini cell scraper (VWR International, Monroeville, PA, USA). The microtiter plate was sonicated (Ultrasonic bath MXB14, Grant Instruments, Royston, England) for 5 min to detach and collect sessile cells. Then, 100 μL of BPW were pipetted into each well. Serial 10-fold dilutions of the sample in BPW were prepared and 10 μL were dropped onto the surface of a tryptone soy agar (TSA) (Scharlab, S.B) plate. After overnight incubation at 30 °C, colonies were enumerated in a stereoscopic magnifier (Nikon SMZ645, Tokyo, Japan).

3.4.2. Biofilm Assessment by Crystal Violet Staining

The microtiter plate was left air drying for 45 min in the laminar flow hood. Biofilm was stained by adding 220 μL of 0.1% crystal violet (bioMérieux, France) solution to each well for 15 min at room temperature. After stain removal, the wells were washed three times with sterile distilled water and left air drying for 30 min in the laminar flow hood. Then, 220 μL of detaining solution (ethanol: acetone 80:20 *v/v*) were added to each well 15 min. The microtiter plate was then shaken (Ultrasonic bath MXB14, Grant Instruments, Royston, England) for 5 min and the crystal violet OD (cvOD) was measured in SpectraMax 340PC (Molecular Devices, Silicon Valley, San Jose, CA, USA). Each absorbance value was corrected by subtracting the average absorbance readings of the blank control wells. Adherence capability of the tested strains was based on the cvOD exhibited by bacterial biofilms, according to Stepanović et al. [40] classification. The cut-off cvOD (cvODc) was defined as three standard deviations above the negative control mean cvOD. Strains were classified as no biofilm producers (cvOD ≤ cvODc), weak biofilm producers (cvODc < cvOD ≤ 2 × cvODc), moderate biofilm producers (2 × cvODc < cvOD ≤ 4 × cvODc), and strong biofilm producers (4 × cvODc < cvOD).

3.5. Biocides Activity Testing Assay

Based on serogrouping and biofilm formation parameters, *L. monocytogenes* strains RM1, RM3 and RM5 and *L. monocytogenes* reference strains CECT 4031 and CECT 935 were selected to be further assessed.

Biocides activity testing was performed according to European standard EN 1276:2009 [64], using the quantitative suspension test for bactericidal activity evaluation of chemical disinfectants used in food and industrial areas. To simulate clean conditions, in all tests, 0.03 g/L of bovine serum albumin (Merck KGaA) was used as an interfering substance. Contact time (5 min) and temperature (20 °C) were established according to the obligatory test conditions specified in EN 1276:2009. For all strains, experimental conditions were previously validated. Biocide activity was assessed using *Escherichia coli* DSMZ 682, *Pseudomonas aeruginosa* ATCC 15442, *Staphylococcus aureus* CECT 239, and *Enterococcus hirae* ATCC 10541D-5, to validate dilution-neutralization, absence of lethal effect in test conditions, including neutralizer toxicity, and efficacy of neutralizing solutions.

Commonly used biocides in food contact surfaces and equipment sanitization in food-producing establishments were selected for further testing. Commercial sanitizers (HigiaBlue, Portugal) containing oxidizing compounds (OxC), ethanol-based compounds (EthC) and quaternary ammonium compounds (QaC) were tested. Benzalkonium chloride (BaC; Merck KGaA) was also evaluated. Table 5 exhibits the tested concentrations for each biocide (diluted in hard water) and respective neutralizers.

Table 5. Tested biocides, concentration range and neutralizers used in biocide activity testing assay (EN 1276:2009).

Biocide	Tested Concentrations			Neutralizer
Oxidizing compound (OxC)	50 ppm	100 ppm	150 ppm	Polysorbate 80, 30 g/L + lecithin, 3 g/L + sodium thiosulphate 10 g/L
Quaternary ammonium compound (QaC)	50 ppm	100 ppm	150 ppm	Polysorbate 80, 30 g/L + sodium dodecyl sulphate, 4 g/L + lecithin, 3 g/L
Ethanol-based compound (EthC)	50%	70%	100%	Polysorbate 80, 30 g/L + saponin, 30 g/L + lecithin, 3 g/L
Benzalkonium chloride (BaC)	Planktonic cells		Biofilm	
	0.8–150 ppm		0.2–500 ppm	Polysorbate 80, 30 g/L + sodium dodecyl sulphate, 4 g/L + lecithin, 3 g/L

All measurements were performed in duplicate and all experiments were performed twice.

3.5.1. Activity towards *L. monocytogenes* Planktonic Suspension

L. monocytogenes strains were incubated in BHI agar (Scharlab, S.B.) at 37 °C for 18 h. Then, 10 mL of the bacterial suspension were prepared to have an OD_{600nm} of 0.15–0.5, corresponding to a concentration of approximately 1.5–5×10^8 cfu/mL. To each tube containing 1 mL of interfering substance, 1 mL of the bacterial suspension was added, and the mixture was vortexed. After 2 min, 8 mL of one of the desired biocide test concentration were added, incubating for 5 min at 20 °C. Then, 1 mL was collected and mixed with 1 mL of hard water and 8 mL of the appropriate neutralizer. After neutralization (5 min at 20 °C), 1 mL was incorporated in TSA (Scharlab, S.B.) in duplicate. After overnight incubation at 37 °C, colonies were enumerated.

3.5.2. Activity towards *L. monocytogenes* 5-day-old Biofilms

For the biocide activity testing on *L. monocytogenes* 5-day-old biofilms, to each well containing the biofilm, 20 µL of interfering substance and 20 µL of tryptone salt solution (Scharlab, S.B.) were added. After 2 min, 160 µL of one of the biocide test concentrations was added and incubated for 5 min at 20 °C. After medium removal, the wells were washed with 40 µL of hard water and 160 µL of the appropriate neutralizer. After neutralization (5 min at 20 °C), the medium was removed and the wells were washed with sterile distilled water, which was also removed. The biofilm quantification was performed according to the procedure described in Section 3.4.1. for biofilm detachment, dilution and colony enumeration.

LD_{90} was then calculated for both planktonic and biofilm assays in order to determine the biocide concentration that reduced 90% of VCC.

3.6. Data Analyses

All quantitative data are presented as mean values with standard deviation (SD) from three independent experiments. Using BioNumerics software package version 6.10 (Applied Maths, Sint-Martens-Latem, Belgium), a dendrogram was constructed based on PFGE patterns of the 17 *L. monocytogenes* strains, with an optimization setting of 1.5% and a band-position tolerance of 1.5% for *Asc*I and *Apa*I restriction. Cluster analysis was performed using the unweighted pair group method (UPGMA) with arithmetic averages and band-based Dice correlation coefficient.

To assess *L. monocytogenes* biofilm formation parameters, Pearson's correlation analyses were used to evaluate the interdependency of cvOD and VCC. To relate biofilm formation parameters and *L. monocytogenes* serogroups, one-way analysis of variance (ANOVA) followed by Tukey's test were performed.

To evaluate the susceptibility of selected *L. monocytogenes* strains to biocides, LD_{90} values were obtained by adjusting experimental data of mortality obtained in biocide testing assays to a polynomial equation or to a linear regression adjusted to a scatter plot of mortality versus biocide concentration in MS Excel 2016 software (Microsoft Corporation, Redmond, USA). Two-way ANOVA was used to compare BaC LD_{90} values in planktonic and biofilm forms. To compare *L. monocytogenes* biofilms QaC LD_{90} values and also BaC LD_{90} values, one-way ANOVA followed by Tukey's test were performed. Pearson's correlation coefficient was also used to relate biofilm formation parameters and QaC and BaC LD_{90}. When $p < 0.05$, a statistically significant difference was considered.

4. Conclusions

Overall, this study provided an assessment of *L. monocytogenes* isolates from a RTE meat-based food industry, using phenotypic and genetic characterization. The use of molecular and subtyping techniques is an important tool to understand the routes and sources of contamination. Our results reveal that *L. monocytogenes* contamination of finished products seems to be related to food-contact surfaces, but also to raw materials. Moreover, some of the obtained pulsotypes revealed high homology (>90%) but were not temporally matched, being collected with months of interval. These results might point out to a common source of contamination and are consistent with the hypothesis that there are stable clonal groups of *L. monocytogenes*, which persist over time, in foods and food-related environments.

All of the studied *L. monocytogenes* strains demonstrated biofilm-forming ability at 30 °C, revealing to be weak biofilm producers. Strains in biofilms were not susceptible to one of the used commercial sanitizers in the industrial premises, QaC, within the recommended concentration range. Similar results were obtained when testing a pure substance biocide, benzalkonium chloride (BaC) in *L. monocytogenes* biofilms. In contrast, *L. monocytogenes* planktonic forms were susceptible to BaC tested concentrations. A positive association was found between biofilm formation parameters and LD_{90} values for QaC and BaC.

Taken together, our results suggest that preventive measures need improvement in the assessed food industry. It also reinforces the necessity of an appropriate selection and application of biocides in food premises, to prevent *L. monocytogenes* biofilm formation and biocide resistance development over time.

Author Contributions: Conceptualization, A.R.H.; methodology, J.C.A., C.d.S.A., A.L.J., software, J.C.A., validation, A.R.H., A.L.J., formal analysis, J.C.A., A.R.H.; investigation, J.C.A., C.d.S.A., A.L.J.; resources, A.S.B., A.L.J.; data curation, A.R.H., A.L.J.; writing—original draft preparation, J.C.A.; writing—review and editing, A.S.B., A.R.H.; supervision, A.S.B., A.R.H.; project administration, A.R.H.; funding acquisition, A.R.H.; All authors have read and agreed to the published version of the manuscript.

Funding: This research was supported by Project UIDP/CVT/00276/2020 (funded by FCT).

Acknowledgments: The authors gratefully acknowledge Maria Helena Fernandes, Maria José Fernandes and Maria Paula Silva for the technical support and Carla Carneiro for the discussion time. We acknowledge the logistic support of CIISA - Centro de Investigação Interdisciplinar em Sanidade Animal, Faculdade de Medicina Veterinária, Universidade de Lisboa, Avenida da Universidade Técnica, 1300-477 Lisboa, Portugal.

Conflicts of Interest: The authors declare no conflict of interest.

References

1. Orsi, R.H.; den Bakker, H.C.; Wiedmann, M. *Listeria monocytogenes* lineages: Genomics, evolution, ecology, and phenotypic characteristics. *Int. J. Med. Microbiol.* **2011**, *301*, 79–96. [CrossRef]

2. Gandhi, M.; Chikindas, M.L. Listeria: A foodborne pathogen that knows how to survive. *Int. J. Food Microbiol.* **2007**, *113*, 1–15. [CrossRef]

3. Russo, P.; Hadjilouka, A.; Beneduce, L.; Capozzi, V.; Paramithiotis, S.; Drosinos, E.; Spano, G. Effect of different conditions on *Listeria monocytogenes* biofilm formation and removal. *Czech J. Food Sci.* **2018**, *36*, 208–214. [CrossRef]

4. Paudyal, R.; Barnes, R.H.; Karatzas, K.A.G. A novel approach in acidic disinfection through inhibition of acid resistance mechanisms; Maleic acid-mediated inhibition of glutamate decarboxylase activity enhances acid sensitivity of *Listeria monocytogenes*. *Food Microbiol.* **2018**, *69*, 96–104. [CrossRef] [PubMed]

5. Bonsaglia, E.C.R.; Silva, N.C.C.; Fernandes, J.A.; Araújo Junior, J.P.; Tsunemi, M.H.; Rall, V.L.M. Production of biofilm by *Listeria monocytogenes* in different materials and temperatures. *Food Control.* **2014**, *35*, 386–391. [CrossRef]

6.   Bolocan, A.S.; Nicolau, A.I.; Alvarez-Ordóñez, A.; Borda, D.; Oniciuc, E.A.; Stessl, B.; Gurgu, L.; Wagner, M.; Jordan, K. Dynamics of *Listeria monocytogenes* colonisation in a newly-opened meat processing facility. *Meat Sci.* **2016**, *113*, 26–34. [CrossRef] [PubMed]

7.   Amajoud, N.; Leclercq, A.; Soriano, J.M.; Bracq-Dieye, H.; Maadoudi, M.E.; Senhaji, N.S.; Kounnoun, A.; Moura, A.; Lecuit, M.; Abrini, J. Prevalence of *Listeria* spp. and characterization of *Listeria monocytogenes* isolated from food products in Tetouan, Morocco. *Food Control.* **2018**, *84*, 436–441. [CrossRef]

8.   Henriques, A.R.; Fraqueza, M.J. Listeria monocytogenes and ready-to-eat meat-based food products: Incidence and control. In *Listeria monocytogenes: Incidence, Growth Behavior and Control*, 1st ed.; Viccario, T., Ed.; Nova Science Publishers Inc.: New York, NY, USA, 2015; pp. 71–103.

9.   Olszewska, M.A.; Zhao, T.; Doyle, M.P. Inactivation and induction of sublethal injury of *Listeria monocytogenes* in biofilm treated with various sanitizers. *Food Control.* **2016**, *70*, 371–379. [CrossRef]

10.   Ortiz, S.; López, V.; Martínez-Suárez, J.V. Control of *Listeria monocytogenes* contamination in an Iberian pork processing plant and selection of benzalkonium chloride-resistant strains. *Food Microbiol.* **2014**, *39*, 81–88. [CrossRef]

11.   Ibusquiza, P.S.; Herrera, J.J.; Vázquez-Sánchez, D.; Parada, A.; Cabo, M.L. A new and efficient method to obtain benzalkonium chloride adapted cells of *Listeria monocytogenes*. *J. Microbiol. Methods* **2012**, *91*, 57–61. [CrossRef]

12.   Nilsson, R.E.; Ross, T.; Bowman, J.P. Variability in biofilm production by *Listeria monocytogenes* correlated to strain origin and growth conditions. *Int. J. Food Microbiol.* **2011**, *150*, 14–24. [CrossRef] [PubMed]

13.   Papaioannou, E.; Giaouris, E.D.; Berillis, P.; Boziaris, I.S. Dynamics of biofilm formation by *Listeria monocytogenes* on stainless steel under mono-species and mixed-culture simulated fish processing conditions and chemical disinfection challenges. *Int. J. Food Microbiol.* **2018**, *267*, 9–19. [CrossRef] [PubMed]

14.   Wang, W.; Zhou, X.; Suo, Y.; Deng, X.; Cheng, M.; Shi, C.; Shi, X. Prevalence, serotype diversity, biofilm-forming ability and eradication of *Listeria monocytogenes* isolated from diverse foods in Shanghai, China. *Food Control.* **2017**, *73*, 1068–1073. [CrossRef]

15.   Doijad, S.P.; Barbuddhe, S.B.; Garg, S.; Poharkar, K.V.; Kalorey, D.R.; Kurkure, N.V.; Rawool, D.B.; Chakraborty, T. Biofilm-forming abilities of *Listeria monocytogenes* serotypes isolated from different sources. *PLoS ONE* **2015**, *10*, e0137046. [CrossRef]

16.   Borucki, M.K.; Peppin, J.D.; White, D.; Loge, F.; Call, D.R. Variation in biofilm formation among strains of *Listeria monocytogenes*. *Appl. Environ. Microbiol.* **2003**, *69*, 7336–7342. [CrossRef] [PubMed]

17.   Mazaheri, T.; Ripolles-Avila, C.; Hascoët, A.S.; Rodríguez-Jerez, J.J. Effect of an enzymatic treatment on the removal of mature *Listeria monocytogenes* biofilms: A quantitative and qualitative study. *Food Control.* **2020**, *114*, 107266. [CrossRef]

18.   Sadekuzzaman, M.; Mizan, M.F.R.; Kim, H.S.; Yang, S.; Ha, S.D. Activity of thyme and tea tree essential oils against selected foodborne pathogens in biofilms on abiotic surfaces. *LWT* **2018**, *89*, 134–139. [CrossRef]

19.   Puga, C.H.; Orgaz, B.; SanJose, C. *Listeria monocytogenes* impact on mature or old Pseudomonas fluorescens biofilms during growth at 4 and 20 °C. *Front. Microbiol.* **2016**, *7*, 134. [CrossRef]

20.   Tamburro, M.; Ripabelli, G.; Vitullo, M.; Dallman, T.J.; Pontello, M.; Amar, C.F.; Sammarco, M.L. Gene expression in *Listeria monocytogenes* exposed to sublethal concentration of benzalkonium chloride. *Comp. Immunol. Microbiol. Infect. Dis.* **2015**, *40*, 31–39. [CrossRef]

21.   Henriques, A.R.; Fraqueza, M.J. Biofilm-forming ability and biocide susceptibility of *Listeria monocytogenes* strains isolated from the ready-to-eat meat-based food products food chain. *LWT* **2017**, *81*, 180–187. [CrossRef]

22.   Kocot, A.M.; Olszewska, M.A. Biofilm formation and microscopic analysis of biofilms formed by *Listeria monocytogenes* in a food processing context. *LWT* **2017**, *84*, 47–57. [CrossRef]

23.   Piercey, M.J.; Elis, T.C.; Macintosh, A.J.; Truelstrup, H.L. Variations in biofilm formation, desiccation resistance and benzalkonium chloride susceptibility among *Listeria monocytogenes* strains isolated in Canada. *Int. J. Food Microbiol.* **2017**, *257*, 254–261. [CrossRef] [PubMed]

24.   Xu, D.; Li, Y.; Zahid, M.S.; Yamasaki, S.; Shi, L.; Li, J.R.; Yan, H. Benzalkonium chloride and heavy-metal tolerance in *Listeria monocytogenes* from retail foods. *Int. J. Food Microbiol.* **2014**, *190*, 24–30. [CrossRef] [PubMed]

25.   Barroso, I.; Maia, V.; Cabrita, P.; Martínez-Suárez, J.V.; Brito, L. The benzalkonium chloride resistant or sensitive phenotype of *Listeria monocytogenes* planktonic cells did not dictate the susceptibility of its biofilm counterparts. *Food Res. Int.* **2019**, *123*, 373–382. [CrossRef] [PubMed]

26. Mena, C.; Almeida, G.; Sá-Carneiro, L.; Teixeira, P.; Hogg, T.; Gibbs, P.A. Incidence of *Listeria monocytogenes* in different food products commercialized in Portugal. *Food Microbiol.* **2004**, *21*, 213–216. [CrossRef]

27. Kérouanton, A.; Marault, M.; Petit, L.; Grout, J.; Dao, T.T.; Brisabois, A. Evaluation of a multiplex PCR assay as an alternative method for *Listeria monocytogenes* serotyping. *J. Microbiol. Methods* **2010**, *80*, 134–137. [CrossRef]

28. Lotfollahi, L.; Chaharbalesh, A.; Ahangarzadeh, R.M.; Hasani, A. Prevalence, antimicrobial susceptibility and multiplex PCR-serotyping of *Listeria monocytogenes* isolated from humans, foods and livestock in Iran. *Microb. Pathog.* **2017**, *107*, 425–429. [CrossRef]

29. Montero, D.; Bodero, M.; Riveros, G.; Lapierre, L.; Gaggero, A.; Vidal, R.M.; Vidal, M. Molecular epidemiology and genetic diversity of *Listeria monocytogenes* isolates from a wide variety of ready-to-eat foods and their relationship to clinical strains from listeriosis outbreaks in Chile. *Front. Microbiol.* **2015**, *6*, 384. [CrossRef]

30. Sheng, J.; Tao, T.; Zhu, X.; Bie, X.; Lv, F.; Zhao, H.; Lu, Z. A multiplex PCR detection method for milk based on novel primers specific for *Listeria monocytogenes* 1/2a serotype. *Food Control.* **2018**, *86*, 183–190. [CrossRef]

31. Rodríguez-López, P.; Ibusquiza, P.S.; Mosquera-Fernández, M.; López-Cabo, M. *Listeria monocytogenes*-carrying consortia in food industry. Composition, subtyping and numerical characterisation of mono-species biofilm dynamics on stainless steel. *Int. J. Food Microbiol.* **2015**, *206*, 84–95. [CrossRef]

32. Torresi, M.; Acciari, V.A.; Zennaro, G.; Prencipe, V.; Migliorati, G. Comparison of Multiple-locus variable number tandem repeat analysis and pulsed-field gel electrophoresis in molecular subtyping of *Listeria monocytogenes* isolates from Italian cheese. *Vet. Ital.* **2015**, *51*, 191–198.

33. Nyarko, E.B.; Donnelly, C.W. *Listeria monocytogenes*: Strain heterogeneity, methods, and challenges of subtyping. *J. Food Sci.* **2015**, *80*, 2868–2878. [CrossRef] [PubMed]

34. Kurpas, M.; Wieczorek, K.; Osek, J. Ready-to-eat meat products as a source of *Listeria monocytogenes*. *J. Vet. Res.* **2018**, *62*, 49–55. [CrossRef] [PubMed]

35. Buchanan, R.L.; Gorris, L.G.M.; Hayman, M.M.; Jackson, T.C.; Whiting, R.C. A review of *Listeria monocytogenes*: An update on outbreaks, virulence, dose-response, ecology, and risk assessments. *Food Control.* **2017**, *75*, 1–13. [CrossRef]

36. Malley, T.J.V.; Butts, J.; Wiedmann, M. Seek and destroy process: *Listeria monocytogenes* process controls in the ready-to-eat meat and poultry industry. *J. Food Prot.* **2015**, *78*, 436–445. [CrossRef]

37. Doumith, M.; Buchrieser, C.; Glaser, P.; Jacquet, C.; Martin, P. Differentiation of the major *Listeria monocytogenes* serovars by multiplex PCR. *J. Clin. Microbiol.* **2004**, *42*, 3819–3822. [CrossRef]

38. Datta, A.R.; Burall, L.S. Serotype to genotype: The changing landscape of listeriosis outbreak investigations. *Food Microbiol.* **2018**, *75*, 18–27. [CrossRef]

39. Su, X.; Zhang, J.; Shi, W.; Yang, X.; Li, Y.; Pan, H.; Kuang, D.; Xu, D.; Shi, X.; Meng, J. Molecular characterization and antimicrobial susceptibility of *Listeria monocytogenes* isolated from foods and humans. *Food Control.* **2016**, *70*, 96–102. [CrossRef]

40. Stepanović, S.; Cirković, I.; Ranin, L.; Svabić-Vlahović, M. Biofilm formation by *Salmonella* spp. and *Listeria monocytogenes* on plastic surface. *Lett. Appl. Microbiol.* **2004**, *38*, 428–432. [CrossRef]

41. Meloni, D.; Consolati, S.G.; Mazza, R.; Mureddu, A.; Fois, F.; Piras, F.; Mazzette, R. Presence and molecular characterization of the major serovars of *Listeria monocytogenes* in ten Sardinian fermented sausage processing plants. *Meat Sci.* **2014**, *97*, 443–450. [CrossRef]

42. Kwasny, S.M.; Opperman, T.J. Static biofilm cultures of gram-positive pathogens grown in a microtiter format used for anti-biofilm drug discovery. *Curr. Protoc. Pharmacol.* **2010**, *13*, 8–23. [CrossRef] [PubMed]

43. Di Bonaventura, G.; Piccolomini, R.; Paludi, D.; D'Orio, V.; Vergara, A.; Conter, M.; Ianieri, A. Influence of temperature on biofilm formation by *Listeria monocytogenes* on various food-contact surfaces: Relationship with motility and cell surface hydrophobicity. *J. Appl. Microbiol.* **2008**, *104*, 1552–1561. [CrossRef] [PubMed]

44. Reis-Teixeira, F.B.; Alves, V.F.; Martinis, E.C.P. Growth, viability and architecture of biofilms of *Listeria monocytogenes* formed on abiotic surfaces. *Braz. J. Microbiol.* **2017**, *48*, 587–591. [CrossRef] [PubMed]

45. Norwood, D.E.; Gilmour, A. The growth and resistance to sodium hypochlorite of *Listeria monocytogenes* in a steady-state multispecies biofilm. *J. Appl. Microbiol.* **2000**, *88*, 512–520. [CrossRef]

46. Aarnisalo, K.; Lundén, J.; Korkeala, H.; Wirtanen, G. Susceptibility of *Listeria monocytogenes* strains to disinfectants and chlorinated alkaline cleaners at cold temperatures. *LWT* **2007**, *40*, 1041–1048. [CrossRef]

47. Ibusquiza, P.S.; Herrera, J.J.R.; Cabo, M.L. Resistance to benzalkonium chloride, peracetic acid and nisin during formation of mature biofilms by *Listeria monocytogenes*. *Food Microbiol.* **2011**, *28*, 418–425. [CrossRef]

48. Sidhu, M.S.; Heir, E.; Sorum, H.; Holck, A. Genetic linkage between resistance to quaternary ammonium compounds and β-Lactam antibiotics in food-related *Staphylococcus* spp. *Microb. Drug Res.* **2001**, *7*, 363–371. [CrossRef] [PubMed]

49. Romanova, N.A.; Gawande, P.V.; Brovko, L.Y.; Griffiths, M.W. Rapid methods to assess sanitizing efficacy of benzalkonium chloride to *Listeria monocytogenes* biofilms. *J. Microbiol. Methods* **2007**, *71*, 231–237. [CrossRef]

50. Rodríguez-López, P.; Rodríguez-Herrera, J.J.; Vázquez-Shánchez, D.; López-Cabo, M. Current knowledge on *Listeria monocytogenes* biofilms in food-related environments: Incidence, resistance to biocides, ecology and biocontrol. *Foods* **2018**, *7*, 85. [CrossRef]

51. Nocker, A.; Sossa, K.E.; Camper, A.K. Molecular monitoring of disinfection efficacy using propidium monoazide in combination with quantitative PCR. *J. Microbiol. Methods* **2007**, *70*, 252–260. [CrossRef]

52. Pan, Y.; Breidt, F.; Kathariou, S. Resistance of *Listeria monocytogenes* biofilms to sanitizing agents in a simulated food processing environment. *Appl. Environ. Microbiol.* **2006**, *72*, 7711–7717. [CrossRef] [PubMed]

53. Katharios-Lanwermeyer, S.; Rakic-Martinez, M.; Elhanafi, D.; Ratani, S.; Tiedje, J.M.; Kathariou, S. Coselection of cadmium and benzalkonium chloride resistance in conjugative transfers from nonpathogenic *Listeria* spp. to other listeriae. *Appl. Environ. Microbiol.* **2012**, *78*, 7549–7556. [CrossRef]

54. Nakamura, H.; Takakura, K.; Sone, Y.; Itano, Y.; Nishikawa, Y. Biofilm formation and resistance to benzalkonium chloride in *Listeria monocytogenes* isolated from a fish processing plant. *J. Food Prot.* **2013**, *76*, 1179–1186. [CrossRef]

55. Moltz, A.G.; Martin, S.E. Formation of biofilms by *Listeria monocytogenes* under various growth conditions. *J. Food Prot.* **2005**, *68*, 92–97. [CrossRef] [PubMed]

56. Kastbjerg, V.G.; Laesen, M.H.; Ingmer, H. Influence of sublethal concentrations of common disinfectants on expression of virulence genes in *Listeria monocytogenes*. *Appl. Environ. Microbiol.* **2010**, *76*, 303–309. [CrossRef] [PubMed]

57. Srey, S.; Jahid, I.K.; Ha, S.D. Biofilm formation in food industries: A food safety concern. *Food Control.* **2013**, *31*, 572–585. [CrossRef]

58. Schlafer, S.; Meyer, R.L. Confocal microscopy imaging of the biofilm matrix. *J. Microbiol. Methods* **2017**, *138*, 50–59. [CrossRef]

59. Gómez, D.; Azón, E.; Marco, N.; Carramiñana, J.J.; Rota, C.; Ariño, A.; Yanguela, J. Antimicrobial resistance of *Listeria monocytogenes* and *Listeria innocua* from meat products and meat-processing environment. *Food Microbiol.* **2014**, *42*, 61–65. [CrossRef] [PubMed]

60. Jahid, I.K.; Ha, S.D. The paradox of mixed-species biofilms in the context of food safety. *Compr. Rev. Food Sci. Food Saf.* **2014**, *13*, 990–1011. [CrossRef]

61. Metselaar, K.; Ibusquiza, P.S.; Ortiz, C.A.R.; Krieg, M.; Zwietering, M.H.; den Besten, H.M.; Abee, T. Performance of stress resistant variants of *Listeria monocytogenes* in mixed species biofilms with *Lactobacillus plantarum*. *Int. J. Food Microbiol.* **2015**, *213*, 24–30. [CrossRef]

62. International Organization for Standardization (ISO). *1290-1:1996. Microbiology of Food and Animal Feeding Stuffs—Horizontal Method for the Detection and Enumeration of Listeria monocytogenes*; Part 1: Detection Method; International Organization for Standardization: Geneva, Switzerland, 1996.

63. Graves, L.M.; Swaminathan, B. PulseNet standardized protocol for subtyping *Listeria monocytogenes* by macrorestriction and pulsed-field gel electrophoresis. *Int. J. Food Microbiol.* **2001**, *65*, 55–62. [CrossRef]

64. European standard (EN). *1276:2009. Chemical Disinfectants and Antiseptics—Quantitative Suspension Test for the Evaluation of Bactericidal Activity of Chemical Disinfectants and Antiseptics Used in Food, Industrial, Domestic and Institutional Areas—Test Method and Requirements (Phase 2, Step 1)*; International Organization for Standardization: Geneva, Switzerland, 2009.

Brief Report

Antimicrobial Drug-Resistant Gram-Negative Saprophytic Bacteria Isolated from Ambient, Near-Shore Sediments of an Urbanized Estuary: Absence of β-Lactamase Drug-Resistance Genes

Charles F. Moritz [1], Robert E. Snyder [1], Lee W. Riley [1], Devin W. Immke [2] and Ben K. Greenfield [2,*]

[1] School of Public Health, University of California, Berkeley, CA 94720, USA; cfmoritz@gmail.com (C.F.M.); robert.snyder@berkeley.edu (R.E.S.); lwriley@berkeley.edu (L.W.R.)

[2] Department of Environmental Sciences, Southern Illinois University, Edwardsville, IL 62026, USA; dimmke@siue.edu

* Correspondence: begreen@siue.edu

Received: 11 June 2020; Accepted: 7 July 2020; Published: 10 July 2020

Abstract: We assessed the prevalence of antimicrobial resistance and screened for clinically relevant β-lactamase resistance determinants in Gram-negative bacteria from a large urbanized estuary. In contrast to the broad literature documenting potentially hazardous resistance determinants near wastewater treatment discharge points and other local sources of aquatic pollution, we employed a probabilistic survey design to examine ambient, near-shore sediments. We plated environmental samples from 40 intertidal and shallow subtidal areas around San Francisco Bay (California, USA) on drug-supplemented MacConkey agar, and we tested isolates for antimicrobial resistance and presence of clinically relevant β-lactamase resistance determinants. Of the 74 isolates identified, the most frequently recovered taxa were *Vibrio* spp. (40%), *Shewanella* spp. (36%), *Pseudomonas* spp. (11%), and *Aeromonas* spp. (4%). Of the 55 isolates tested for antimicrobial resistance, the *Vibrio* spp. showed the most notable resistance profiles. Most (96%) were resistant to ampicillin, and two isolates showed multidrug-resistant phenotypes: *V. alginolyticus* (cefotaxime, ampicillin, gentamicin, cefoxitin) and *V. fluvialis* (cefotaxime, ampicillin, cefoxitin). Targeted testing for class 1 integrons and presence of β-lactam-resistance gene variants TEM, SHV, OXA, CTX-M, and *Klebsiella pneumonia* carbapenemase (KPC) did not reveal any isolates harboring these resistance determinants. Thus, while drug-resistant, Gram-negative bacteria were recovered from ambient sediments, neither clinically relevant strains nor mobile β-lactam resistance determinants were found. This suggests that Gram-negative bacteria in this well-managed, urbanized estuary are unlikely to constitute a major human exposure hazard at this time.

Keywords: antibiotic resistance; aquatic contamination; probabilistic sampling; San Francisco Estuary; coast; *Pseudomonas*; *Shewanella algae*; *Vibrio parahaemolyticus*

---

1. Introduction

The development of antimicrobial resistance in Gram-negative bacteria is a serious and growing global concern. Anthropogenic selection of highly resistant bacteria is driven by the overuse of antimicrobial agents in healthcare and agriculture as well as their mismanagement during waste disposal [1–6]. This selective process has dramatically affected global health; drug-resistant infections have become widespread globally [7–10] and were recently estimated at over 2 million infections in the United States annually [11]. Environmental and saprophytic bacteria are important as indicators and reservoirs of antibiotic resistance determinants that may be shared by human bacterial pathogens [12–19].

The β-lactams are currently the most widely used class of antimicrobial agents for treatment of bacterial infections in humans [20]. Gram-negative bacteria (GNB) have evolved to develop resistance to β-lactams by producing β-lactamase enzymes that hydrolyze β-lactams. Indeed, 2771 unique β-lactamase enzymes were discovered as of 2018 [21]. Extended-spectrum β-lactamases (ESBLs) such as TEM-, SHV-, OXA-, and CTX-M-type β-lactamases have become widespread in clinical and environmental settings, threatening the utility of broader-spectrum β-lactam drugs [21,22]. More recently, resistance to carbapenem drugs in GNB of the family Enterobacteriaceae, through production of *Klebsiella pneumonia* carbapenemases (KPCs), has become an imminent public health threat [9,14,23,24]. Genes encoding these β-lactamases are often located on mobile genetic elements that mediate their transfer between bacteria of the same or different species. This mechanism may contribute to dissemination of resistance determinants from the natural environment to healthcare settings [25].

Coastal and river waters located in populated areas with limited or overextended water and sanitation infrastructure harbor high rates of drug-resistant bacteria [6,17,26,27], but the extent to which this is true in areas with reliable secondary and tertiary wastewater treatment facilities is not as well characterized. San Francisco Bay (CA, USA) is located in a highly populated and urbanized region with extensive wastewater treatment infrastructure [28,29]. San Francisco Bay also has a legacy of environmental contamination that has resulted in elevated concentrations of a broad range of pollutants [30]. This includes fecal contamination observed at ponds managed as bird habitats and sloughs [31], and occasionally at swimming beaches [32,33]. Here, we assessed the prevalence of antimicrobial resistance in GNB in near-shore sediments collected from San Francisco Bay, an estuarine environment with ambient urban pollution. We determined resistant strain taxa and tested resistant isolates for class 1 integrons and presence of β-lactam-resistance gene variants TEM, SHV, OXA, CTX-M, and KPC. To our knowledge, this is the first probabilistic spatial survey of an estuary's sediment for clinically relevant genetic resistance elements in GNB.

2. Results

None of the 40 collection sites were immediately adjacent to treated wastewater discharge locations (Figure 1). Bacteria colonies grew on unsupplemented plates from all 40 sites, but none presented as positive for lactose utilization, indicating that lac+ colonies (e.g., *Escherichia coli*) were absent from all samples. From the 40 sites, bacterial isolates that grew in the presence of ampicillin, gentamicin, imipenem, and cefotaxime were found at 34 (85%), 27 (67.5%), 15 (37.5%), and 9 (22.5%) sites, respectively (Table 1). From the initial antibiotic-containing MacConkey agar plates, 174 isolates were obtained and subjected to further analyses. Bacteria isolated from plates containing ampicillin were the most prevalent (87 isolates from 32 sites), followed by gentamicin (39 isolates from 13 sites), imipenem (37 from 15 sites), and cefotaxime (11 isolates from 8 sites) (Table 1).

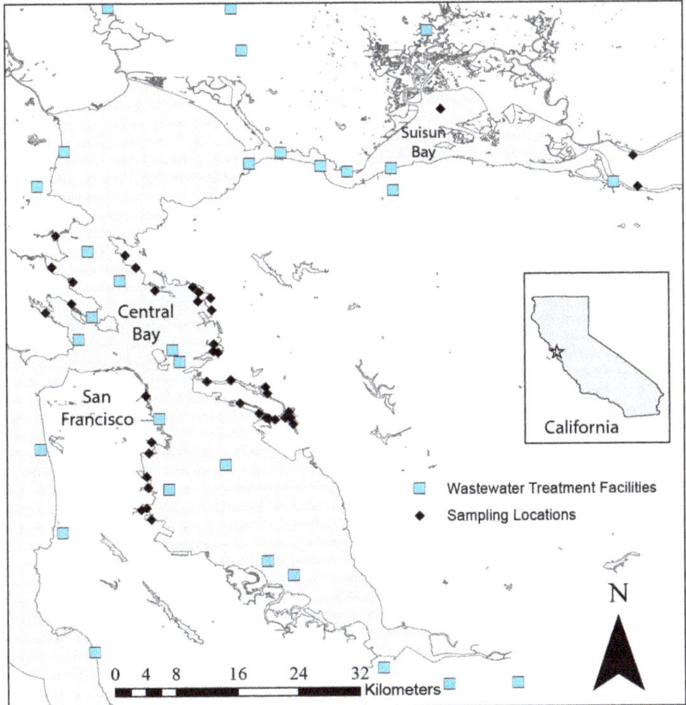

Figure 1. San Francisco Bay. Black diamonds (♦) indicate sediment collection location. Light blue squares (■) indicate wastewater treatment discharge locations in the region. Inset: Location within California, USA.

Table 1. Number of morphologically distinct bacterial colonies isolated from estuarine sediments in the San Francisco Bay Area, 2015, by antibiotic used for screening.

Antibiotic [a]	Sites with Growth (N, %)	CFU/g [b]	Sites with Isolates Obtained (N)	Morphologically Distinct Isolates Obtained (N) [c]
No antimicrobial agent	40 (100%)	3513		
Ampicillin	34 (85%)	1280	32	87
Cefotaxime	9 (22.5%)	16	8	11
Imipenem	15 (37.5%)	106	15	37
Gentamicin	27 (67.5%)	196	13	39

[a] Concentrations of antibiotic embedded in MacConkey agar plates: ampicillin, 16 µg mL^{-1}; imipenem, 1 µg mL^{-1}; cefotaxime, 1 µg mL^{-1}; and gentamicin, 10 µg mL^{-1}. [b] Colonies were counted on MacConkey agar plates and multiplied by the dilution factor to approximate the number of CFU/g sediment in each sediment sample. [c] Number of bacteria isolated from all antibiotic screening plates.

Seventy-two different Gram-negative bacterial isolates were identified by their 16S rRNA sequences. They included 1 *Acinetobacter* sp. (1.4%), 3 *Aeromonas* spp. (4.2%), 1 *Castellaniella* sp. (1.4%), 1 *Gallaecimonas* sp. (1.4%), 8 *Pseudomonas* spp. (11%), 1 *Rhizobium* sp. (1.4%), 26 *Shewanella* spp. (36.1%), 2 *Stenotrophomonas* spp. (2.8%), and 29 *Vibrio* spp. (40.3%) (Table 2). Fifty-three of the identified isolates were tested for their susceptibility to seven different antimicrobial agents (Table 3). Among 23 *Vibrio* spp. isolates, 22 (95.7%) were resistant to ampicillin. This included one isolate (*V. alginolylticus*) resistant to ampicillin and gentamicin and two isolates (8.7%) that displayed multidrug-resistant (MDR) phenotypes: *V. alginolyticus* (cefotaxime (CTX), ampicillin (AMP), gentamicin (GEN), and cefoxitin (FOX)) and *V. fluvialis* (CTX, AMP, amoxicillin–clavulanic acid (AMC), FOX). Among the 26

Shewanella spp. isolates, none were resistant to any of the drugs tested, except for three isolates that had intermediate resistance to imipenem. Due to a lack of Clinical and Library Standards Institute (CLSI) interpretive guidelines for the disc-diffusion test, we were unable to test *Pseudomonas* spp. isolates for phenotypic resistance.

Table 2. Identity of bacterial species recovered from San Francisco Bay sediment, 2015, by antibiotic used to select for resistance in initial MacConkey agar plate.

Antibiotic	Species	Isolates (N)	Antibiotic	Species	Isolates (N)
Ampicillin (16 µg mL^{-1})	Total	21	Imipenem (1 µg mL^{-1})	Total	35
	Vibrio alginolyticus	6		*Aeromonas australiensis*	1
	Vibrio parahaemolyticus	6		*Aeromonas hydrophila*	1
	Vibrio alginolyticus/parahaemolyticus [a]	7		*Aeromonas veronii*	1
	Vibrio alginolyticus/azureus [a]	2		*Castellaniella defragrans*	1
Cefotaxime (1 µg mL^{-1})	Total	11		*Pseudomonas* sp. [b]	1
	Acinetobacter venetianus	1		*Shewanella algae*	7
	Gallaecimonas xiamenensis	1		*Shewanella algae/haliotis* [a]	11
	Pseudomonas fluorescens	2		*Shewanella loihica*	8
	Pseudomonas oleovorans	1		*Stenotrophomonas maltophilia*	2
	Pseudomonas putida	3		*Vibrio diazotrophicus*	1
	Pseudomonas stutzeri	1		*Vibrio fluvialis*	1
	Rhizobium sp. [b]	1	Gentamicin (10 µg mL^{-1})	*Vibrio parahaemolyticus*	5
	Vibrio fluvialis	1			

[a] Unable to discriminate between two species after 16S sequence analysis. [b] Species not determined.

Table 3. Species and antibiotic resistance profiles of bacteria recovered from estuarine sediments in San Francisco Bay, 2015, from drug-supplemented media. Plate: Drug supplementation on plate (see Methods).

Species	Isolates (N)	Plate	Resistance (Disc Diffusion) [a]	Intermediate Resistance (Disc Diffusion) [a]
Acinetobacter venetianus	1	CTX	CTX	
Aeromonas australiensis	1	IPM	AMC	
Aeromonas hydrophila	1	IPM	FOX	AMC
Aeromonas veronii	1	IPM	None	
Shewanella algae	4	IPM	None	
Shewanella algae	3	IPM		IPM
Shewanella algae/halitosis [b]	11	IPM	None	
Shewanella loihica	8	IPM	None	
Vibrio alginolyticus	3	AMP	AMP	
Vibrio alginolyticus	1	AMP	AMP, CTX, GEN, FOX	
Vibrio alginolyticus	1	AMP	AMP, GEN	
Vibrio alginolyticus/parahaemolyticus [b]	4	AMP	AMP	
Vibrio diazotrophicus	1	IPM	None	
Vibrio fluvialis	1	CTX	AMP, CTX, AMC	FOX
Vibrio fluvialis	1	IPM	AMP	AMC
Vibrio parahaemolyticus	5	AMP	AMP	
Vibrio parahaemolyticus	1	AMP	AMP	CIP
Vibrio parahaemolyticus	5	GEN	AMP	

[a] AMP, ampicillin; CTX, cefotaxime; IPM, imipenem; AMC, amoxicillin–clavulanic acid; GEN, gentamicin; FOX, cefoxitin; CIP, ciprofloxacin. [b] Unable to discriminate between two species after 16S sequence analysis.

Of the 174 isolates that grew on drug-supplemented MacConkey agar plates, 174, 37, 98, and 11 isolates were tested for the presence of genes that encode class 1 integrons, carbapenemase (KPC), ESBLs (TEM, OXA, SHV), and CTX-M-type ESBLs, respectively. All PCR reactions were negative for these resistance genes.

3. Discussion

From 40 near-shore sites in the Bay Area, we isolated 18 distinct species of Gram-negative saprophytic bacteria (Table 2) on drug-supplemented plates. Although no recognized pathogenic GNB species were identified, many culturable isolates exhibited resistance to clinically used antimicrobial agents. Most studies assessing the presence of drug resistance in environmental bacteria thoroughly characterize a small number of sites, typically near known point-source pollutant effluent locations [1,17,18,26,34,35]. There have also been some comparative surveys across multiple water bodies [6,15]. In contrast to these designs, our sampling scheme extensively sampled a near-shore environment under ambient urban influence. In particular, our sampling sites were probabilistically chosen from intertidal and shallow subtidal areas around a large, urbanized estuary [36]. In this regionally representative sampling program, environmental bacteria were successfully isolated from every sampling site.

While species of several genera identified here (e.g., *Aeromonas* spp., *Pseudomonas* spp., *Shewanella* spp., and *Vibrio* spp.) have been described as opportunistic pathogens, they are all commonly found in marine-sediment environments, and their presence is rarely considered a public health risk [37–41]. Nevertheless, the genera *Aeromonas*, *Pseudomonas*, and *Shewanella* have been implicated as natural progenitors of, and reservoirs for, resistance genes such as CTX-M-, GES-, VIM-, and OXA-type ESBLs and carbapenemases that can be horizontally transferred into more pathogenic bacteria [25,42–44]. High rates of ampicillin resistance in *Vibrio* spp. have been well documented [45], consistent with the resistance rate of 96% found in this study. We also found five *Vibrio* isolates (22%) that displayed other resistance phenotypes. However, none of these harbored any of the common clinical resistance genes we tested for, including TEM, SHV, and OXA. The majority of bacteria that grew under selective pressure for imipenem resistance were *Shewanella* spp. (26 isolates; 74%); however, only three of these *Shewanella* spp. isolates (12%) exhibited intermediate resistance to carbapenems, with none being resistant. *Shewanella* spp. have been reported elsewhere to have reduced susceptibility to carbapenems, and the genus has also been identified as a natural progenitor of several OXA-type carbapenemases [42,46], yet none were found in our study.

Due to an absence of CLSI guidelines for nonclinically relevant bacteria, the antimicrobial susceptibilities of the isolated *Pseudomonas* spp. (including *Pseudomonas fluorescens*, *P. oleovorans*, *P. putida*, and *P. stutzeri*) were not tested by the disc-diffusion method. However, further investigation into the antimicrobial resistance profiles and genes for these isolates is warranted because the genus has been observed to harbor genes that mediate resistance to antimicrobial agents. Environmentally occurring *Pseudomonas* spp. harboring carbapenemases and ESBLs, namely VIM, IMP, and several CTX-M variants, have been widely reported in the past decade [19,43,47,48]. The CTX-M variants that we tested for in isolated *Pseudomonas* spp. were those known to be circulating in the region and were previously found in *P. putida* and *P. teessidea* in retail spinach [16].

A notable result of this study was the absence of drug-resistant bacteria from the Enterobacteriaceae family as well as the absence of fecal indicator bacteria [19]. A number of similar studies found an abundance of such bacteria, but these studies were conducted in water bodies and under conditions that would suggest a priori high levels of fecal contamination [4,17,40]. The frequency of fecal contamination in San Francisco Bay beaches is variable but generally low, and Bay beaches are typically safe for human recreation, with most beaches considered safe for swimming, especially during dry weather [32,33]. Carbapenem-resistant Enterobacteriaceae (CRE) present serious public health risk, and they were a major target of the present study; however, no such bacteria were isolated from the areas we tested.

There were several limitations in this study. Certain species may have been inhibited by the stress of the freeze–thaw step in combination with drug-supplemented MacConkey agar. Further, the techniques described here were culture-dependent, and PCR analysis was restricted to class 1 integrons and β-lactam resistance, which precluded the identification of other integrons or potentially relevant resistance mechanisms. Nevertheless, given the clinical importance of class 1 integrons [49], the observation of other associated resistance genes such as trimethoprim–sulfa or aminoglycoside would be unlikely in their absence. That said, the observed absence of β-lactam resistance mechanisms does not consider the full range of possible resistance genes. In the future, metagenomic study of DNA present in San Francisco Bay sediment samples or other whole-resistome screening approaches could reveal other clinically or environmentally relevant mechanisms [6,35].

Importantly, this study probabilistically sampled from 37 of the sites [36] in order to assess for regional patterns, rather than focusing only on areas of anthropogenic contamination. Studies that target wastewater treatment plants, hospital effluents, or animal livestock runoff could yield a higher prevalence of antimicrobial resistance among clinically relevant bacteria. In our study, the absence of clinically relevant drug-resistant GNB harboring β-lactamases and related resistance determinants suggests that GNB from ambient sediments in this well-managed, urbanized estuary may not constitute a major human exposure hazard at this time. These findings may be related to secondary and tertiary treatment operations and control measures for all wastewaters that drain into the Bay [28], in combination with the large dilution factor due to tidal exchange, resulting in low ambient sediment bacterial pollution in this estuary. These hypotheses could be tested in the future by evaluating resistance profiles and mechanisms in bacteria obtained from point sources and adjacent locations, including wastewater discharge effluents [1,4,18,34,35]. However, our study represents just one line of evidence, and routine water monitoring does periodically detect elevated fecal coliforms at some beaches [32]. Resistance to β-lactams continues to spread globally in GNB while, in parallel, novel resistance genes in environmental bacteria continue to be described. Therefore, routine environmental surveillance is needed to assess for the presence of potentially harmful bacteria or drug-resistance genes.

4. Materials and Methods

4.1. Sample Collection and Processing

Thirty-seven near-shore sites were sampled from Central San Francisco Bay (Central Bay), and three from Suisun Bay, both sub-basins of San Francisco Bay (Figure 1). The Central Bay sites were selected using a generalized random-tessellation stratified methodology, which is a probabilistic but spatially balanced method developed to identify locations for the sampling of natural resources [50]. The Suisun Bay samples were convenience samples, employing collection methods identical to those of the Central Bay sites. Although all sites were near-shore, a variety of habitats were included in the spatial sample, including both open water and narrow channels, sites adjacent to densely populated areas (e.g., San Francisco, CA, USA; Oakland, CA, USA), and sites proximate to more sparsely populated areas (e.g., Marin County and Suisun Bay) (Figure 1) [36].

Coastal Conservation and Research (Moss Landing, CA, USA) sampled all sites between 27 July and 14 September, 2015, as part of the Regional Monitoring Program (RMP)'s Bay Margins Sediment Study [51,52]. Sediments were collected by boat, with personnel using a modified VanVeen grab (0.1 m^2 sampling area), from which 15 mL of surface sediment was scraped into a 50-mL conical tube (Fischer Scientific, Hampton, NH, USA). Sediment samples were combined with 20 mL of a preservative solution (15% glycerol in phosphate-buffered saline solution, PBS) and stored on dry ice (−78.5 °C) for transportation to the laboratory, after which they were immediately stored at −20 °C until analysis.

4.2. Gram-Negative Bacteria Isolation

Samples were thawed prior to analysis and diluted 10-fold with PBS. We selected for bacteria with reduced drug susceptibility by incubating 100 μL of this PBS–sediment solution on MacConkey agar (Difco Laboratories Inc., Detroit, MI, USA), supplemented with one of four antibiotics: ampicillin (16 μg mL^{-1}), gentamicin (10 μg mL^{-1}), imipenem (1 μg mL^{-1}), or cefotaxime (1 μg mL^{-1}). An additional plate without any antibiotics was used to assess baseline growth. Plates were incubated at 37 °C for 24 h and assessed for growth. In the absence of any growth at this stage, plates were incubated for another 24 h. All plates were examined for well-formed colonies, and the total number of CFUs was recorded (Table 1). Up to four colonies were selected from each antibiotic plate for further analysis. In an attempt to increase the diversity of species isolated, we tried to choose morphologically distinct isolates within a plate, based on visual observation. The selected colonies were streaked for isolation on MacConkey agar and incubated again at 37 °C for 24 h. An isolated colony from each of these plates was then streaked for isolation on Luria–Bertani (LB) agar (Difco Laboratories Inc., Detroit, MI, USA) and incubated at 37 °C for 24 h. Finally, an isolated colony from each LB plate was subcultured in 4 mL of LB broth (Difco Laboratories Inc., Detroit, MI, USA) at 37 °C for 24 h. A 1-mL aliquot of this culture was saved in a 15% glycerol stock, and a separate 1-mL aliquot was used to extract DNA: Bacteria were concentrated by centrifugation (60 s, 14,000 RPM), resuspended in water, and placed in a boiling water bath for 10 min; excess cell debris was collected by centrifugation (30 s, 14,000 RPM); and the supernatant containing the DNA was pipetted to a separate tube and stored at −20 °C before analysis.

4.3. Bacterial Species Identification and Drug-Susceptibility Tests

Bacteria were identified by 16S rRNA sequencing. PCR was first carried out with the primers 16s8F/16s806R18 (94 °C for 5 min, then 30 cycles of 94 °C for 30 s, 62 °C for 30 s, and 72 °C for 90 s) as previously described by Raphael et al. [16]. The amplified DNA products (approximately 800 bp) were sequenced on an Applied Biosystems 3730 DNA analyzer (Applied Biosystems, Foster City, CA, USA) at the University of California, Berkeley, DNA Sequencing Facility. Antimicrobial susceptibility profiles were assessed with a disc-diffusion assay according to the CLSI interpretive guidelines [53].

All isolates were tested for the presence of genes encoding class 1 integrons by PCR following a procedure previously described by Raphael et al. [16]. Class 1 integrons were chosen because of their clinical relevance and prevalence and widespread distribution in Gram-negative bacteria, both globally [49] and in the San Francisco Bay region [10]. All isolates obtained from MacConkey agar plates containing ampicillin (16 µg mL^{-1}), and those that were resistant to ampicillin by the disc-diffusion test, were examined by PCR for the presence of the following extended spectrum β-lactamase variants: TEM (including TEM-1 and TEM-2), SHV (including SHV-1), and OXA (OXA-1, OXA-4, and OXA-30). For this, we employed multiplex primers and reaction conditions described by Dallenne et al. [54]. Isolates obtained from MacConkey agar plates containing cefotaxime (1 µg mL^{-1}), and those that were resistant to cefotaxime by the disc-diffusion test, were tested for CTX-M genes using a set of multiplex primers and conditions for CTX-M variants (CTX-M-1, CTX-M-3, and CTX-M-15) as described by Adams-Sapper et al. [22]. Isolates obtained from plates supplemented with imipenem were tested for the variants of the carbapenemase gene KPC using primers and conditions described by Dallenne et al. [54]. Additionally, all bacteria with multidrug-resistant phenotypes were tested for variants of all the above-mentioned genes: TEM, SHV, OXA, CTX-M, and KPC.

Author Contributions: Conceptualization, L.W.R. and B.K.G.; methodology, C.F.M., R.E.S., L.W.R., and B.K.G.; formal analysis, C.F.M. and B.K.G.; investigation, C.F.M.; resources, L.W.R.; data curation, C.F.M. and B.K.G.; writing—original draft preparation, C.F.M.; writing—review and editing, C.F.M., R.E.S., L.W.R., D.W.I., and B.K.G.; supervision, L.W.R. and B.K.G.; and project administration, C.F.M., D.W.I., and B.F.G. All authors have read and agreed to the published version of the manuscript.

Funding: This research received no external funding.

Acknowledgments: Sediment samples were provided by Russell Fairey and Marco Sigala of Moss Landing Marine Laboratories as part of the Regional Monitoring Program for Water Quality in San Francisco Bay. We thank Don Yee of the San Francisco Estuary Institute as well as Sheila Adams-Sapper and Nicole Tarlton from UC Berkeley for their valuable study guidance, in addition to Megan Danielson for her assistance with Figure 1.

Conflicts of Interest: The authors declare no conflict of interest.

References

1. Korzeniewska, E.; Harnisz, M. Extended-Spectrum Beta-Lactamase (ESBL)-Positive Enterobacteriaceae in Municipal Sewage and Their Emission to the Environment. *J. Environ. Manage.* **2013**, *128*, 904–911. [CrossRef] [PubMed]
2. Goñi-Urriza, M.; Capdepuy, M.; Arpin, C.; Raymond, N.; Pierre Caumette, C.Q. Impact of an Urban Effluent on Antibiotic Resistance of Riverine Enterobacteriaceae and *Aeromonas Spp.* *Appl. Environ. Microbiol.* **2000**, *66*, 125–132. [CrossRef] [PubMed]
3. Tello, A.; Austin, B.; Telfer, T.C. Selective Pressure of Antibiotic Pollution on Bacteria of Importance to Public Health. *Environ. Health Perspect.* **2012**, *120*, 1100–1106. [CrossRef] [PubMed]
4. Amos, G.C.A.; Hawkey, P.M.; Gaze, W.H.; Wellington, E.M. Waste Water Effluent Contributes to the Dissemination of CTX-M-15 in the Natural Environment. *J. Antimicrob. Chemother.* **2014**, *69*, 1785–1791. [CrossRef]
5. Fleming-Dutra, K.E.; Hersh, A.L.; Shapiro, D.J.; Bartoces, M.; Enns, E.A.; File, T.M.; Finkelstein, J.A.; Gerber, J.S.; Hyun, D.Y.; Linder, J.A.; et al. Prevalence of Inappropriate Antibiotic Prescriptions among Us Ambulatory Care Visits, 2010–2011. *JAMA-J. Am. Med. Assoc.* **2016**, *315*, 1864–1873. [CrossRef]
6. Zhu, Y.G.; Zhao, Y.; Li, B.; Huang, C.L.; Zhang, S.Y.; Yu, S.; Chen, Y.S.; Zhang, T.; Gillings, M.R.; Su, J.Q. Continental-Scale Pollution of Estuaries with Antibiotic Resistance Genes. *Nat. Microbiol.* **2017**, *2*, 16270. [CrossRef]
7. Laxminarayan, R.; Duse, A.; Wattal, C.; Zaidi, A.K.M.; Wertheim, H.F.L.; Sumpradit, N.; Vlieghe, E.; Hara, G.L.; Gould, I.M.; Goossens, H.; et al. Antibiotic Resistance-the Need for Global Solutions. *Lancet Infect. Dis.* **2013**, 1057–1098. [CrossRef]
8. Nagvekar, V.; Sawant, S.; Amey, S. Prevalence of Multi Drug Resistant Gram-Negative Bacteria Cases at Admission in Multispecialty Hospital. *J. Glob. Antimicrob. Resist.* **2020**. [CrossRef]
9. Fasciana, T.; Gentile, B.; Aquilina, M.; Ciammaruconi, A.; Mascarella, C.; Anselmo, A.; Fortunato, A.; Fillo, S.; Petralito, G.; Lista, F.; et al. Co-Existence of Virulence Factors and Antibiotic Resistance in New *Klebsiella Pneumoniae* Clones Emerging in South of Italy. *BMC Infect. Dis.* **2019**, *19*, 928. [CrossRef]
10. Adams-Sapper, S.; Sergeevna-Selezneva, J.; Tartof, S.; Raphael, E.; Diep, B.A.; Perdreau-Remington, F.; Riley, L.W. Globally Dispersed Mobile Drug-Resistance Genes in Gram-Negative Bacterial Isolates from Patients with Bloodstream Infections in a US Urban General Hospital. *J. Med. Microbiol.* **2012**, *61*, 968–974. [CrossRef]
11. Center for Disease Control and Prevention. *Antibiotic Resistance Threats in the United States;* Center for Disease Control and Prevention: Atlanta, GA, USA, 2013.
12. Berglund, B. Environmental Dissemination of Antibiotic Resistance Genes and Correlation to Anthropogenic Contamination with Antibiotics. *Infect. Ecol. Epidemiol.* **2015**, *5*, 28564. [CrossRef] [PubMed]
13. Davies, J.; Davies, D. Origins and Evolution of Antibiotic Resistance. *Microbiol. Mol. Biol. Rev.* **2010**, *74*, 417–433. [CrossRef]
14. Woodford, N.; Wareham, D.W.; Guerra, B.; Teale, C. Carbapenemase-Producing Enterobacteriaceae and Non-Enterobacteriaceae from Animals and the Environment: An Emerging Public Health Risk of Our Own Making? *J. Antimicrob. Chemother.* **2014**, *69*, 287–291. [CrossRef]
15. Czekalski, N.; Sigdel, R.; Birtel, J.; Matthews, B.; Bürgmann, H. Does Human Activity Impact the Natural Antibiotic Resistance Background? Abundance of Antibiotic Resistance Genes in 21 Swiss Lakes. *Environ. Int.* **2015**, *81*, 45–55. [CrossRef] [PubMed]
16. Raphael, E.; Wong, L.K.; Riley, L.W. Extended-Spectrum Beta-Lactamase Gene Sequences in Gram-Negative Saprophytes on Retail Organic and Nonorganic Spinach. *Appl. Environ. Microbiol.* **2011**, *77*, 1601–1607. [CrossRef]
17. Matyar, F.; Kaya, A.; Dinçer, S. Antibacterial Agents and Heavy Metal Resistance in Gram-Negative Bacteria Isolated from Seawater, Shrimp and Sediment in Iskenderun Bay, Turkey. *Sci. Total Environ.* **2008**, *407*, 279–285. [CrossRef]
18. Griffin, D.W.; Banks, K.; Gregg, K.; Shedler, S.; Walker, B.K. Antibiotic Resistance in Marine Microbial Communities Proximal to a Florida Sewage Outfall System. *Antibiotics* **2020**, 118. [CrossRef] [PubMed]

19. Tacão, M.; Correia, A.; Henriques, I.S. Low Prevalence of Carbapenem-Resistant Bacteria in River Water: Resistance Is Mostly Related to Intrinsic Mechanisms. *Microb. Drug Resist.* **2015**, *21*, 497–506. [CrossRef] [PubMed]

20. Bush, K.; Bradford, P.A. β-Lactams and β-Lactamase Inhibitors: An Overview. *Cold Spring Harb. Perspect. Med.* **2016**, *6*, a025247. [CrossRef]

21. Bush, K. Past and Present Perspectives on β-Lactamases. *Antimicrob. Agents Chemother.* **2018**, *62*, e01076-18. [CrossRef]

22. Adams-Sapper, S.; Diep, B.A.; Perdreau-Remington, F.; Riley, L.W. Clonal Composition and Community Clustering of Drug-Susceptible and -Resistant *Escherichia Coli* Isolates from Bloodstream Infections. *Antimicrob. Agents Chemother.* **2013**, *57*, 490–497. [CrossRef]

23. Papp-Wallace, K.M.; Endimiani, A.; Taracila, M.A.; Bonomo, R.A. Carbapenems: Past, Present, and Future. *Antimicrob. Agents Chemother.* **2011**, *55*, 4943–4960. [CrossRef] [PubMed]

24. Nordmann, P.; Dortet, L.; Poirel, L. Carbapenem Resistance in Enterobacteriaceae: Here Is the Storm! *Trends Mol. Med.* **2012**, *18*, 263–272. [CrossRef] [PubMed]

25. Scotta, C.; Juan, C.; Cabot, G.; Oliver, A.; Lalucat, J.; Bennasar, A.; Albertí, S. Environmental Microbiota Represents a Natural Reservoir for Dissemination of Clinically Relevant Metallo-β-Lactamases. *Antimicrob. Agents Chemother.* **2011**, *55*, 5376–5379. [CrossRef] [PubMed]

26. Montezzi, L.F.; Campana, E.H.; Corrêa, L.L.; Justo, L.H.; Paschoal, R.P.; Da Silva, I.L.V.D.; Souza, M.D.C.M.; Drolshagen, M.; Picão, R.C. Occurrence of Carbapenemase-Producing Bacteria in Coastal Recreational Waters. *Int. J. Antimicrob. Agents* **2015**, *45*, 174–177. [CrossRef]

27. Lu, S.Y.; Zhang, Y.L.; Geng, S.N.; Li, T.Y.; Ye, Z.M.; Zhang, D.S.; Zou, F.; Zhou, H.W. High Diversity of Extended-Spectrum Beta-Lactamase-Producing Bacteria in an Urban River Sediment Habitat. *Appl. Environ. Microbiol.* **2010**, *76*, 5972–5976. [CrossRef] [PubMed]

28. Novick, E.; Senn, D.B. *External Nutrient Loads to San Francisco Bay*; San Francisco Estuary Institute: Richmond, CA, USA, 2014.

29. Seto, E.Y.; Konnan, J.; Olivieri, A.W.; Danielson, R.E.; Gray, D.M.D. A Quantitative Microbial Risk Assessment of Wastewater Treatment Plant Blending: Case Study in San Francisco Bay. *Environ. Sci. Water Res. Technol.* **2015**, *2*, 134–145. [CrossRef]

30. Trowbridge, P.R.; Davis, J.A.; Mumley, T.; Taberski, K.; Feger, N.; Valiela, L.; Ervin, J.; Arsem, N.; Olivieri, A.; Carroll, P.; et al. The Regional Monitoring Program for Water Quality in San Francisco Bay, California, USA: Science in Support of Managing Water Quality. *Reg. Stud. Mar. Sci.* **2016**, *4*, 21–33. [CrossRef]

31. Shellenbarger, G.G.; Athearn, N.D.; Takekawa, J.Y.; Boehm, A.B. Fecal Indicator Bacteria and *Salmonella* in Ponds Managed as Bird Habitat, San Francisco Bay, California, USA. *Water Res.* **2008**, *42*, 2921–2930. [CrossRef]

32. Heal the Bay. *2018–2019 Beach Report Card*; Heal the Bay: Santa Monica, CA, USA, 2019.

33. San Francisco Estuary Partnership. *The State of the Estuary 2015*; San Francisco Estuary Partnership: Oakland, CA, USA, 2015.

34. Picão, R.C.; Cardoso, J.P.; Campana, E.H.; Nicoletti, A.G.; Petrolini, F.V.B.; Assis, D.M.; Juliano, L.; Gales, A.C. The Route of Antimicrobial Resistance from the Hospital Effluent to the Environment: Focus on the Occurrence of KPC-Producing *Aeromonas Spp.* and Enterobacteriaceae in Sewage. *Diagn. Microbiol. Infect. Dis.* **2013**, *76*, 80–85. [CrossRef]

35. Wang, Z.; Zhang, X.-X.; Huang, K.; Miao, Y.; Shi, P.; Liu, B.; Long, C.; Li, A. Metagenomic Profiling of Antibiotic Resistance Genes and Mobile Genetic Elements in a Tannery Wastewater Treatment Plant. *PLoS ONE* **2013**, *8*, e76079. [CrossRef] [PubMed]

36. Yee, D.; Wong, A.; Shimabuku, I.; Trowbridge, P. *Characterization of Sediment Contamination in Central Bay Margin Areas*; San Francisco Estuary Institute: Richmond, CA, USA, 2017.

37. Baker-Austin, C.; Stockley, L.; Rangdale, R.; Martinez-Urtaza, J. Environmental Occurrence and Clinical Impact of *Vibrio Vulnificus* and *Vibrio Parahaemolyticus*: A European Perspective. *Environ. Microbiol. Rep.* **2010**, 7–18. [CrossRef] [PubMed]

38. Janda, J.M.; Abbott, S.L. The Genus *Shewanella*: From the Briny Depths below to Human Pathogen. *Crit. Rev. Microbiol.* **2014**, *40*, 293–312. [CrossRef] [PubMed]

39. Holt, H.M.; Gahrn-Hansen, B.; Bruun, B. *Shewanella Algae* and *Shewanella Putrefaciens*: Clinical and Microbiological Characteristics. *Clin. Microbiol. Infect.* **2005**, *11*, 347–352. [CrossRef]

40. Maravić, A.; Skočibušić, M.; Šamanić, I.; Fredotović, Ž.; Cvjetan, S.; Jutronić, M.; Puizina, J. *Aeromonas Spp.* Simultaneously Harbouring BlaCTX-M-15, BlaSHV-12, BlaPER-1 and BlaFOX-2, in Wild-Growing Mediterranean Mussel (*Mytilus Galloprovincialis*) from Adriatic Sea, Croatia. *Int. J. Food Microbiol.* **2013**, *166*, 301–308. [CrossRef]

41. Parkins, M.D.; Gregson, D.B.; Pitout, J.D.D.; Ross, T.; Laupland, K.B. Population-Based Study of the Epidemiology and the Risk Factors for *Pseudomonas Aeruginosa* Bloodstream Infection. *Infection* **2010**, *38*, 25–32. [CrossRef]

42. Poirel, L.; Héritier, C.; Nordmann, P. Chromosome-Encoded Ambler Class D β-Lactamase of *Shewanella Oneidensis* as a Progenitor of Carbapenem-Hydrolyzing Oxacillinase. *Antimicrob. Agents Chemother.* **2004**, *48*, 348–351. [CrossRef]

43. Pellegrini, C.; Mercuri, P.S.; Celenza, G.; Galleni, M.; Segatore, B.; Sacchetti, E.; Volpe, R.; Amicosante, G.; Perilli, M. Identification of BlaIMP-22 in *Pseudomonas Spp.* in Urban Wastewater and Nosocomial Environments: Biochemical Characterization of a New IMP Metallo-Enzyme Variant and Its Genetic Location. *J. Antimicrob. Chemother.* **2009**, *63*, 901–908. [CrossRef]

44. Rossolini, G.M.; Condemi, M.A.; Pantanella, F.; Docquier, J.D.; Amicosante, G.; Thaller, M.C. Metallo-β-Lactamase Producers in Environmental Microbiota: New Molecular Class B Enzyme in *Janthinobacterium Lividum*. *Antimicrob. Agents Chemother.* **2001**, *45*, 837–844. [CrossRef]

45. Shaw, K.S.; Rosenberg Goldstein, R.E.; He, X.; Jacobs, J.M.; Crump, B.C.; Sapkota, A.R. Antimicrobial Susceptibility of *Vibrio Vulnificus* and *Vibrio Parahaemolyticus* Recovered from Recreational and Commercial Areas of Chesapeake Bay and Maryland Coastal Bays. *PLoS ONE* **2014**, *9*, e89616. [CrossRef] [PubMed]

46. Héritier, C.; Poirel, L.; Nordmann, P. Genetic and Biochemical Characterization of a Chromosome-Encoded Carbapenem-Hydrolyzing Ambler Class D β-Lactamase from *Shewanella Algae*. *Antimicrob. Agents Chemother.* **2004**, *48*, 1670–1675. [CrossRef] [PubMed]

47. Girlich, D.; Poirel, L.; Nordmann, P. Novel Ambler Class A Carbapenem-Hydrolyzing β-Lactamase from a *Pseudomonas Fluorescens* Isolate from the Seine River, Paris, France. *Antimicrob. Agents Chemother.* **2010**, *54*, 328–332. [CrossRef] [PubMed]

48. Quinteira, S.; Ferreira, H.; Peixe, L. First Isolation of BlaVIM-2 in an Environmental Isolate of *Pseudomonas Pseudoalcaligenes*. *Antimicrob. Agents Chemother.* **2005**, *49*, 2140–2141. [CrossRef] [PubMed]

49. Deng, Y.; Bao, X.; Ji, L.; Chen, L.; Liu, J.; Miao, J.; Chen, D.; Bian, H.; Li, Y.; Yu, G. Resistance Integrons: Class 1, 2 and 3 Integrons. *Ann. Clin. Microbiol. Antimicrob.* **2015**, *14*, 45. [CrossRef]

50. Stevens, D.L., Jr.; Olsen, A.R. Spatially Balanced Sampling of Natural Resources. *J. Am. Stat. Assoc.* **2004**, *99*, 262–278. [CrossRef]

51. SFEI. *Regional Monitoring Program for Water Quality in San Francisco Bay 2015 Bay Margins Sediment Study Cruise Plan*; San Francisco Estuary Institute: Richmond, CA, USA, 2015.

52. Coastal Contamination & Research. *Contaminant Concentrations in Central Bay Margins Sediment Cruise Report*; Coastal Contamination & Research: Moss Landing, CA, USA, 2015.

53. CLSI. *M100-S25 Performance Standards for Antimicrobial Susceptibility Testing; Twenty-Fifth Informational Supplement*; Clinical and Laboratory Standards Institute: Wayne, PA, USA, 2015.

54. Dallenne, C.; da Costa, A.; Decré, D.; Favier, C.; Arlet, G. Development of a Set of Multiplex PCR Assays for the Detection of Genes Encoding Important β-Lactamases in Enterobacteriaceae. *J. Antimicrob. Chemother.* **2010**, *65*, 490–495. [CrossRef]

MDPI

St. Alban-Anlage 66

4052 Basel

Switzerland

Tel. +41 61 683 77 34

Fax +41 61 302 89 18

www.mdpi.com

Antibiotics Editorial Office

E-mail: antibiotics@mdpi.com

www.mdpi.com/journal/antibiotics